International Conference on Design, Mechanical and Material Engineering (D2ME 2016)

MATEC Web of Conferences Volume 82 (2016)

Auckland, New Zealand
8-10 September 2016

Editors:

B. B. Sahari
R. Das
Q. Qin

ISBN: 978-1-5108-3169-8

Printed from e-media with permission by:

Curran Associates, Inc.
57 Morehouse Lane
Red Hook, NY 12571

Some format issues inherent in the e-media version may also appear in this print version.

Printed by Curran Associates, Inc. (2016)

For additional information, please contact EDP Sciences – Web of Conferences at the address below.

EDP Sciences – Web of Conferences
17, Avenue du Hoggar
Parc d'Activité de Courtabœuf
BP 112
F-91944 Les Ulis Cedex A
France

Phone: +33 (0) 1 69 18 75 75
Fax: +33 (0) 1 69 28 84 91

contact-edps@webofconferences.org

Additional copies of this publication are available from:

Curran Associates, Inc.
57 Morehouse Lane
Red Hook, NY 12571 USA
Phone: 845-758-0400
Fax: 845-758-2633
Email: curran@proceedings.com
Web: www.proceedings.com

TABLE OF CONTENTS

D2ME 2016 Committee

General Chair

Dr. Raj Das, University of Auckland, New Zealand

Advisory Committee Chair

Prof. Kai Cheng, Brunel University, UK

Conference Committee Co-Chairs

Prof. Zhan Chen, AUT University, Auckland, New Zealand

Prof. Josua Meyer, University of Pretoria, South Africa

Program Committee Chairs

Prof. Qinghua Qin, The Australian National University, Australia

Prof. Shinn-Liang Chang, National Formosa University, Taiwan

Prof. Barkawi b. Sahari, Universiti Putra Malaysia, Malaysia

Prof. Liyong Tong, The University of Sydney, Australia

International Technical Committees

Prof. Wong Shaw Voon, Universiti Putra Malaysia, Malaysia (Chair)

Prof. Li Chang, The University of Sydney, Australia

Prof. Gangadhara Prusty, UNSW AUSTRALIA, Australia

Dr. Khalid Arif, Massey University, New Zealand

Dr.Chi Kit Au, The University of Waikato, New Zealand

Dr. Johan Verbeek, The University of Waikato, New Zealand

Dr. Adrian Lowe, The Australian National University, Australia

Prof. Chong Soo Lee, Pohang University of Science and Technology, South Korea

Dr. Xiaowen Yuan, Massey University, New Zealand

PREFACE

The 2016 International Conference on Design, Mechanical and Material Engineering (D2ME 2016) was held at the Waipuna Hotel and Conference Centre, by the beautiful Panmure basin in Auckland, New Zealand during September 8-10, 2016. The conference was organized by Hong Kong Society of Mechanical Engineers (HKSME). During the conference, delegates from South Africa, Germany, Italy, Japan, China, Taiwan, Malaysia, Brunei, Australia and New Zealand gathered together and presented their research in a variety of D2ME topics. The organizer (HKSME) would like to thank all of the presenters who have come far to NZ and have contributed to the enjoyable and valuable discussions during the conference.

Peer-reviewing of all submitted papers was conducted before the papers were selected for presentation in the conference. For each paper included in this proceedings, the paper was peer-reviewed by 2-3 experts from the conference technical committee following the normal refereeing procedure. The conference proceedings containing 39 papers are divided into 3 parts: Design Engineering, Mechanical Engineering and Materials Engineering. The organizer (HKSME) would like to thank the contributions made by the technical committee. They gave their times and reviewed all the papers in their respective fields of expertise.

Prof. Zhan Chen
Auckland University of Technology, NZ

Statement of Peer review

In submitting conference proceedings to MATEC Web of Conferences, I certify to the Publisher that I adhere to the **Policy on Publishing Integrity** of the journal in order to safeguard good scientific practice in publishing.

1. All articles have been subjected to peer review administered by the proceedings editors.
2. Reviews have been conducted by expert referees, who have been requested to provide unbiased and constructive comments aimed, whenever possible, at improving the work.
3. Proceedings editors have taken all reasonable steps to ensure the quality of the materials they publish and their decision to accept or reject a paper for publication has been based only on the merits of the work and the relevance to the journal.

Title, date and place of the conference

2016 International Conference on Design, Mechanical and Material Engineering (D2ME 2016)

Sept. 8-10, 2016 Auckland, New Zealand

Proceedings editor(s):

Prof. Barkawi b. Sahari, Prof. Qinghua Qin, Dr. Raj Das

Date and editor's signature

MATEC Web of Conferences **82**, 01001 (2016) DOI: 10.1051/matecconf/20168201001
D2ME 2016

Correct use of Membrane Elements in Structural Analysis

Timothy Rothman[1] and Liming Ji[2 a]

[1]*Intel Corporation, 1900 Prairie City Road, Folsom, California, 95623, USA*
[2]*Intel Corporation, Raycom Infotech Park A, No.2 KeXueYuan South Road, ZhongGuanCun, Beijing China 100190*

Abstract. Structural analysis of consumer electronic devices such as phones and tablets involves Finite Element Analysis (FEA). Dynamic loading conditions such as device dropping and bending dictate accurate FEA models to reduce design risk in many areas. The solid elements typically used in structural analysis do not have integration points on the surface. The outer surface is of most interest because that is where the cracks start. Analysts employ a post processing trick through using membranes to bring accurate stress/strain results to the surface. This paper explains numerical issues with implementation of membranes and recommends a methodology for accurate structural analysis.

1 Introduction

Structural analysis of the complete system is done with different loadings such as bending or drop test [1]. Shear locking and hourglassing are usually understood and modelling techniques to avoid related errors are regularly used [2]. Finite Element Analysis (FEA) is used to solve this complex problem. During a typical design cycle at least 3 FEA iterations are needed. During a crash event, a number of structural components must sustain abnormal loadings in order to meet stringent integrity requirements [3]. Dynamic analysis to analyze drop test is not a trivial activity, and careful selection of element types and meshing strategies is needed. Membrane elements are used to accurately describe surfaces stresses related to underlying geometry that is not accurate at the surface when using solid elements alone [4]. These membrane elements need to be "glued" to the model, so *TIE or *EMBEDD [5] options are used. Careful use of membrane elements through mesh density, and method of attaching are explored.

2 Structural analysis using FEA

Structural analysis in the context of this paper is mechanical analysis of the complete system to different loadings such as bending or drop test [1]. Finite Element Analysis (FEA) or the Finite Element Method is used to discretize the domain and solve the 4 mathematical sets of equations of elasticity. It is usually necessary to characterize materials at appropriate loading rates to ensure the measured results will extend properly to the geometry of interest. In the design of phones, tablets and laptops correct structural design is of paramount concern and design cycles do not allow multiple

[a] Corresponding author : liming.ji@intel.com

iterations. The philosophy is to make to the system correct by design the first time through mechanical analysis and allow iterations to be related to minor manufacturing or cosmetic issues. In a typical design cycle at least 3 design iterations are needed to work though design risk areas and ensure the proposed fixes are manufacturable and can be implemented in the limited physical space available for phones and tablets. Figure 1 show an example of structural design issues resolved using FEA.

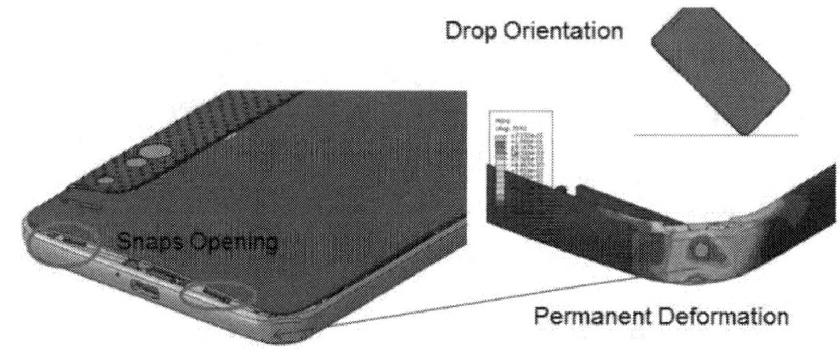

Figure 1. Drop Test at 1.5m FEA of prototype phone. Issues resolved during design process.

The major limitations to getting accurate, timely results are mesh accuracy, boundary or loading assumptions, run time relative to computational resources and schedule and material characterization accuracy. The focus of this paper is on the mesh accuracy. Mesh accuracy obviously includes how accurately the mesh matches the underlying geometry and if the mesh is too coarse, the area at risk may be missed entirely. Mesh accuracy also includes a much more subtle aspect of element choice, implementation and which elements to be used for which analysis.

3 Element choices in Abaqus

Commercial FEA tools like Simulia Abaqus provide analysts many choices of elements. The nodes are where the displacements are directly calculated, and the integration points are where additional information such as strains and stresses are calculated. Shown in Figure 2 are examples of 3D solid element each having 3 degrees of freedom (displacements) for each node. Many analysts rely on tribal knowledge on element choice due to the overwhelming choices available.

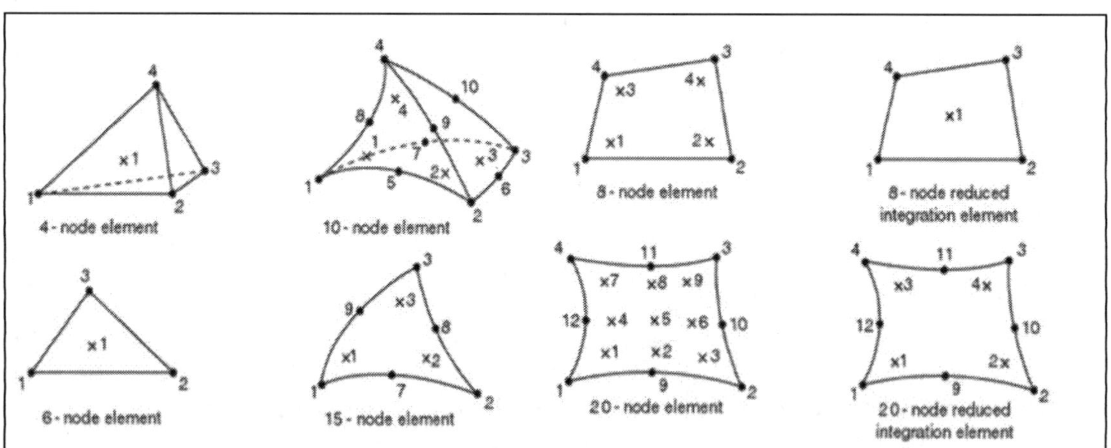

Figure 2. 3D Solid (Continuum) Element Types Available in Abaqus. x denotes integration point locations.

4 Sources of numerical inaccuracies in FEA due to element choice

There are many known sources of inaccuracy due to element density and choice. Element density is a fairly apparent source of error, but element choice is often misunderstood or the subject is avoided by the novice analyst. Poor choice of element density and type based on the geometry and loading often leads to erroneous results. In fact the true art of FEA lies in correct choice of the element.

4.1 Shear locking and hourglassing

It is well established that shear locking and hourglassing can dramatically affect the accuracy of the results [2]. Shear locking is observed with first order fully integrated elements in bending where the element mathematically cannot follow the actual shape caused by the loading. Many commercial FEA codes offer reduced integration element which is intended to resolve the shear locking phenomenon, while keeping computational time at a minimum. Hourglassing is the effect of a single integration point structure being excessively flexible unable to accurately represent the load handing ability of the structure. Even second order reduced integration elements suffer from hourglassing when only 1 layer of elements is used. Abaqus also provides in compatible mode elements that overcome the effects of shear locking and are practically not affected by hourglassing.

4.2 Implementation of membrane elements

The key structural components of the majority of transportation vehicles are designed as thin-walled components. During a crash event, a number of structural components must sustain abnormal loadings in order to meet stringent integrity requirements [3]. The loading and structural requirements for consumer electronics is similar, except transportation vehicles implement a "crush zone" and although the need for thin members is necessary to minimize consumer device thickness, the space for crush zones is typically not be available in handheld electronics.

Special element choices need to be considered for these high load, thin structures. The outer surface is of most interest because that is where the impact occurs and where the cracks start. The solid elements typically used in structural analysis do not have integration points on the surface (specifically C3D8I, C3D10M, and C3D10I). Analysts therefore use a post processing methodology to bring accurate stress/strain results to the surface. This is done by adding membrane elements to the model. [4] Membranes are used as the outer surface of every element in the model. The membrane in some cases is made directly from the outer faces of existing elements and no new nodes are created resulting in no new degrees of freedom and no additional run time needed for calculations (incremental run time need only for storing the results). In addition a small membrane called a patch is added for post processing in specific areas (i.e. such emulating a strain gage in an area of high solder stress under BGAs). These patches need to be "glued" to the model, so *TIE or *EMBEDD options are used.

4.3 Case studies

4.3.1 Simple models

In order to correctly assess accuracy models are simplified to ones which have analytical solutions. Figure 3 shows two simplified models chosen. One is a 200x40x1mm beam in pure bending and held at one end (all degrees of freedom) and a moment applied at the free end. The second simplified model is a plate 100x100x1mm bending due to uniform pressure held on 4 sides. Both models use Young's Modulus of 25GPa with poisson's ratio of 0.3 and element size of 0.5mm.

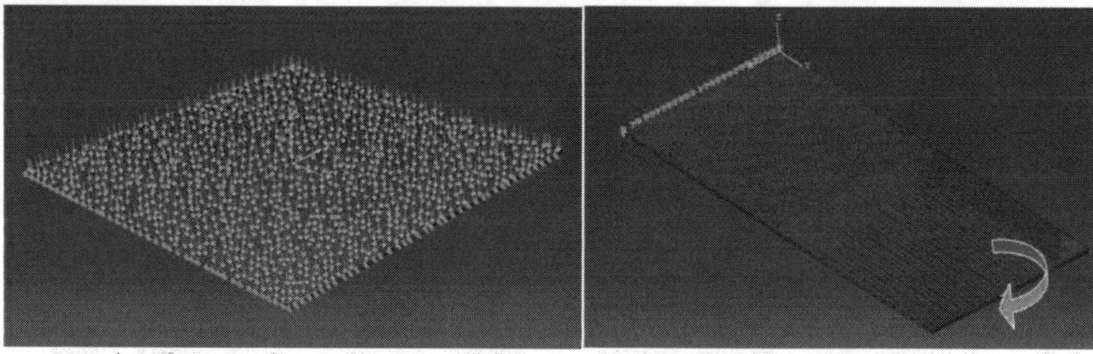

Simple Plate under pressure model: 100*100*1mm Simple Beam under bending model: 200*40*1mm

Figure 3. Simplified models. Orange denotes holding conditions.

Tables 1 and 2 indicate that the size of the membrane is NOT relevant to the results. Also the use of either *TIE or *EMBED has no impact on the patch used for post processing. There is only a minor influence of inaccuracy of the C3D10M element.

Table 1. Deformation and stress of Simple Plate under pressure indicating minimal inaccuracy using membranes.

	Max def. U3 (mm)	Difference	Stress (Mpa) S11 - Average of int. pt.	Difference
Analytical solution	**1.77200**		**28.74000**	
S4R	1.7892	0.97%	28.8948	0.54%
C3D8I, 2 layers	1.78497	0.73%	28.8948	0.54%
C3D8I, 2 layers + MEM	1.78373	0.66%	28.8266	0.30%
C3D10M, one layer+MEM	1.85727	4.81%	30.0394	4.52%
C3D10I, one layer+MEM	1.78717	0.86%	28.8608	0.42%
C3D8I, 2 layer +MEM +0.8mm Patch -TIE			28.8282	0.31%
C3D8I, 2 layer +MEM +1.52mm Patch *TIE			28.8215	0.28%
C3D8I, 2 layer +MEM +1.52mm Patch *EMBED			28.8173	0.27%
Measured at Patch	**Measured at Membrane**			

Table 2. Stress of Simple Beam under bending indicating minimal inaccuracy using membranes.

	Curve - S22(Mpa)	Difference
Analytical solution	**15**	
C3D8I+MEM	14.9909	-0.06%
C3D8I, one layer+MEM	14.991	-0.06%
C3D8R+ENH+MEM	14.9909	-0.06%
C3D10I+MEM	14.9905	-0.06%
C3D10M + MEM	15.5	3.33%
C3D8I + 0.8mm Patch -*TIE	14.997	-0.02%
C3D8I + 0.8mm Patch - *EMBEDDED	14.9975	-0.02%
Measured at Patch	**Measured at Membrane**	

4.3.2 Complex models

Simple models yield surprisingly accurate results, but the simplified state of bending can give a false sense of confidence because the same methodology a more complex model will give inaccurate results. In consumer electronics the BGA stress on the PCB is of key interest. Ensuring solder joint damage does not occur is critical. So a BGA package is added to a 100x60x1mm plate using C3D8I elements under pressure and patches are used in each BGA package corner. The stress plots in Figure 4 when using a 1.52mmx1.52mm patch seem similar between *TIE and *EMBEDDED, but closer inspection of the stress on the patches as shown in Table 3 suggests major differences. The 1.52mmx1.52mm patch is then discretized with 15x15 elements on each patch instead of a single element and Figure 5 and Table 4 show perfect correlation of results when using *TIE and *EMBEDDED.

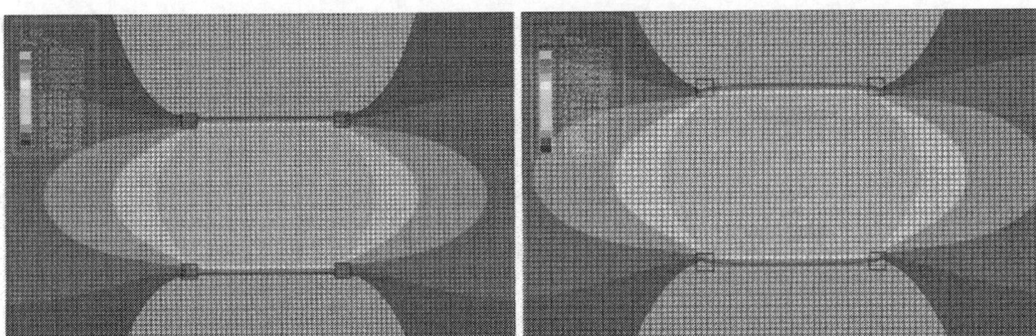

Figure 4. Stress plots of complex model with BGA package using 1.52mmx1.52mm patch shown on the same scale suggesting the stress fields are the same

Table 3. Stress comparison of complex model with BGA package on 1.52mmx1.52mm patch. *TIE and *EMBEDDED do NOT match.

Corner	*TIE	*EMBEDDED
1	7.67974	9.08999
2	7.66387	9.08827
3	-7.60217	-9.84414
4	-7.60342	-9.84611

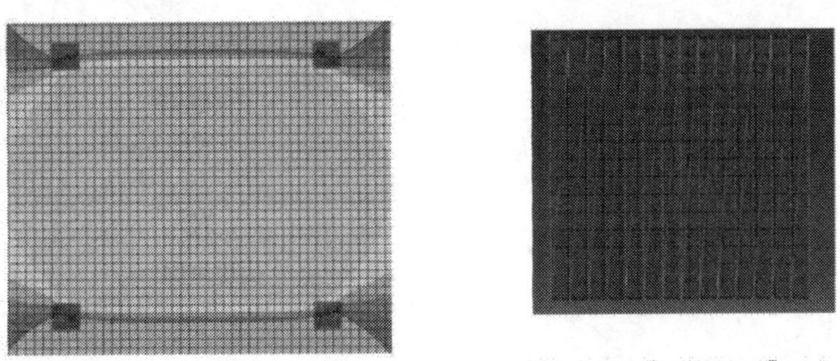

Figure 5. Stress plots of complex model with BGA package using 1.52mmx1.52mm patch discretized

Table 4. Stress comparison of complex model with BGA package on 1.52mmx1.52mm patch discretized with 15x15 elements. *TIE and *EMBEDDED are in agreement.

Corner	*TIE	*EMBEDDED
1	9.60765	9.60726
2	9.59861	9.5982
3	-9.98875	-9.98875
4	-9.99104	-9.99104

Cautionary note for using tetrahedral elements such as C3D10M. The C3D10M has middle nodes and in cases such as shown in Figure 1 where the corresponding membrane is M3D3 used as a coating on the underlying tetrahedral, the M3D3 may not yield accurate results since it has no midside nodes.

5 Conclusions

Care must be taken in ensuring the membrane skin or local post processing patch are the correct size and element type to match the underlying element bending behaviour. An accurate mesh is needed for both the solid and the membrane and the membrane should be used to bring the correct results to the surface, not create a new set of inaccurate results. If the membrane is bigger than the under solid element, the output may not be correct in bending for both *TIE and *EMBEDDED, the under beam is curved but big membrane element cant capture this deflection. See Figure 6.

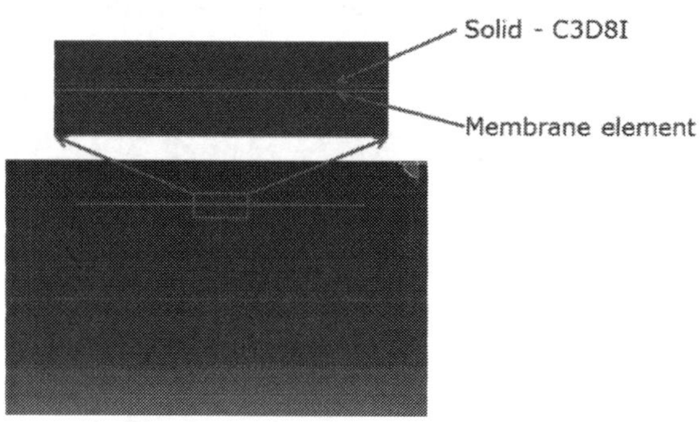

Figure 6. Exploded view of Membrane element placed just below surface of underling solid

References

1. JEDEC Standard JESD22-B111, *Board level drop test method of components for handheld electronic products* (2003)
2. Eric Quili Sun, Shear Locking and Hourglassing in MSC Nastan, ABAQUS, and ANSYS, *MSC.Software's Virtual Product Development Conference 2006* Web Publication (2006)
3. T. Deihl , D. Carroll, Utilizing ABAQUS' 10-Node Modified Tet for Analyzing Impact Problems Involving Thin-Walled Structures, *Proceedings Abaqus Users' Conference 2000, Newport RI* (2000)
4. W. Abramowicz, Thin-walled Structures as Impact Energy Absorbers, *Thin-Walled Structures* 91–107 (2000)
5. Abaqus Analysis User's Guide

Appendix

*EMBEDDED - Abaqus searches for the geometric relationships between nodes of the embedded elements and the host elements. If a node of an embedded element lies within a host element, the translational degrees of freedom at the node are eliminated and the node becomes an "embedded node." *The translational degrees of freedom of the embedded node are constrained to the interpolated values of the corresponding degrees of freedom of the host element*. Embedded elements are allowed to have rotational degrees of freedom, but these rotations are not constrained by the embedding [5].

*TIE - *Constrains each of the nodes on the slave surface to have the same motion as the point on the master surface to which it is closest [5].*

Both TIE and EMBEDDED may get incorrect result if the membrane patch is much bigger than the under solid element.

MATEC Web of Conferences **82**, 01002 (2016) DOI: 10.1051/matecconf/20168201002
D2ME 2016

Development of Hybrid Type Flexible Pneumatic Cylinder for Considering Less Air Consumption

Hiroaki Tamaki[1], Shujiro Dohta[1], Tetsuya Akagi[1,a], Wataru Kobayashi[1] and Yasuko Matsui[1]

[1]*Department of Intelligent Mechanical Engineering, Okayama University of Science 1-1, Ridai-cho, Kita-ku, Okayama, 700-0005, Japan*

Abstract. Inexpensive rehabilitation devices that can be used at home are required because of a lack of PT and welfare workers. In the previous study, the low-cost portable rehabilitation device using a flexible spherical actuator that consists of flexible pneumatic cylinder was proposed and tested. However, a compact and high power compressor that supplies air pressure to pneumatic actuator has not been developed yet. In particular, the heat generated by compressing air prevents to miniaturize it. To realize a home rehabilitation, the small-sized compressors or less air consuming flexible actuators are required. In this study, a hybrid type flexible pneumatic cylinder driven by electric motors and air pressure is proposed and tested. The concept, the construction and the operating principle of the proposed actuator were described. The position control using the tested actuator is also carried out.

1 Introduction

In an aging society, it is required to develop a system to aid in nursing care [1-4] and to support the activities of daily life for the elderly and the disabled [5]. Rehabilitation devices help the elderly who is injured temporally to recover their physical ability for keeping Quality of Life (QOL). In the previous study [6], to realize a home rehabilitation using a low-cost device, a flexible spherical actuator using a flexible pneumatic cylinder was proposed and tested. A portable rehabilitation device using the actuator that can be used while handling it with hands was also tested. However, in order to drive pneumatic actuator at home, a compact and high power compressor is necessary. Unfortunately, compact compressors that can generate continuous flow rate for a long period have not been developed yet. It is related to the heat problem caused by compressing air. To realize a home rehabilitation, small-sized compressors or less air consuming flexible actuators are required. In this study, aiming to develop the flexible pneumatic cylinder that consumes less air, a hybrid type flexible pneumatic cylinder utilizing an electric motor is proposed and tested. The position control using the tested cylinder is also carried out.

2 Construction of hybrid type flexible pneumatic cylinder

Figure 1 shows the flexible pneumatic cylinder developed in our previous study [7]. The cylinder consists of a flexible tube as a cylinder and gasket, one steel ball as a cylinder head and a slide stage

[a] Corresponding author : akagi@are.ous.ac.jp

that can move along the outside of the cylinder tube. The steel ball in the tube is pinched by two pairs of brass rollers from both sides of the ball. The operating principle of the cylinder is as follows. When the supply pressure is applied to one side of the cylinder, the inner steel ball is pushed. At the same time, the steel ball pushes the brass rollers and then the slide stage moves toward the opposite side while it deforms the tube. The slide stage is moved forward and backward even if the cylinder bends.

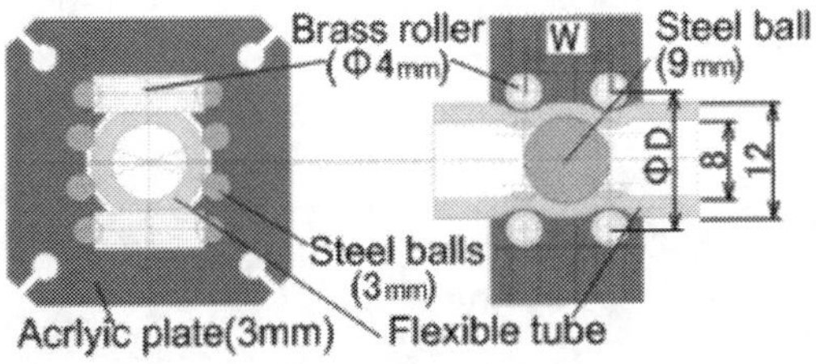

Figure 1. Flexible pneumatic cylinder.

To construct a flexible pneumatic cylinder driven by air and electric motor, the wire winding type motor is applied to the cylinder. Figures 2 and 3 show the view and construction of prototype of hybrid type flexible pneumatic cylinder, respectevely. Compared with a typical flexible pneumatic cylinder, it has a steal ball connected with wire as a cylinder head, a tee connecter for air supply, a tube end connecter for sealing and a wire winding motor. The wire winding motor consists of an acrylic wire spool with diameter of 10 mm and a small-sized geared motor (Seeed Studio Co. Ltd., MOT102A2B). The steel ball with the outer diameter of 9 mm has a hole with the inner diameter of 2 mm as shown in Fig.4. To keep a seal, there are two rubber balls with the outer diameter of 4.6 mm on both side of the steel ball. They are penetrated by a screw connected with a wire hook. The wire is connected to the motor and passed through the tee connecter, thin tubes, and the tube end connecter as shown in Fig.4. The tube end connecter consists of a tube connector, an acrylic cover, and a rubber ball with a thin hole. By adjusting the pushing force acting on the rubber ball using screw in the connecter, it can keep a seal even if the wire slides.

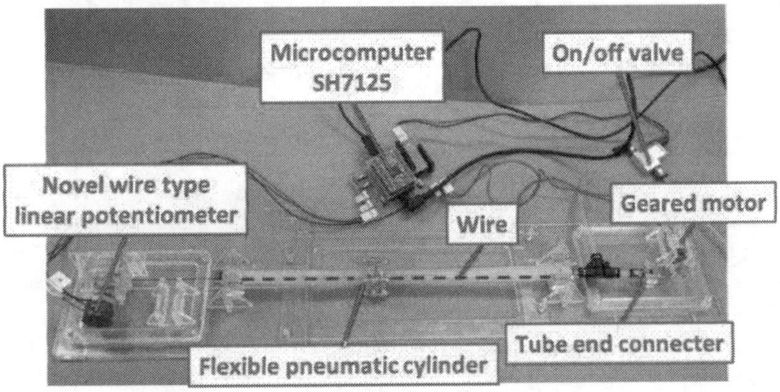

Figure 2. View of the tested hybrid type flexible pneumatic cylinder.

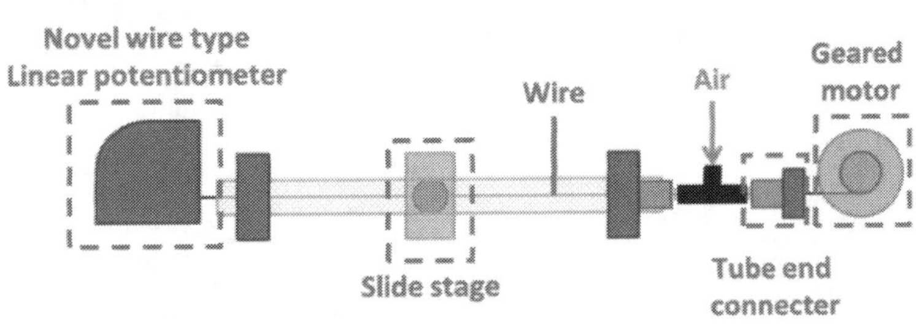

Figure 3. Construction of the tested hybrid type flexible pneumatic cylinder.

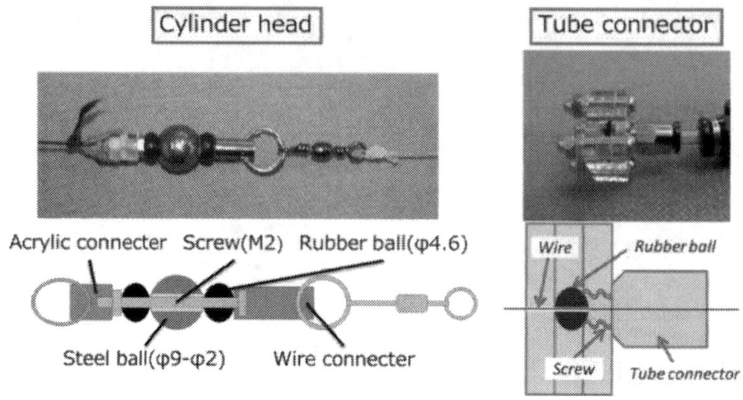

Figure 4. Cylinder head with the wire and inner schematic diagram of the tube end connecter.

3 Operating principle and experimental results

The operating principle of the cylinder is as follows. To drive the slide stage toward motor side, the motor winds up the wire. To drive toward opposite side, the wire is released by motor and an on/off valve is driven to supply the pressurized air to the steel ball. By pushing and pulling the steel ball, the ball pushes the rollers in slide stage while deforming the tube. Then, the slide stage moves. Thus, this method can reduce the consumed air for driving to push toward the motor side. As a result, the consumed pressurized air could be a half compared with the typical pneumatic cylinder. This is a natural phenomenon because the electric energy in partially used instead of pneumatic energy. You may ask what the advantage of the hybrid actuator is. The answer will be given in the following section.

Figure 5 shows the transient response of the controlled displacement of the cylinder for various desired position. In the experiment, the cylinder displacement was measured by a novel wire type linear potentiometer connected with the ball from the opposite side. Figure 6 shows the construction of the tested wire type linear potentiometer [8]. The tested potentiometer consists of a helical potentiometer (BOURNS Co. Ltd., 3590S-A26-104L) that can measure 10 times rotational angle, a clockwork wire spool with diameter of 22 mm and a stainless steel wire with diameter of 0.4 mm. Both shafts of the potentiometer and the wire spool are connected each other. From a rotational angle of the helical potentiometer and the diameter of the wire spool, the maximum length for measurement of approximately 0.7 m can be expected. The resolution of the potentiometer using 10 bit A/D converter is approximately 0.74 mm. The resolution can be improved by using 12 bit A/D converter. The cost of the linear potentiometer except the stainless steel wire is inexpensive, that is 8 US dollars. As a control scheme, simple on/off control scheme was used. In Fig.5, a broken and solid lines shows the desired and controlled displacement. It can be confirmed that the displacement of the tested cylinder can be controlled using both electric and pneumatic power. However, the speed of slide stage in the case when the motor winds the wire is slow, it is 10 mm/s.

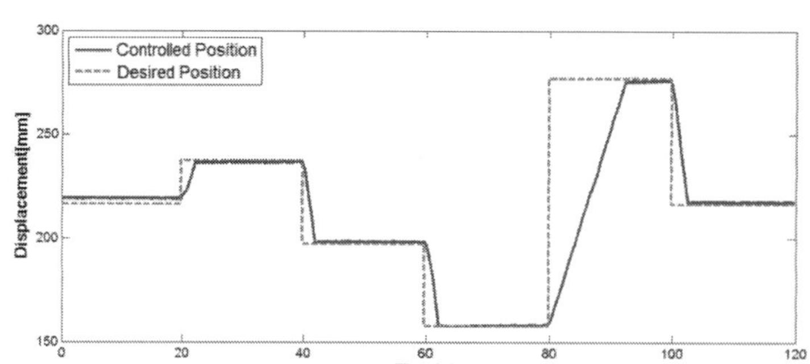

Figure 5. Transient response of the displacement of the tested cylinder.

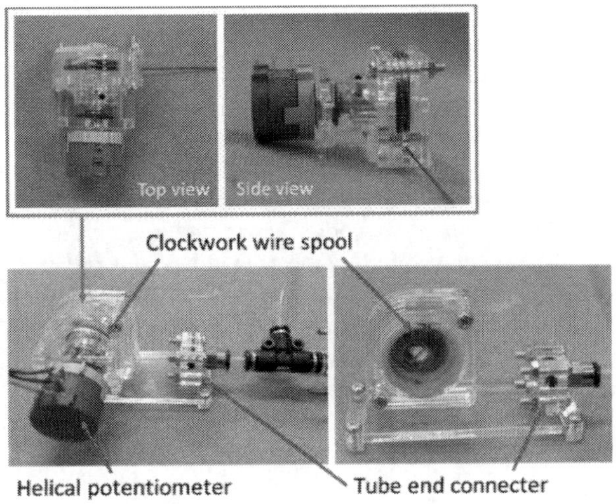

Figure 6. Low-cost wire type linear potentiometer

4 Improvement of hybrid type flexible pneumatic cylinder

The tested cylinder mentioned above needs a compressor to drive it for a long time such as one hour. If the pressurized air is not exhausted, the cylinder can be used for a long time without a compressor. Therefore, the improved type hybrid pneumatic cylinder without consuming pressurized air is proposed and tested. Figure 7 shows the improved hybrid type flexible pneumatic cylinder. Compared with the previous one, the motor was changed from one motor to two motors to increase the winding torque. Two output shafts of two motors are connected with each other through the wire spool. In addition, there is a PET bottle tank with volume of 500 cc and the initial pressure of 400 kPa, that is connected to the input port of motor side in direct. The operating principle of the improved cylinder is almost same as the previous. To drive the slide stage toward motor side, the motor winds up the wire. To drive toward opposite side, the wire is only released by motor. Then, the steel ball is pushed by always applied pressurized air of the tank. By this method, in the condition when the cylinder is always applied by a constant pressure, the slide stage can control the position by only using the motor. It means that the consumed pressurized air is externally decreased, that is zero in ideal, compared with driving by only air pressure. In this method, the cylinder can be driven for relative longer period of 1 or 2 hours by using a small Li-ion battery.

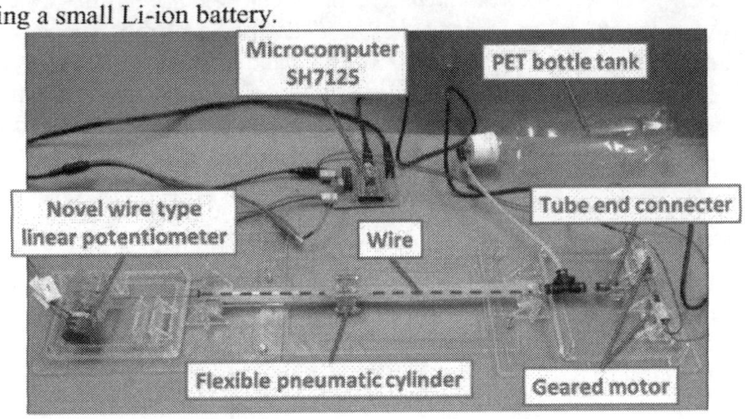

Figure 7. Improved hybrid type flexible pneumatic cylinder with embedded position controller.

The advantage of the proposed hybrid type flexible pneumatic cylinder is as follows. They can consider a similar system using a coil spring instead of air tank. However, the generated force of the coil spring increases according to the displacement. Therefore, the generated force is changed by the position. They can also consider the system using only electric motors. However, it is difficult to keep the wire's tension constant by synchronizing two electric motors. Therefore, the air spring using the pneumatic pressure was used.

Figure 8 shows the transient response of the controlled displacement of the improved cylinder for various desired position. As a control scheme, a simple on/off control scheme of the motor was applied. In the control, electric circuits of two motor connected in series so as to apply same current to both motors. The applied voltage is controlled by the embedded controller (Renesas Co. Ltd. SH7125) through a motor driver IC (Toshiba Co. Ltd. TA7291P). The suppled voltage of 12 V was used. In Fig.8, a broken and solid lines show the desired and controlled displacement, respectively. Compared with the result using the pervious cylinder as shown in Fig.5, the speed of slide stage in the case when the motor winds the wire is increased from 10 mm/s to 22 mm/s. As a result, it can be confirmed that the improved cylinder can be driven for more than one hour without a compressor.

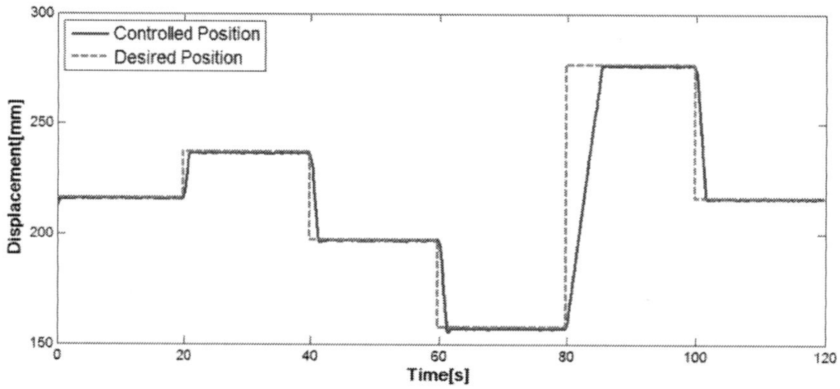

Figure 8. Transient response of the displacement of the improved cylinder.

5 Conclusions and future work

Aiming to develop a home rehabilitation device driven using a small air tank for 1 or 2 hours, the hybrid type flexible pneumatic cylinder for considering less air consumption was proposed and tested. The construction of the cylinder was described. The operating principle of the tested cylinder is also illustrated. The position control of the cylinder was carried out. As a result, the tested cylinder can be driven for more than one hour by using small sized PET tank with initial pressure of 400 kPa. It can be confirmed that the hybrid type flexible pneumatic cylinder with no consumed pressurized air can be realized. As a future work, we develop a more potable rehabilitation device using the hybrid type flexible pneumatic cylinder.

Acknowledgements

This study was supported by the MEXT of Japan through a Financial Assistance Program for QOL Innovative Research.

References

1. M. Ishii, K. Yamamoto, and K. Hyodo, Stand-alone wearable power assist suit -development and availability-, *Journal of Robotics and Mechatronics*, vol. **17**, no. 5, pp. 575-583, (2005).

2. J. Piquion, A. Nayar, A. Ghazaryan, R. Papanna, W. Klimek, and R. Laroia, Robot-assisted gynecological surgery in a community setting, *Journal of Robotic Surgery* 3, 2, pp. 61-64, (2009).
3. T. Noritsugu, M. Takaiwa, D. Sasaki, Development of Power Assist Wear Using Pneumatic Rubber Artificial Muscles, *Journal of Robotics and Mechatronics*, Vol. **21**, No. 5, pp.607-613, (2009).
4. H. Kobayashi, T. Shiban, Y. Ishida, Realization of all 7 motions for the upper limb by a muscle suit, *Journal of Robotics and Mechatronics*, Vol. **16**, pp.504-512, (2004).
5. Y. Nagata, Soft actuators -forefront of development-, NTS Ltd., pp. 291-335, (2004).
6. Y. Matsui, T. Akagi, S. Dohta, M. Aliff, and C. Liu, Development of Portable Rehabilitation Device Using Flexible Spherical Actuator and Embedded Controller, *Lecture Notes in Electrical Engineering* 293, Vol. **1**, Springer, pp. 231-238, (2014).
7. T. Akagi, and S. Dohta, Development of a Rodless Type Flexible Pneumatic Cylinder and Its Application, *Transactions on Robotics and Auto-mation of the JSME (C)*, Vol. **73**, No. 731, pp.2108–2114, (2007) (in Japanese).
8. Y. Matsui, T. Akagi, and S. Dohta, Development of Flexible Displacement Measuring System Using Wire-type Linear Encoder for Flexible Spherical Actuator, *Procedia Computer Science* 76, pp. 113-118, (2015).

Finding Pull Directions for Interference-free Mould Openings

Alan C. Lin[1, a] and Mohammad Effendi[1]

[1]*Department of Mechanical Engineering, National Taiwan University of Science and Technology, 43, Section 4, Keelung Road, Taipei 10607, Taiwan.*

Abstract. This paper proposes an approach to determining pull directions of an undercut region using the visibility map algorithm. A number of pull direction candidates are firstly generated and projected onto a spherical surface using a regular placement algorithm. The normal directions of undercut faces are then extracted and used to eliminate unwanted pull direction candidates. A group of pull directions are finally selected and used to design a side core mechanism in the design of an injection mould. Furthermore, one point at the center of the group is chosen to represent the pull direction of the undercut region.

1 Introduction

Figure 1(*a*) shows a typical cell phone component with holes and undercut areas distributed in various locations. To manufacture the component using injection moulding, several sliders and lifters with appropriate pull directions are needed, as shown in Figure 1(*b*). In the field of mould manufacturing, the determination of pull directions is critical to design sliders and lifters so that the mould can be opened without any interference [1]. A number of studies have been proposed regarding this issue. For example, Nee *et al.* [1] proposed a method based on geometrical characteristic and topological relationship, where the pull direction is obtained from normal directions of undercut faces. However, this method fails for cases with irregular faces. Lin *et al.* [2, 3] use Visibility Map (V-map) algorithms to determine the pull direction of an undercut region based on the STL format. However, the interference between the core and moulding parts shows that their algorithm should be improved. Based on attributes and nodes in an Extended Face Adjacency Attribute Graph (EFAAG) algorithm, Jian and Guoxiang [4] proposed a method for recognizing undercut regions. The V-map is used in their method to find pull directions of an undercut. Chen and Chou [5] introduced Augmented Visibility Map (AVM) to minimize the number of undercut regions for pockets. Chen *et al.* [6] proposed an algorithm to determine pull directions based on V-map using a Gaussian Mapping method. The purpose of their research was to minimize the number of cores based on the Pair of Parting Direction (PPD) algorithm. The pull direction is calculated with the Maximally Covered Point (MCP) algorithm. Liu and Ramani [7] proposed an exact visibility map method based on facet triangle meshing. Each facet of the V-map is projected onto a spherical surface, and then the intersection between all of it is selected as undercut pull direction.

[a] Corresponding author: alin@mail.ntust.edu.tw

DOI: 10.1051/matecconf/20168201003

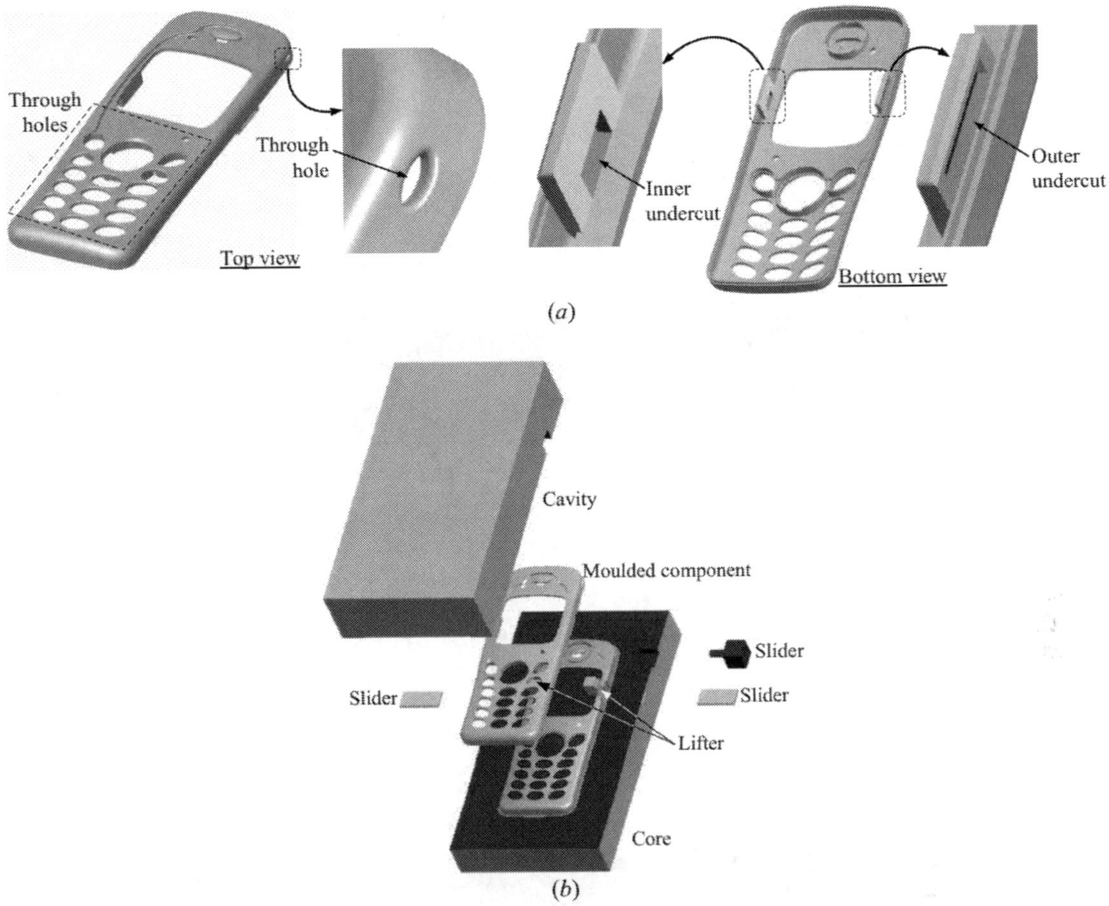

Figure 1. Normal directions of (*a*) an irregular face and (*b*) a regular face.

Most of the above mentioned approaches use regular faces and work only for limited cases. To overcome this problem, this paper proposes a method to determine pull directions of an undercut region with both regular and irregular faces.

2 Determination of Pull Directions

In our proposed method, a number of pull direction candidates are first determined and plotted onto a spherical surface using a regular placement algorithm as V-map representation. Then, the normal direction of both regular and irregular faces are extracted to create a 3D circle line. The V-map region is then built based on the 3D circle line and the normal direction, and used to eliminate unwanted pull direction candidates. A group of pull direction candidates of an undercut is finally selected, as the intersection of all V-maps through the Boolean operation. A center point of the group is then used as the pull direction for the undercut region. Figure 2 shows the proposed algorithm, where all the steps are described in the following sections.

2.1 Calculation of Normal Directions

Faces of an undercut have been classified into regular and irregular faces. An irregular face is a common face for complex manufacturing products such as turbine blades, cell phones, art products based on 3D printing, etc. For such faces the normal direction is perpendicular to the tangent plane at the selected point. A mesh in *x*-, *y*- and *z*- coordinates is then set and used as a sampling tool to obtain

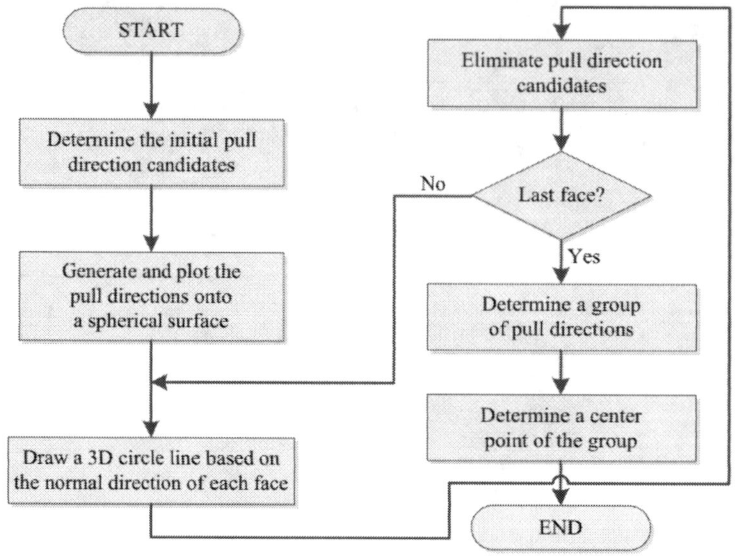

Figure 2. Flowchart of the proposed method.

normal direction, which depends on the coordinates of mesh. For a regular face, *e.g.* a plane, the normal directions are perpendicular to its plane at selected points. Figure 3 highlights an illustration of the normal directions for both of regular and irregular face. To overcome this problem, the related program is employed using C# code and OpenNX API (Application Program Interface).

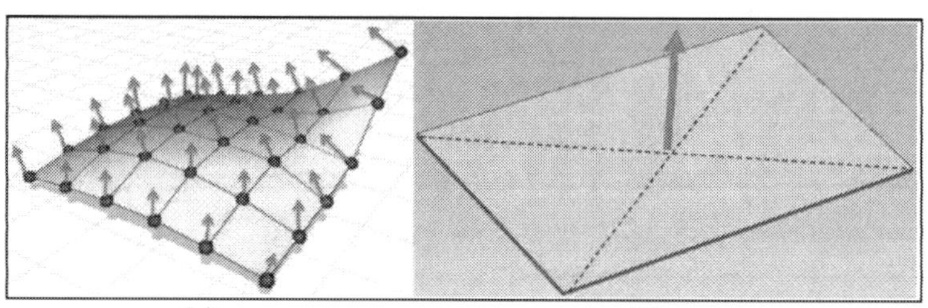

(*a*) (*b*)

Figure 3. Normal directions of (*a*) an irregular face and (*b*) a regular face.

In OpenNX, the AskFaceData function is used to calculate the normal direction for a regular face. The AskFaceData, AskFaceParm, and AskFaceProps functions are used together to find normal direction for irregular faces, as shown in the Figure 4. The normal directions of both regular and irregular faces, are then collected and used to select pull direction candidates for an undercut based on the V-map principle.

2.2 Generation and Projection of Pull Direction Candidates

Deserno [8] proposed two methods to generate and distribute points onto a spherical surface. A distributed point denotes a pull direction candidate of an undercut. The first method called the "random placement method", randomly plots the distributed points onto a spherical surface. Fluctuation and overlapping of distributed points show the weakness of this method, therefore the points are not evenly distributed, as can be seen in the area 1 and 2 in Figure 5(*a*). The second method called the "regular placement method", places each point from the other point at the same distance in two orthogonal directions, as can be seen in the Figure 5(*b*). Using this method, all of the determined points can be evenly distributed onto the spherical surface. Although the second method shows an emerging partial crystallinity problem, our preliminary results show that it gives better performance. We therefore choose the second method.

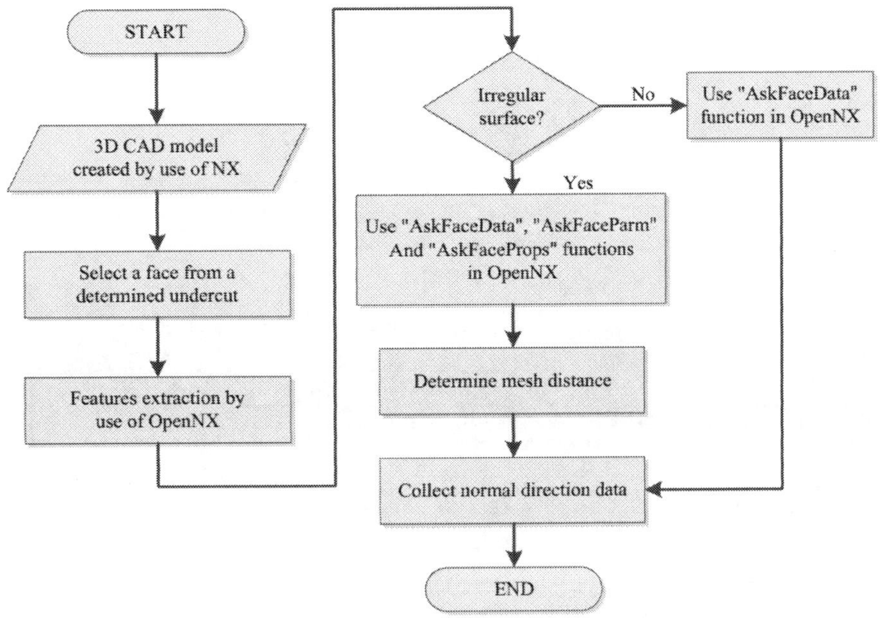

Figure 4. Flowchart to calculate normal directions of undercut's faces.

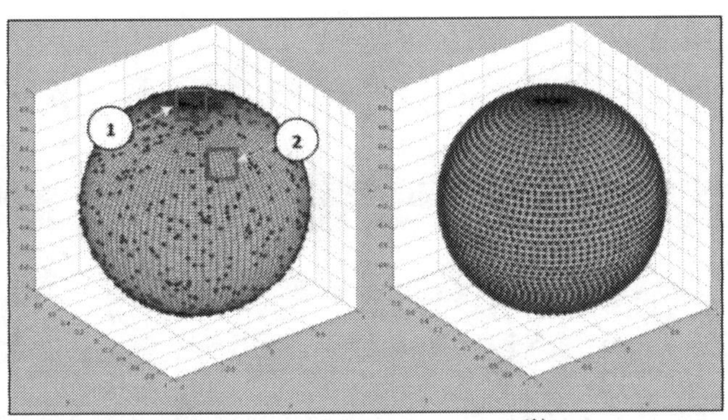

(*a*) (*b*)

Figure 5. Projection of pull direction candidates using (*a*) random placement, and (*b*) regular placement method.

2.3 Visibility Map of an Undercut Region

To illustrate the basic concept of the proposed method, a simple undercut is considered for demonstration. Figure 6 shows an undercut with 5 different regular faces, where the normal direction of each face can be seen in Table 1. Figure 7(*a*) highlights the blue points, which shows the pull direction candidates of an undercut region. The red line shows the 3D circle location, and the yellow point shows the normal direction of a face. Both of the red line and the yellow point represent the location of the face's V-map region.

The normal direction of a face is used as a reference to create a 3D circle line. It represents V-map which is used to eliminate unwanted pull direction candidates. The distance between the 3D circle line and the normal direction point is then calculated and used as a threshold to perform it. The distance between each drawing candidate and the normal direction point is calculated for the next step. If the distance is greater than the threshold, then the candidate will be removed, otherwise it will be kept for the next step. Thereby, the number of pull direction candidates will be gradually reduced, and the center point of leftover candidates is chosen as the undercut's pull direction.

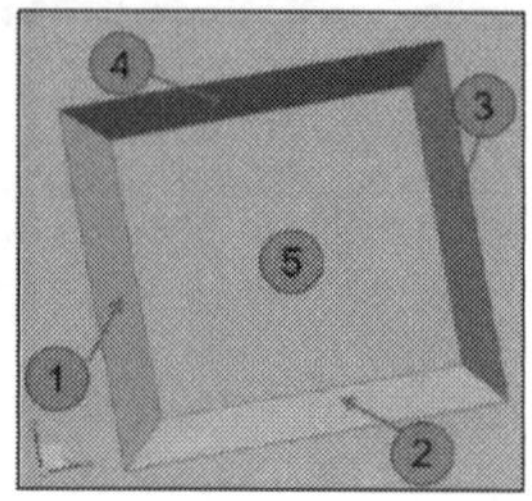

Figure 6. Example of undercut with 5 regular faces.

Table 1. Normal directions of an undercut faces in Figure 6.

Face No.	*X* normal direction	*Y* normal direction	*Z* normal direction
1	0.985	-0.174	0
2	0	-0.174	0.985
3	-0.985	-0.174	0
4	0	-0.174	-0.985
5	0	-1	0

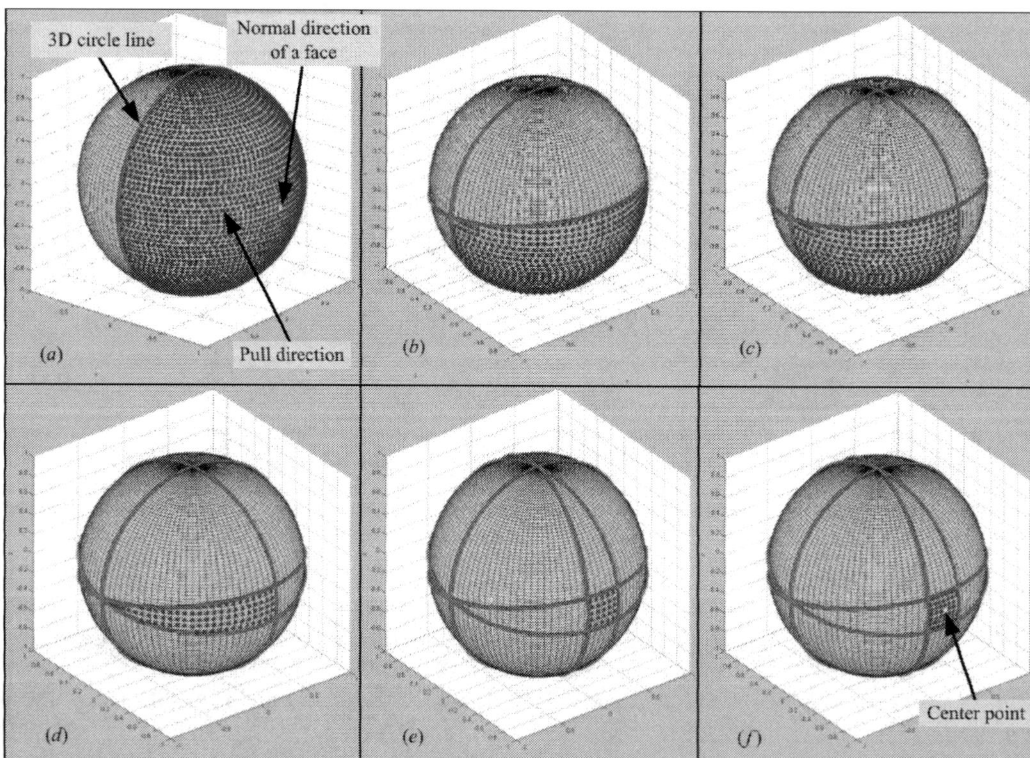

Figure 7. Step by step illustration of the proposed method.

In the demonstration, five thousand pull direction candidates are eliminated with the V-map of the first face, leaving 2,498 candidates as can be seen in the Figure 7(*a*). Similarly, Figure 7(*b*) shows the 1,272 pull direction candidates after using the V-map of the second face. Furthermore, the V-map of the third face is used to eliminate the pull direction into 658 candidates as can be seen in the Figure 7(*c*). In the next step, the V-map of the fourth face is used to reduce the pull direction into 163 candidates as we see in the Figure 7(*d*).

After elimination by the V-map of the last face, 49 pull directions are finally obtained as shown in Figure 7(*e*). These points show an intersection region of all face's V-maps based on a Boolean operation, which is formulated by the following equation:

$$Vmap = \bigcap_{i=1}^{m} Vmap_i \tag{1}$$

where m is the total number of undercut faces. Center point (0.0126 -0.9999 0) is chosen to represent 49 directions, as shown in Figure 7(*f*).

3 System Implementation

Figure 8(*a*) shows an undercut with 13 different irregular faces, and is used to illustrate the applicability and advantages of the proposed method. The mesh size is set at 4 mm x 4 mm x 4 mm, and the total number of normal direction is 103. The initial number of undercut pull direction candidates is set at 10,000 points, which are generated by a regular placement method. After elimination, the total number of drawing candidates is 839. Finally, the point [0.1542 -0.0352 0.9874] is chosen as the center point to represent 839 candidates as can be seen in the Figure 8(*b*).

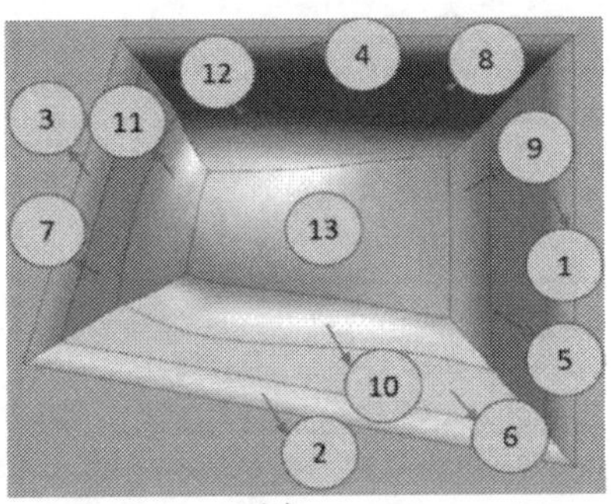

(*a*)

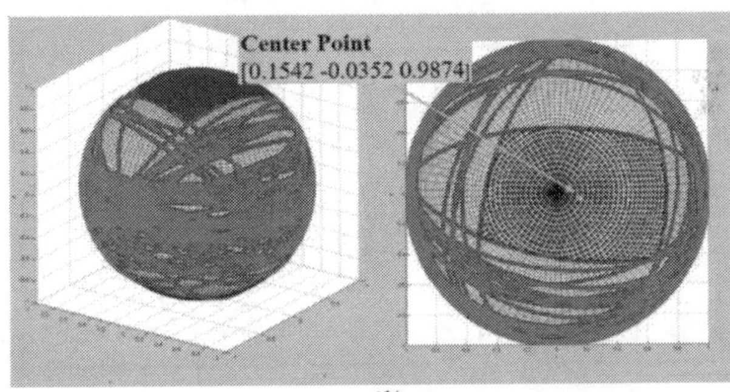

(*b*)

Figure 8. (*a*) Undercut with irregular surfaces, (*b*) pull directions (blue) bounded by 3D circle lines (red).

4 Conclusions

An algorithm for automatic determination of an undercut pull direction for both regular and irregular faces is presented in this paper based on the V-map method. The pull direction candidates are generated by a random placement method, and plotted onto a spherical surface. The normal directions of undercut faces are extracted to create V-map regions, and used to eliminate unwanted pull direction candidates. Eventually, a group of drawing candidates is selected, and a point in the center of the group is chosen to represent the pull direction for the undercut region. The pull direction becomes a vital element in the design of an injection mould.

References

1. A. Nee, M. Fu, J. Fuh, K. Lee, Y. Zhang, Ann. of the CIRP, **46** (1997).
2. A. Lin, A. Son, App. Mech. and Mat., **275-277,** 2606-2610 (2013).
3. N. Quang, A. Lin, App. Mech. and Mat., **275-277,** 2635-2639 (2013).
4. S. Jian, S. Guoqiang, ICSE, 1468-1471 (2010).
5. L. Chen, S. Chou, Jour. of Man. Sys, **14** (1995).
6. L. Chen, S. Chou, T. Woo, Comp.-Aid. Des., **25** (1993).
7. M. Liu, K. Ramani, Sol. and Phys. Mod. Symp., 367-372 (2007).
8. M. Deserno, Technical Report, Department of Chemistry and Biochemistry, UCLA, USA (2004).

MATEC Web of Conferences **82**, 01004 (2016) DOI: 10.1051/matecconf/20168201004
D2ME 2016

Drag Performance of Twist Morphing MAV Wing

N.I. Ismail[a], A.H. Zulkifli, R.J. Talib, H. Zaini and H.Yusoff

Faculty of Mechanical Engineering, Universiti Teknologi MARA, 3500 Permatang Pauh, Pulau Pinang, Malaysia

Abstract. Morphing wing is one of latest evolution found on MAV wing. However, due to few design problems such as limited MAV wing size and complicated morphing mechanism, the understanding of its aerodynamic behaviour was not fully explored. In fact, the basic drag distribution induced by a morphing MAV wing is still remained unknown. Thus, present work is carried out to compare the drag performance between a twist morphing wing with membrane and rigid MAV wing design. A quasi-static aeroelastic analysis by using the Ansys-Fluid Structure Interaction (FSI) method is utilized in current works to predict the drag performance a twist morphing MAV wing design. Based on the drag pattern study, the results exhibits that the morphing wing has a partial similarities in overall drag pattern with the baseline (membrane and rigid) wing. However, based C_D analysis, it shows that TM wing induced higher C_D magnitude (between 25% to 82% higher) than to the baseline wing. In fact, TM wing also induced the largest C_D increment (about 20% to 27%) among the wings. The visualization on vortex structure revealed that TM wing also produce larger tip vortex structure (compared to baseline wings) which presume to promote higher induce drag component and subsequently induce its higher C_D performance.

1 Introduction

Nowadays, Micro Air Vehicle (MAV) is seen as an ideal replacement for unmanned aerial vehicles (UAVs) particularly in intelligence and surveillance mission at confined space areas. With dimension lesser than 15cm, MAV is seen as small scale aircraft that can also fly at low-speed with Reynolds number regime between 10^4–10^5. In order to improve the MAV performance, the biological inspired MAV design such as morphing wing is seen as a new potential to be explored[1]. Most of these biological designs are inspired from flying characteristics of airborne mammals, birds[2] and insects[3]. However, the implementation of the morphing mobility (e.g. twist morphing wing) on such limited MAV wing size wing is very challenging. The morphing MAV wing designer have to consider the limitation in wing size, limited energy, complex morphing mechanism, and high aerodynamic-wing structural interaction[4]–[6]. Therefore, earlier research on such MAV wing had concentrated more on the its manoeuvrability rather than understanding its aerodynamic performance[7], [8]. Thus, the influence of morphing mobility on drag performances was not well explored. Based on this issue, current study is conducted to enhance the understanding on the drag performance of morphing MAV wing.

Here, a study is carried out by comparing the drag performance of a twist morphing MAV wing with a series of baseline MAV wing known as membrane and rigid MAV wing design. Twist

[a] Corresponding author : iswadi558@ppinang.uitm.edu.my

morphing (TM) mobility is chosen in current study due its simple deformation and actuation system. The study is mainly conducted based on quasi-static analysis by using the Fluid Structure Interaction (FSI) method. The result is elucidated through the comparison study on the drag performance between a twist morphing (TM), membrane and rigid wings..

2 FSI Computational Method

FSI simulation based on Ansys software is fully utilized in this work to analyze the drag performance of twist morphing, membrane and rigid wing. The combination of CFX and Static Structural module is used to solve the flow field and wing deformation, respectively. A steady state condition and incompressible turbulent flow boundary conditions is implemented to solve the airflow field. The airflow solution was obtained based on the RANS equations with Shear Stress Turbulent (SST) turbulent model. Meanwhile, Static Structural module is used to solve the linear-static structural problems on found on morphing and membrane wings. The summary of FSI simulation framework is shown in Figure 1.

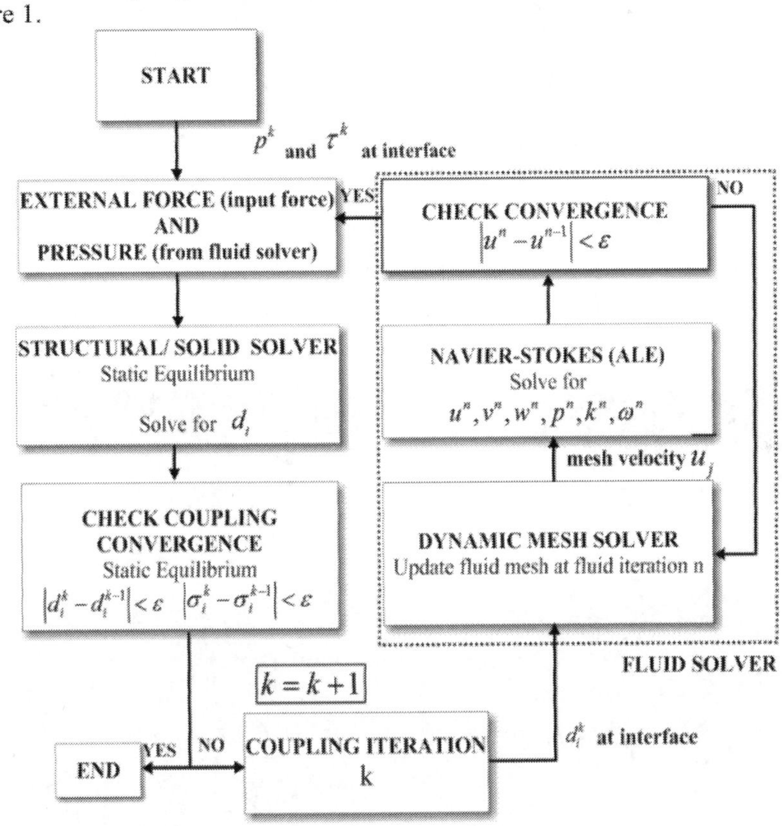

Figure 1. The summary of FSI simulation framework.

3 MAV wing model and boundary conditions

The twist morphing, membrane and rigid MAV wing are modeled based on previous research done at the University of Teknologi MARA[9], [10]. Due to symmetrical wing condition, all wing model are simplified into half wing. In general, all wing have almost indentical Zimmerman wing shape with wingspan, wing aspect ratio, root chord length and built-in wing twist for each wing is approximately 150 mm, 1.25, 150 mm and 0.6°, respectively. However, the TM wing has included the morphing force component (for twist mobility) and membrane skin (also membrane wing). Generally, the TM wing design is identical to the membrane wing design but the wing has additional morphing force component (3N) enforced at 90mm from the leading edge and parallel to wing spanwise axis. The function of morphing force is to induce the twisted mobility on the TM wing. The thickness for wing skeleton and membrane skins is constantly at 1.0mm. The material for wing structure and membrane skin is set to be Perspex[10] and silicone rubber[10], respectively. The detail structure of each wing can be found in [9] and represented in Fig. 2. The mechanism to induce the morphing mobility on the TM wing is shown in Figure 3.

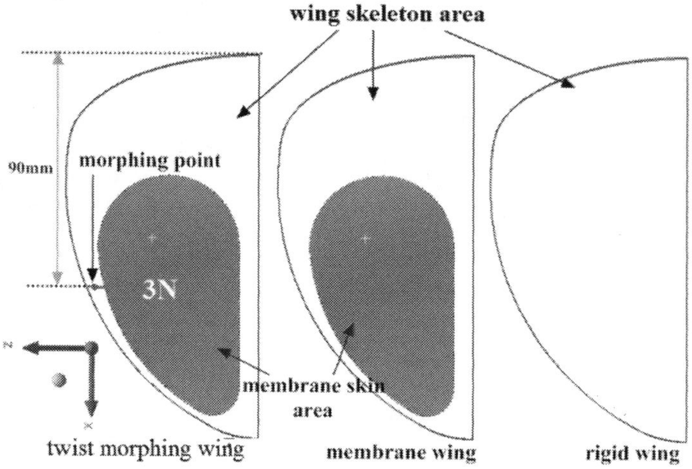

Figure 2. The TM (left), membrane (middle) and rigid wing configuration.

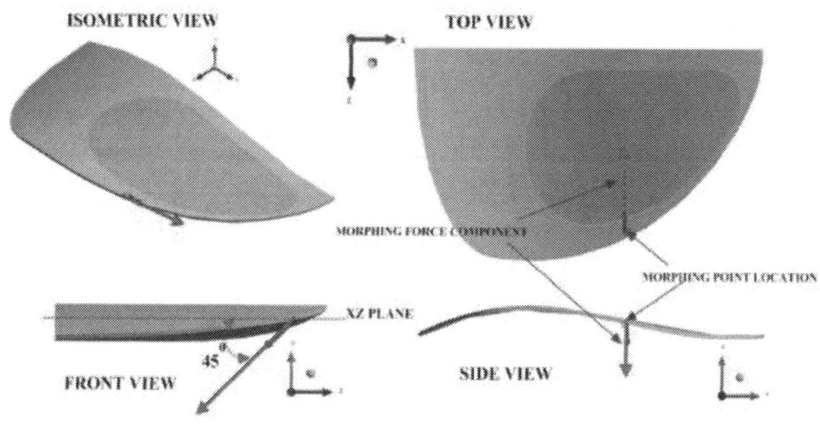

Figure 3. Mechanism principle applied on TM wing.

The size of flow domain surrounding the MAV wing is dimensioned in root chord as shown in Figure 4. The optimized number of elements for current flow domain is 1,000,000 elements with y^+ is obtained below than 1.The inlet, outlet and symmetrical wall location are shown in Figure 4. The airflow speed is imposed at the inlet and set at 9.7m/s which is equivalent to Re=100,000 at chord (define as the maximum speed for MAV). The angle of attack (AOA) increment starts form -7° to 30° with 2° interval. The turbulence intensity of 5% with automatic wall function is also imposed to solve the flow viscous effect.

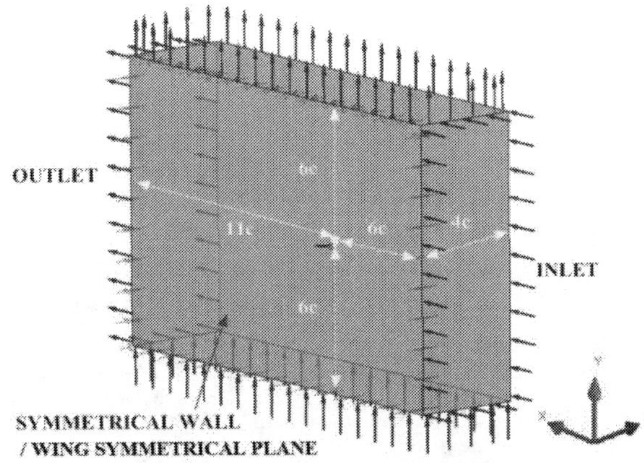

Figure 4. flow domain dimensioned in the wing root chord.

4 Drag coefficient results

The drag coefficient (C_D) results for all wings are shown in Fig. 4. The result shows that all wings had exhibited almost a constant C_D magnitude at the early AOA stages between -15° to 2°. However C_D magnitude for all wings had drastically increase after the AOA increase higher than 2°. The increment of C_D magnitude was slight dented at stall angle (between AOA= 22° to 26°) before the C_D value increase further almost linearly up to AOA=35°. In general, the C_D magnitude produced by the baseline wing is almost identical. However, the membrane wing managed to induced slightly (2% to 3%) higher C_D magnitude compared to the rigid wing.

Based on the detail analysis on the magnitude of C_D, the result shows that TM wing able to produce lower C_D magnitude (compared to baseline wing) at the early AOA stages (between -15° to -8°). At this AOA stage, TM wing exhibited about 7% to 18% lower C_D magnitude compared to baseline wing. The minimum C_D magnitude found on TM wing is 0.0363 exhibited at AOA=-6°. However, the minimum C_D magnitude for TM wing is induced at earlier angle compared to the baseline wing (AOA=-2°). As the AOA increase (AOA>-2°), the C_D value for TM wing had also significantly increase. Analytically, TM wing managed to produce about 25% to 82% higher C_D magnitude than the baseline wing at AOA between -2° to 22°. TM wing stalled at AOA= 22° with C_D = 0.557 which is about 47% higher than the baseline wing produced. The stall angle for the baseline wing delayed at AOA=26°. Based on depth C_D analysis, TM wing exhibited a drastic increment in C_D magnitude especially at AOA between -2° to 12°. At this AOA stage, the C_D magnitude for TM wing increase about 20% to 27% for every 2° AOA increment. Despite of high C_D increment found in TM wing, the baseline just induced slightly lower C_D increment than the TM wing at the same AOA range (-2° to 12°). The baseline wing exhibited about 14% to 24% of C_D increment for every 2° AOA rise. However, as the AOA increase higher than 12° (up to 22°), the increment of C_D magnitude for both TM and baseline wings had reduced below than 10%.

Based on this result, it clearly shows that TM wing had induced higher C_D distribution compared to the baseline wing particularly at every AOA cases higher than -2°. Based on previous study shown in references [9], [10], this malevolent performance is partially contributed by the enlargement of tip vortex structure induced by TM wing. Larger tip vortex structure contributed into higher induce drag component in which consequently contributed into higher drag performance[11]. Thus, a preliminary investigation on the tip vortex structure found on TM and baseline wings is carried out to justify the drag distribution findings.

Figure 6 shows the vortex structure found on TM and baseline wings at AOA= 5° and 10°. The vortex structure result is visualized based on Q criterion magnitude equal to 0.04 and the tip vortex structures are recognized based on its location that induced at the wing tip as shown in Figure 5. The airflow moves from top to the bottom of the page. Based on this figure, it clearly shows that each

 DOI: 10.1051/matecconf/20168201004

wing produce tip vortex structure that flows downstream behind the wing tips. Based on relative comparison (estimated by the relative length and diameter), the results clearly shows that TM wing induce larger tip vortex structure compared to the baseline wings. In fact, the size of vortex structure for TM (and also baseline) wing continue to increase as the AOA increase from 5° to 10°. However, the size of tip vortex structure for rigid and membrane wings are almost identical based on estimated length and diameter. Thus, the results explain the reason behind the enormous C_D distribution found on TM wing. Larger tip vortex structure found on TM wing induce its induce drag component which subsequently contribute in amplifying its overall C_D performance. On the other hand, C_D performance for rigid and membrane wings are almost identical due to the similarities (in terms of size) of tip vortex structure found on both wings.

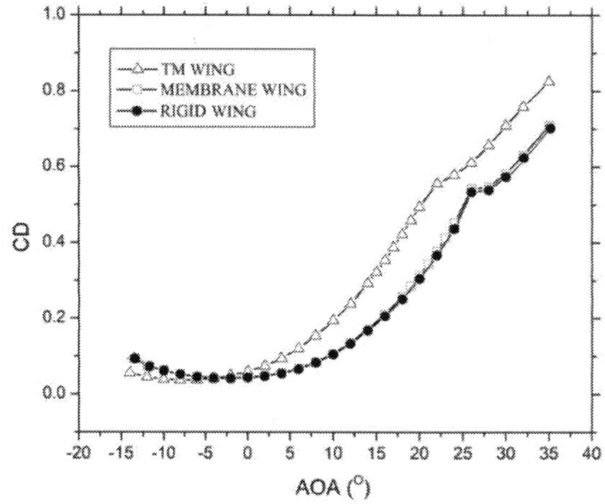

Figure 5. Drag performance for twist morphing, membrane and rigid wings.

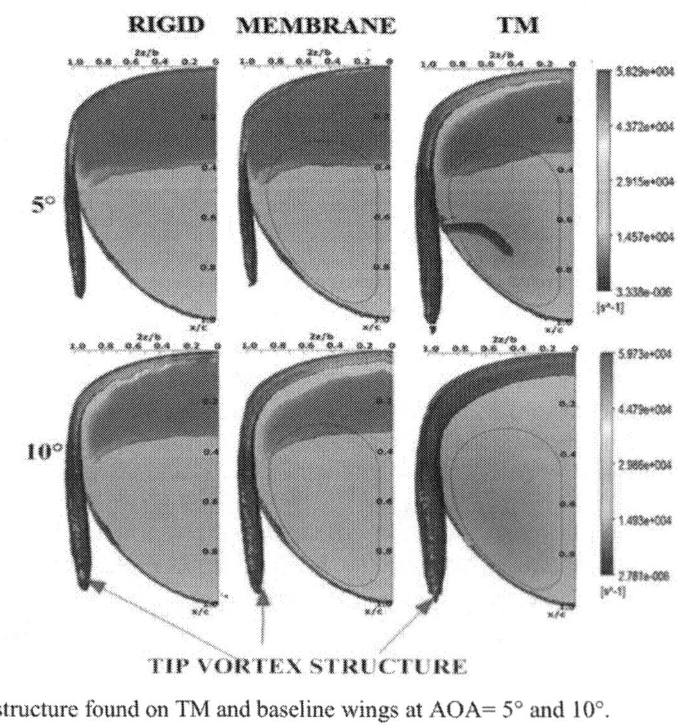

Figure 6. The vortex structure found on TM and baseline wings at AOA= 5° and 10°.

5 Conclusion

Understanding the basic drag performance of morphing MAV wing is very crucial in order to enhance its application in MAV area. Thus, current study is carried out to compare the drag performance of a twist morphing MAV wing with membrane and rigid wings. Based on initial drag pattern study, the results indicates that the wings have a partial similarities in overall drag pattern. However, based on the detail C_D analysis, the results clearly shows that TM wing induced higher C_D magnitude (between 25% to 82%) compared to the baseline wing in most AOA cases. In fact, TM wing induced the largest C_D increment among the wings (about 20% to 27%) at AOA between -2° to 12°. The visualization on vortex structure also revealed that TM wing also produce larger tip vortex structure compared to baseline wings. Larger tip vortex structure found on TM wing presume to promote higher induce drag component and subsequently induce its higher C_D performance.

Acknowledgments

The authors acknowledge technical and financial support from Universiti Teknologi MARA and the Government of Malaysia via the sponsorship by the Malaysia Ministry of Higher Education's Fundamental Research Grant Scheme (FRGS) (600-RMI/FRGS 5/3 (152/2014)).

References

1. M. Abdulrahim, H. Garcia, and R. Lind, J. Aircr., **42**, 131 (2005)
2. A. H. Supekar, *Design, Analysis and Development of A Morphable Wing Structure For Unmanned Aerial Vehicle Performance Augmentation*, , (University of Texas, 2007)
3. S. A. Combes and T. L. Daniel, J. Exp. Biol., **206**, 2989, (2003)
4. Z. Chen, *Micro Air Vehicle Design for Aerodynamic Performance and Flight Stability*, (University of Sheffield, 2013)
5. R. Banerjee and P. Seshaiyer, *Int. J. Aerosp. Light. Struct.*, **3**, 1, (2013)
6. Y. J. Abudaram, S. Rohde, J. P. Hubner, and P. Ifju, *Composite Materials and Joining Technologies for Composites*, (Springer New York, 2013)
7. R. Pecora, F. Amoroso, and L. Lecce, J. Aircr., **49**, 1666 (2012)
8. S. Vasista, L. Tong, and K. C. Wong, J. Aircr., **49**, 11 (2012)
9. N. I. Ismail, A. H. Zulkifl, M. Z. Abdullah, M. H. Basri, N. S. Abdullah, A. H. Zulkifli, M. Hisyam Basri, and N. Shah Abdullah, Chinese J. Aeronaut., **26**, 1093 (2013)
10. N. I. Ismail, a. H. Zulkifli, M. Z. Abdullah, M. H. Basri, and N. S. Abdullah, Chinese J. Aeronaut., **27**, 475 (2014)
11. B. K. Stanford, P. Ifju, R. Albertani, and W. Shyy, *Prog. Aerosp. Sci.*,**44**, 258 (2008)

Improvement of Flexible Linear Stepping Actuator Driven by Pneumatic Balloons and Brakes

Yuya Eguchi, Tetsuya Akagi [a], Shujiro Dohta and Wataru Kobayashi

Department of Intelligent Mechanical Engineering, Okayama University of Science, Ridai-cho, Kita-ku, Okayama, 700-0005, Japan

Abstract. The development of soft actuator for the rehabilitation device and power assisting devices has been required based on aging society. However, a flexible actuator that can generate both larger force and longer displacement has not been developed. It is also difficult to realize a flexible displacement sensor with long stroke while deforming its form according to the actuator's shape. In the previous study, the flexible actuator with larger force and longer stroke that can adjust its stroke by giving stepping motion using pneumatic balloons and brakes was proposed and tested. However, the slipping of the pneumatic brake prevents the larger generated force. In this paper, the improved pneumatic brake using a pneumatic balloon and a mechanical chuck is described. The fundamental performance of the actuator using the improved brake is also described.

1 Introduction

Under the situation of aging society and decreasing birth rate in Japan [1], the development of soft actuator for the rehabilitation device and power assisting devices [2-5] has been required. A rubber artificial muscle is well-known as a soft actuator. Because it generates larger force even if its mass is small. However, the muscle cannot work with longer stroke. The maximum displacement of the muscle is less than one fourth of its original length [6]. In addition, the generated force of the muscle decreases according to the increase of the displacement of the muscle. As a long stroke flexible actuator, a flexible pneumatic cylinder was proposed and tested in the previous study [7]. The cylinder can be used even if it bends. Stroke of the cylinder is able to be set longer according to user's requirement. As applications of the developed cylinder, a flexible robot arm for wrist rehabilitation using three flexible pneumatic cylinders was proposed and tested [8]. A portable rehabilitation device using a spherical actuator that consist of two flexible cylinders was also developed [9]. From these applications, the validity of flexible pneumatic cylinder for application as a rehabilitation device was confirmed [10, 11]. However, the generated force of the cylinder is not so large. It depends on sectional area of the cylinder. Therefore, it is strongly required to develop a flexible actuator that can generate both larger force and longer stroke. In addition, it is difficult to develop a flexible displacement sensor with long stroke while deforming its form along to the shape of flexible actuator. In ideal, the flexible actuator with larger force and longer stroke that can adjust its stroke without

[a] Corresponding author : akagi@are.ous.ac.jp

sensor is desired. To satisfy such a greedy demand, in the previous study, the flexible linear stepping actuator was proposed and tested [12]. It consists of two pneumatic balloons and brakes. The performance of the tested actuator was also investigated. However, the generated force of the tested actuator is not so large. Because holding force of the pneumatic brake is small. In this paper, the improvement of the brake will be described. The performance of whole actuator is also described.

2 Flexible pneumatic cylinder and previous linear stepping actuator

In our previous study, the novel flexible pneumatic cylinder as shown in Fig.1 was developed [7]. The cylinder consists of a steel ball with the outer diameter of 9 mm as a cylinder head, a flexible tube (SMC Co. Ltd., TSU1208) as a cylinder, and a slide stage. The slide stage has two brass rollers set on the inner bore of the stage to press and deform the tube. The roller has an outer diameter of 4 mm. The steel ball is held by two slide stages from both sides of the ball. The operating principle is as follows. When the supply pressure is applied to one side of the cylinder, the tube is moved forward or backward. The cylinder can move with a longer stroke of more than 10 m. It can also move even if the cylinder bends. However, compared with the theoretical generated force of 25 N at supply pressure of 500 kPa, the frictional force of the cylinder is relatively large. It is about 10 N. Therefore, the cylinder can generate the force of about 15 N under the condition that the supply pressure is 500 kPa. Generally, the generated force of the cylinder depends on the sectional area of the cylinder. However, even if the cylinder tube with a larger sectional area is used, the frictional force increases according to increasing the diameter of the cylinder. This is because the contacting area of the sliding mechanism between the inner ball and rollers is increased. In addition, it is difficult to realize a displacement sensor that can measure a longer stroke while changing its shape according to deformation of the cylinder.

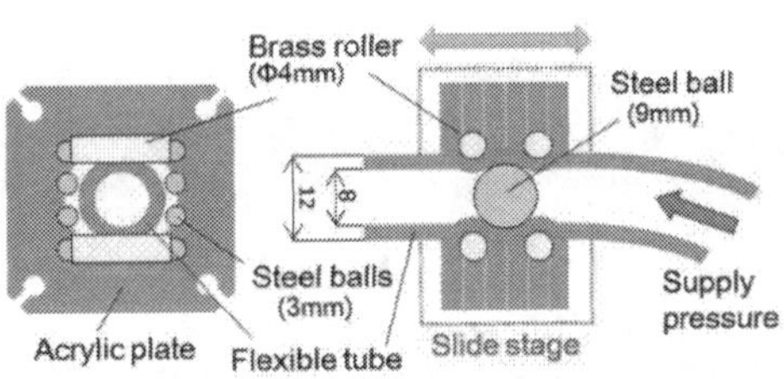

Figure 1. Construction of flexible pneumatic cylinder.

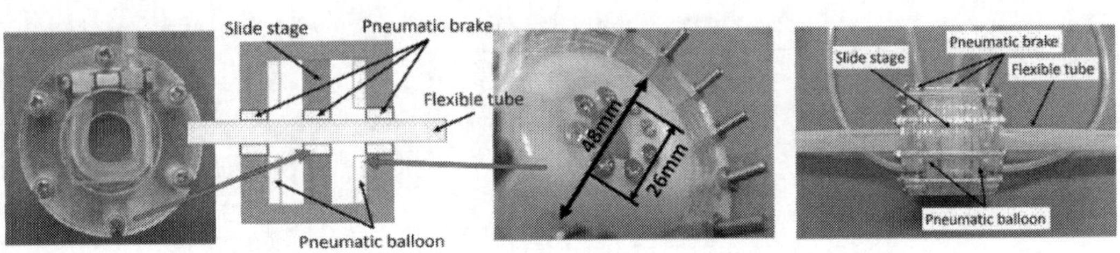

Figure 2. Construction and view of tested flexible linear stepping actuator.

Therefore, in the previous study, a novel pneumatic actuator that can generate larger force and work with a longer stroke was proposed [8]. It can also adjust the displacement without sensor by giving the stepping motion. We called it "a flexible linear stepping actuator". Figure 2 shows the construction and view of the tested actuator. The actuator consists of two doughnut-shaped balloons such as a diaphragm, two pneumatic driven brakes set on both sides of actuator, a moving round stage with a pneumatic driven brake and a flexible tube as a rod. The pneumatic brake consists of silicone rubber tube with outer diameter of 4 mm surrounded by an acrylic cover. The operating principle of the brake is as follows. When the supply pressure is applied to the silicone rubber tube in the brake, the tube in

the brake expands. By using the acrylic cover, the rubber tube expands toward the center of the flexible tube and it holds the flexible tube. The moving stage with a brake located on the center of the actuator is sandwiched by two balloons. The doughnut-shaped balloon that consists of a silicone rubber film with thickness of 0.5 mm has the outer diameter of 48 mm and the inner diameter of 26 mm. The theoretical generated force calculated from the sectional area of the balloon is about 639 N when the supply pressure of 500 kPa is applied. The size of the actuator is not so large. It has the length of 50 mm and the outer diameter of 60 mm. The mass of the actuator without flexible tube (rod) is about 180 g.
The operating principle of the tested actuator is shown in Fig.3. The operating principle is as follows: First, as shown in Fig.3 (1), the both side brakes are driven for 0.2 s in order to hold the tube. In the condition, the left side balloon is driven for 0.15 s and the moving stage moves toward right as shown in Fig.3 (2). When the moving stage reaches at the right side wall by giving the input pressure for 0.15 s, the brake on the stage is activated to hold the tube as shown in Fig.3 (3). After that, while the left side balloon and both side brakes being released, the right side balloon is driven for 0.15 s. So the brake on the moving stage keeps holding the tube, the flexible tube is pushed toward left as shown in Fig.3 (4). By repeating this procedure, the tube can move toward left every a certain stroke. The stroke is as same as the width of the balloon chamber; that is 5mm.

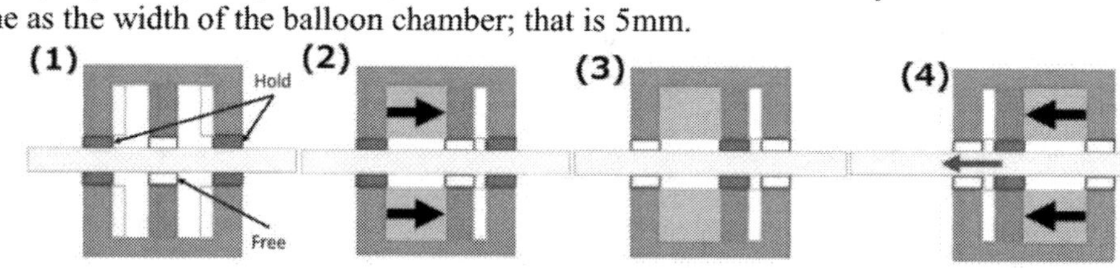

Figure 3. Operating principle of tested flexible linear stepping actuator.

3 Control system and experimental result of previous actuator

Figure 4 on the left shows the schematic diagram of the control system of the tested actuator using an embedded controller. Figure 4 on the right shows the time chart of the driving valves. The system consists of the tested actuator, the embedded controller (Renesas Co. Ltd., SH7125) and four on/off control valves (Koganei Co. Ltd., G010E1). In the system, valves 1 and 2 are used to drive both side balloons, respectively. Valve 3 can drive the brake on the moving stage. Valve 4 is used to drive both side brakes. These valves are driven and controlled by the embedded controller through transistors.

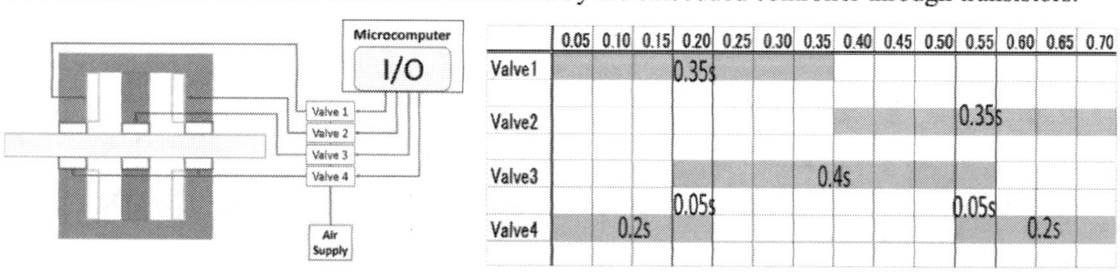

Figure 4. Schematic diagram of control system of tested linear stepping actuator and time chart of driving valves.

Figure 5 shows the transient view of the movement of the tested actuator when the moving pattern of the brakes and balloons are given as shown in Fig. 4 on the right. From the time chart of driving valves, it can be expected that the actuator can push the tube with a step stroke of 5 mm every 0.7 s. This means the moving speed of the actuator is 7.14 mm/s. From the experiment as shown in Fig.5, the average speed of the tested actuator was about 5.4 mm/s. In the experiment, the generated force of the actuator with this moving pattern was also investigated. As a result, the maximum generated force is about 28N. This value is much smaller than the theoretical value using the area of the balloon. It is caused that the total generated force of the actuator depends on the holding force of the pneumatic brake and the generated force of the balloon. In this actuator, as the generated force of the balloon is too enough, the net generated force is equal to holding force of the brake. The theoretical holding force of about 71 N for radial direction of the brake is calculated from the contacting area of the silicone rubber tube in the brake and the supply pressure of 500 kPa. Therefore, with consideration of slip and frictional condition between brake and tube, the generated force of the actuator is less than 71 N. However, this holding force of the brake is too small. Therefore, it is necessary to improve the holding force of the brake.

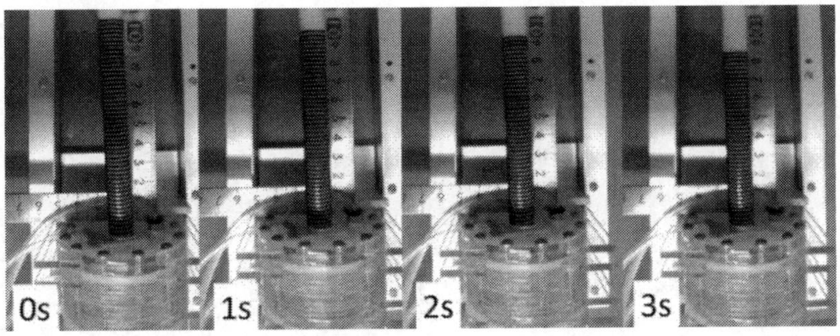

Figure 5. Transient view of movement of the tested actuator.

4 Improved pneumatic brake and linear stepping actuator

Figure 6 shows the schematic diagram of the improved pneumatic brake in order to increase the holding force. The improved pneumatic brake consists of a doughnut-shaped balloon with the inner diameter of 24 mm and the outer diameter of 36 mm, a ring-shaped plastic plate and a mechanical chuck. The mechanical chuck is made of 8 plastic claws set on the stage with the inclined angle of 45 degrees. In the tested chuck, two types of claws were used. One is the claw with width of 6 mm. Another is the claw with width of 3 mm. These two types of claws were used, because the generated braking force was larger than the case using the same size of claws from the preliminary experimental result. In addition, number of the claws was also investigated such as 4, 8 and 16 claws. As a result of these preliminary experimental results, it can be found that the alternative arrangement of two types of claw around the flexible tube is suitable to increase the braking force. The inclined angle of 45 deg. was calculated as an optimal value theoretically. In the case when the generated force F of the balloon for the brake is applied to the mechanical chuck, the holing force F_b obtained by the surrounding pushing force is given as following equation.

$$F_b = \frac{1}{2} F \sin 2\alpha \qquad (1)$$

Where, α means the inclined angle of the braking stage. From Eq.1, the maximum holding force is given in the case that the inclined angle is 45 deg. Then, from the sectional area and supplied pressure of 500 kPa, the calculated generated force of the balloon in the brake is about 283 N. Therefore, the maximum generated force of the brake is estimated as less than 141 N. The operating principle of the brake is as follows. When the supply pressure is applied to the balloon, the balloon expands and pushes the ring-shaped plate. The plate pushes eight claws. The claws slide on the inclined stage and hold the flexible tube. By using the balloon with relatively larger sectional area, the holding force will be increased compared with the previous brake.

Figure 7 shows the schematic diagram of the improved linear stepping actuator using the improved pneumatic brake. The fundamental construction of the actuator is almost same as the previous one. Compared with the previous actuator, the actuator has two improved brakes. One is installed between two balloons as a slide stage. Another is set on the outside of the sliding mechanism that consists of two balloons and the slide stage. The size of the actuator without flexible tube becomes larger than previous one. The whole length of the actuator is 102 mm and the outer diameter is same, that is 60 mm. The total mass is about 360g.

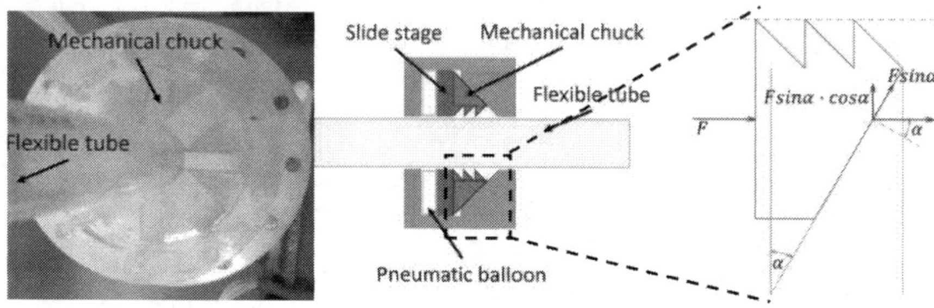

Figure 6. Schematic diagram of improved pneumatic brake using balloon and mechanical chuck.

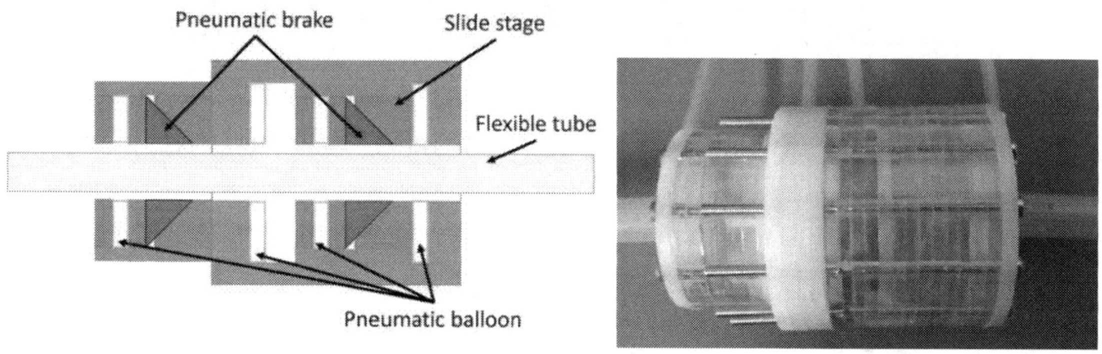

Figure 7. Schematic diagram and view of improved linear stepping actuator.

Figure 8 shows the transient view of movement of the tested actuator when it is lifting up the load of 30N. In the experiment, the same driving pattern of valves as shown in Fig.4 was used. From Fig.8, it can be seen that the actuator was lifting up the load of 30 N with average speed of 5 mm/s. In addition, from the result of the static generated force, it can be found that the maximum generated force of 90 N can be obtained. This force is about three times larger than that of previous actuator.

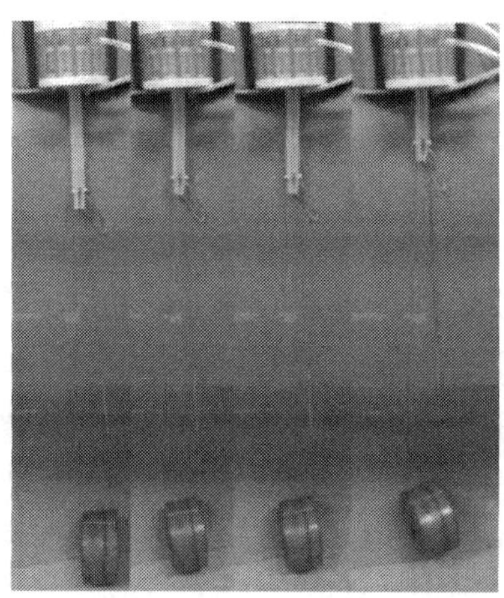

Figure 8. Transient view of the improved actuator when it is lifting the load of 30 N.

5 Conclusions

As a flexible actuator with larger force and longer stroke, the previous developed flexible linear stepping actuator driven by pneumatic balloons and brakes was introduced. In order to increase the generated force of the actuator, the improved pneumatic brake that consists of the balloon and mechanical chuck. The linear stepping actuator using the improved brake was also constructed. The moving test lifting up the constant load was carried out. As a result, it can be found that the tested actuator can lift the load of 30 N with speed of 5 mm/s in the experiment. It can be confirmed that the actuator is useful to apply as a rehabilitation device because of larger generated force and constant step stroke of 5 mm.

Acknowledgements

This research was supported by the Ministry of Education, Culture, Sports, Science and Technology of Japan through a Financial Assistance for QOL Innovative Research (2012-2016) and JSPS KAKENHI Grant Number 16K060202.

References

1. Ministry of Internal Affairs and Communications, Statistics Bureau, Statistics, Population Estimates, Result of the Population Estimates, Monthly Report. Result of the Population Estimates (online), Available: http://www.stat.go.jp/english/data/jinsui/tsuki/ (2015)
2. M. Ishii, K. Yamamoto, K. Hyodo, Stand-Alone Wearable Power Assist Suit -Development and Availability-. *Journal of Robotics and Mechatronics*, Vol.**17**, No.5, 575–583 (2005)
3. J. Piquion, A. Nayar, A. Ghazaryan, R. Papann, W. Klimek, and R. Laroia, Robot-assisted gynecological surgery in a community setting. *Journal of Robotics and Surgery*, Vol.**3**, Issue 2, 61–64 (2009)
4. T. Noritsugu, M. Takaiwa, D. Sasaki, Development of Power Assist Wear Using Pneumatic Rubber Artificial Muscles, *Journal of Robotics and Mechatronics*, Vol. **21**, No. 5, 607-613(2009)
5. H. Kobayashi, T. Shiban, Y. Ishida, Realization of all 7 motions for the upper limb by a muscle suit, *Journal of Robotics and Mechatronics*, Vol. 16, 504-512(2004)
6. Y.Nagata ed., Soft Actuators –Forefront of Development-, NTS Ltd.(2004) (in Japanese)
7. T. Akagi, S. Dohta, Development of a Rodless Type Flexible Pneumatic Cylinder and Its Application, *Transactions on Robotics and Automation of the JSME (C)*, Vol. **73**, No. 731, 2108–2114(2007) (in Japanese)
8. Mohd Aliff, S. Dohta, T.Akagi, H. Li, Development of a simple-structured pneumatic robot arm and its control using low-cost embedded controller, *Journal of Procedia Engineering*, Vol.**41** 134 – 142(2012)
9. S. Dohta, T. Akagi, C. Liu and A. Ando, Development and Control of Flexible Spherical Actuator Using Flexible Pneumatic Cylinders, *International Journal of Advanced Mechatronic Systems*, Vol. **5**, No.3,184- 192(2013)
10. Mohd Aliff, S. Dohta, T. Akagi, T. Morimoto, Control of Flexible Pneumatic Robot Arm Using Master Device with Pneumatic Brake Mechanism, *JFPS International Journal of Fluid Power System*, Vol.**8**, No.1, 38-43(2015)
11. Y. Matsui, T. Akagi and S. Dohta, Development and Control of Flexible Spherical Actuator for Portable Rehabilitation Device, *International Journal of Materials Science and Engineering*, Vol.**3**, No.1, 49-54(2015)
12. Y. Eguchi, T. Akagi , S, Dohta, Preliminary Design of Flexible Linear Stepping Actuator Driven by Pneumatic Balloons and Brakes, *MATEC Web of Conferences*, Vol.**51**,No.02004, 1-4(2016)

MATEC Web of Conferences **82**, 01006 (2016) DOI: 10.1051/matecconf/20168201006
D2ME 2016

Miniaturization of a Quasi-Servo Valve and Its Application to Positon Control of a Rubber Artificial Muscle with Built-in Sensor

Yoshinori Moriwake[1], Shujiro Dohta[1,a], Tetsuya Akagi[1] and So Shimooka[1]

[1]*Okayama University of Science, 1-1, Ridai-cho, Kita-ku, Okayama, Japan*

Abstract. Nowadays, the care and welfare pneumatic devices to support a nursing care and a self-reliance of the elderly and the disabled are actively researched and developed by many researchers. These wearable devices require many actuators and control valves for multi degrees of freedom. The total weight and volume of the wearable devices increases according to the degree of freedom. Our final goal is to develop a compact wearable actuator with built-in sensor, controller and control valve and to apply it to a wearable assisted device. In our previous study, a small-sized quasi-servo valve which consists of two on/off control valves and an embedded controller was developed. In this study, the quasi-servo valve composing of much smaller-sized (40% in mass, 42% in volume) on/off valves is proposed and tested. In addition, the rubber artificial muscle with an ultrasonic sensor as a built-in displacement sensor is proposed and a position control of the muscle is carried out using the tested tiny valve and built-in sensor. As a result, it was confirmed that the position control of the muscle can be realized using the tested ultrasonic sensor.

1 Introduction

Today, the care and welfare pneumatic devices to support a nursing care and a self-reliance of the elderly and the disabled are actively researched and developed by many researchers [1] to [5]. These wearable devices require many actuators and control valves for multi degrees of freedom and precise control performance of the device. The total weight and volume of the wearable devices increases according to the degree of freedom. Therefore, we aim to develop a compact and low-cost servo valve and a wearable actuator built-in sensor. In our previous studies [6] to [8], a small-sized, light-weight and inexpensive pressure control type quasi-servo valve which consists of two on/off control valves using a low-cost embedded controller and a pressure transducer was proposed and tested. In this paper, a quasi-servo valve by using the much smaller, lighter and inexpensive on/off control valves is produced and tested. In addition, as an application of the tested valve, a position control of a rubber artificial muscle is carried out by using the tested quasi-servo valve and an ultrasonic sensor as a displacement sensor. The ultrasonic sensor is installed in the rubber artificial muscle to develop a compact wearable actuator with built-in sensor. The control performance of the muscle is shown and the usefulness of the tested built-in sensor is discussed.

[a] Corresponding author : dohta@are.ous.ac.jp

2 Quasi-Servo Valve

Figures 1 and 2 show the schematic diagram and the view of the quasi-servo valve, respectively. The valve consists of two on/off control valves (SMC Co. Ltd., S070C-SDG-32) whose both output ports are connected to each other. One valve is used as a switching valve to supply or exhaust, and the other is used as a PWM control valve that can adjust output flow rate like a variable fluid resistance [6]. The size of the on/off valve is 12×33×7mm, and the mass is only 6g. The previous on/off valve (Koganei Co. Ltd., G010HE-1) was 33×20×10mm, and the mass was 15g [6]. The proposed valve is 58% smaller and 60% lighter than the previous valve. The maximum output flow rate is 15 liter/min at the supply pressure of 500kPa.

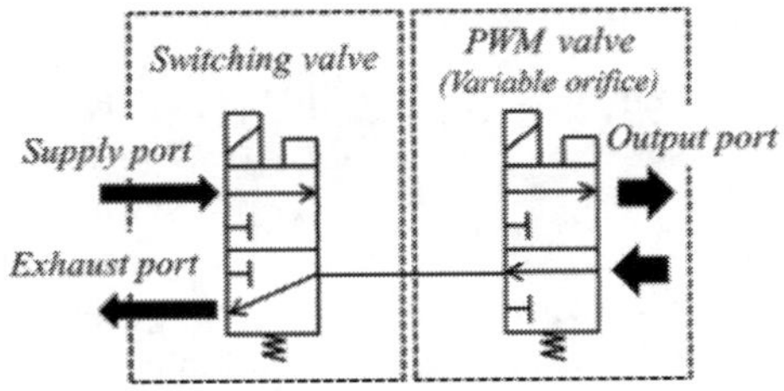

Figure 1. Schematic diagram of quasi-servo valve.

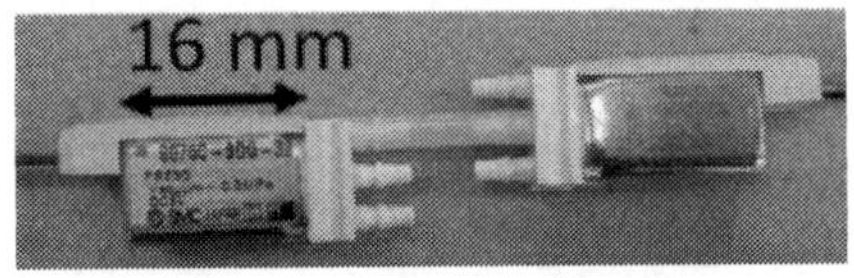

Figure 2. View of quasi-servo valve.

3 Artificial Rubber Muscle with Built-In Displacement Sensor

As a built-in displacement sensor of the rubber artificial muscle, an ultrasonic sensor was selected, because the ultrasonic sensor has many advantages that it can measure the displacement by non-contacting and is a low-cost sensor. Figures 3 and 4 show a schematic diagram of the system and a view of a rubber artificial muscle (FESTO Co. Ltd., MXAM-10-AA) with a built-in sensor and an embedded controller, respectively. The original length of rubber artificial muscle is 250 mm and the inner diameter is 10 mm. The stroke of the muscle is about 50mm in the case when the supply pressure of 500 kPa is applied. The tested ultrasonic sensor is rebuilt by using the sensor on the market (Parallax Inc. Ltd., 28015). The transmitter and receiver on the sensor are changed smaller ones as shown in Fig.8 on the right side (the transmitter: Nippon Ceramic Co. Ltd., T4008A1) and on the left side (the receiver: Nippon Ceramic Co. Ltd., R4008A1), respectively. The measuring principle is well known as follows. The distance is estimated by the time until the ultrasonic wave reaches at the receiver from the transmitter. This time is counted by an embedded controller (Renesas Co. Ltd., SH7125). The distance between the transmitter and the receiver can be obtained by using the velocity of sound and the time interval; counting value.

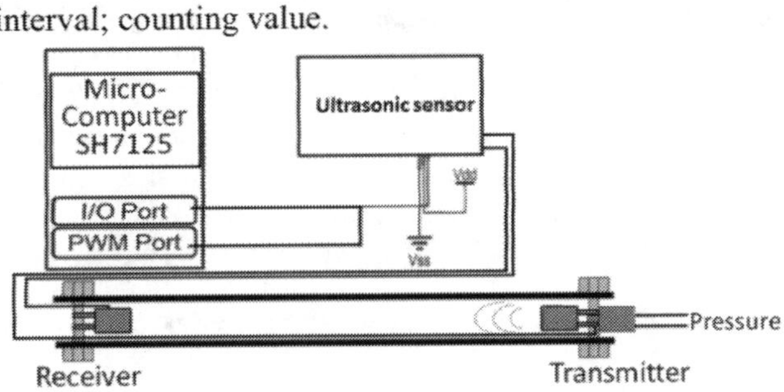

Figure 3. Schematic diagram of artificial muscle with built-in sensor.

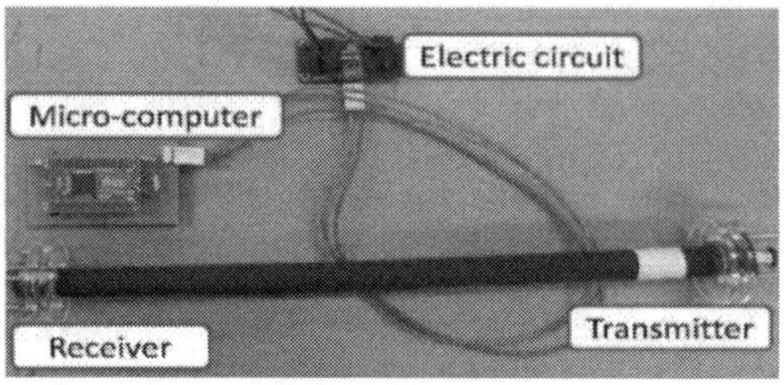

Figure 4. View of rubber artificial muscle with built-in sensor.

Figure 5 shows the relation between the counting value and the measured length of the rubber artificial muscle. The counting value was obtained by the tested built-in ultrasonic sensor and the measured length was obtained by the potentiometer. In the experiment, the artificial muscle was pressurized from 200 to 500 kPa every 50 kPa. From Fig.5, it can be seen that the relation between the sensor output and the length of the muscle is linear even if the rubber muscle is pressurized. This result means that the tested ultrasonic sensor is useful as a built-in displacement sensor of the rubber artificial muscle. From the result, the relation between the length L_m[mm] and the counting value C_m[-] can be expressed by the following equation.

$$L_m = 0.197 C_m + 129 \tag{1}$$

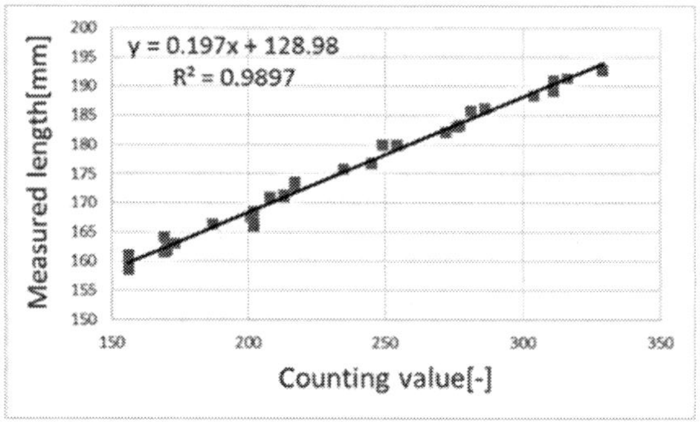

Figure 5. Relation between applied pressure and length of muscle.

4 Position Control of Artificial Rubber Muscle with Built-In Sensor Using Quasi-Servo Valve

Figures 6 and 7 show the schematic diagram and the view of the position control system of the tested rubber artificial muscle with built-in sensor, respectively. The control system consists of the tested quasi-servo valve, the tested rubber artificial muscle and an embedded controller. In addition, the potentiometer is connected with the tested muscle in order to measure the true displacement as a monitor. The position control is done as follows. First, the embedded controller counts the time of the output signal from the tested ultrasonic sensor through I/O port. The embedded controller also calculates the length of the muscle from the counted value mentioned above. The error between the measured displacement and the reference displacement is calculated by the controller and the control input for the quasi-servo valve is also calculated based on a control scheme. Finally, the control input is applied to the quasi-servo valve as PWM signal and on/off signal for two on/off valve in quasi-servo valve.

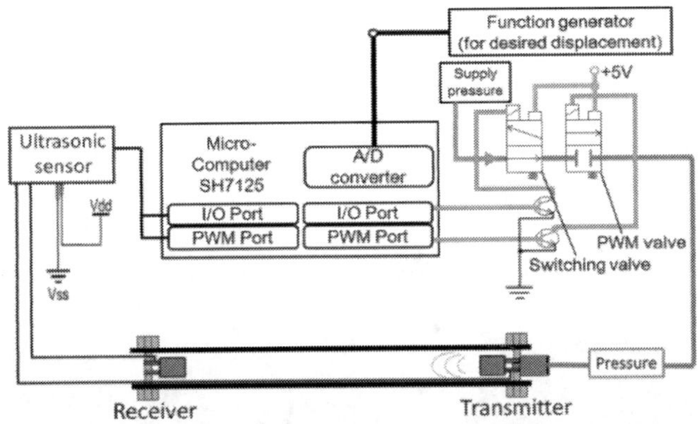

Figure 6. Schematic diagram of position control system.

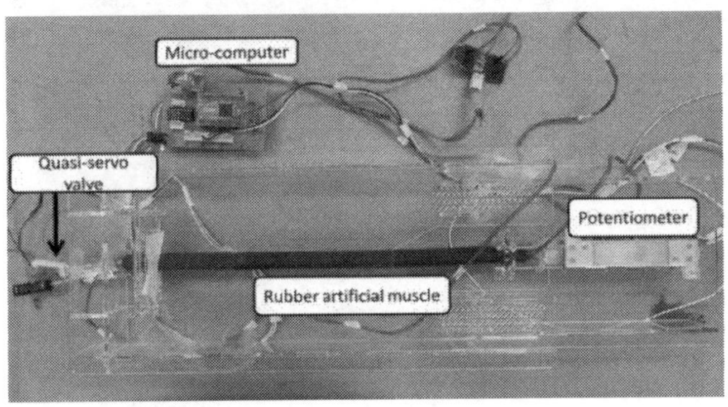

Figure 7. View of the position control system.

Figure 8 shows the transient response of displacement of the rubber artificial muscle with built-in displacement sensor. In the experiment, the desired sinusoidal position with the offset of 25mm, the amplitude of 10mm and the frequency of 0.1Hz was applied to the control system as a reference. As a feedback displacement signal, the output signal from the potentiometer was used. The supply pressure is 500kPa. PD control scheme expressed by following equations was used.

$$u_d = K_p e_c + K_d \Delta e_c \tag{2}$$

$$u = |u_d| + 20 \tag{3}$$

Where u_d, K_p (= 65 %/mm), e_c, K_d (= 35 %/mm), Δe_c and u mean the control input, the proportional gain, the error from the reference, the derivative gain, difference of error and input duty ratio to the PWM valve, respectively. As mentioned above, the switching valve switches to supply if $u_d > 0$, and to exhaust in other case. In Fig.8, the blue, red and green lines show the reference displacement, the true displacement measured by potentiometer and the measured displacement by using ultrasonic sensor, respectively. From Fig.8, it can be observed that the measured displacement by a potentiometer can track the reference displacement and that the measured displacement using the ultrasonic sensor is a little larger than the true displacement while the displacement is increasing: the air is being supplied. This is because the velocity of the sound wave is increased by the air supply as shown in Fig.6, and the length of the muscle is estimated shorter and consequently the contracted displacement becomes larger. Figure 9 shows the transient response of the displacement of the rubber muscle using the signal from ultrasonic sensor as a feedback signal. From Fig.9, it can be found that the measured displacement using ultrasonic sensor can track well the reference displacement. However, the displacement is a little larger than the true displacement while the displacement is increasing and the displacement is much larger than the true displacement when the displacement is decreasing: the air is being exhausted. This phenomenon can be explained as same as mentioned above. But, its difference is larger than the case when the air is supplied. The investigation for this phenomenon and its compensation for the position control are our future works. However, it is concluded that the tested ultrasonic sensor can be used as a built-in displacement sensor for the rubber artificial muscle.

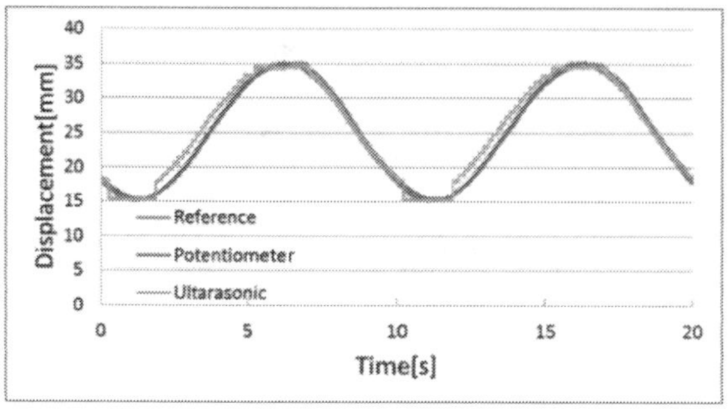

Figure 8. Position control using the potentiometer as a feedback signal.

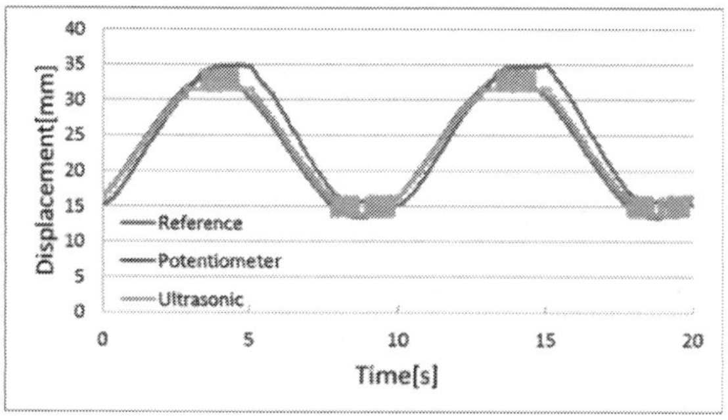

Figure 9. Position control using the ultrasonic sensor as a feedback signal.

5 Conclusions

This study can be summarized as follows. The new small-sized quasi-servo valve using a tiny on/off valve which is 58% smaller and 60% lighter than previous one was produced and tested. In addition, the rubber artificial muscle using the built-in ultrasonic sensor was proposed and tested. The measuring system of the tested sensor using the embedded controller was also proposed and tested. In order to confirm the effectiveness of the tested sensor, the relation between the counting value in the tested sensor and true displacement measured by the potentiometer was investigated. And the position control system that consists of the rubber artificial muscle with built-in displacement sensor, the small-sized quasi-servo valve and an embedded controller was proposed and tested. As a result, it was confirmed that the ultrasonic sensor could be used as a built-in sensor of the rubber artificial muscle.

Acknowledgements

This research was supported by the Ministry of Education, Culture, Sports, Science and Technology of Japan through a Financial Assistance Program for QOL Innovative Research (2012-2016).

References

1. T. Noritsugu, M. Takaiwa and D. Sasaki, *Development of Power Assist Wear Using Pneumatic Rubber Artificial Muscles*, Journal of Robotics and Mechatronics, **21**, 607-613 (2009)
2. N. Hogan and H. I. Krebs, *Interactive robots for neuro-rehabilitation*, Restorative Neurology and Neuroscience, **22**, 349-358 (2004)
3. H. Kobayashi, T. Shiban and Y. Ishida,*Realization of all 7 motions for the upper limb by a muscle suit*, Journal of Robotics and Mechatronics, **16**, 504-512 (2004)
4. M. Ishii, K. Yamamoto and K. Hyodo, *Stand-Alone Wearable Power Suit-Development and Availability*, Journal of Robotics and Mechatronics, **17, 5**, 575-583 (2005)
5. Y. Nagata ed., *Soft Actuators -Forefront of Development*, NTS Ltd., 291-335 (2004).
6. F. Zhao,S. Dohta and T. Akagi, *Development and Analysis of Small-sized Quasi-servo valves for Flexible Bending Actuator*, Tansactions of the JSME (C), **76, 772**, 3665-3671, (2010)
7. Y. Moriwake, T. Akagi, S. Dohta and F. Zhao, *Development of low-cost pressure control type quasi-servo valve using embedded controller*, Journal of Procedia Engineering, **41**, 493-500 (2012)
8. S. Shimooka, S. Dohta, T. Akagi, Y. Moriwake and F. Zhao, *Estimation of Pressure Control Performance in Low-Cost Quasi-Servo Valve Using Embedded Controller*, Lecture Notes in Electrical Engineering 293, Springer, **1**, 359-366 (2014)

MATEC Web of Conferences **82**, 01007 (2016) DOI: 10.1051/matecconf/20168201007
D2ME 2016

Vibration Analysis of Suspension Cable with Attached Masses by Non-linear Spline Function Method

Jian Qin[a], Liang Qiao, Jun Chen, Jiancheng Wan and Yongjun Xia

China Electric Power Research Institute, 100055, Beijing, China

Abstract. The nonlinear strain and stress expressions of suspension cable are established from the basic condition of suspension structure on the Lagrange coordinates and the equilibrium equation of the suspension structure is obtained. The dynamics equations of motion of the suspended cable with attached masses are proposed according to the virtual work principle. Using the spline function as interpolation functions of displacement and spatial position, the spline function method of dynamics equation of suspension cable is formed in which the stiffness matrix is expressed by spline function, and the solution method of stiffness matrix, matrix assembly method based on spline integral, is put forwards which can save cost time efficiency. The vibration frequency of the suspension cable is calculated with different attached masses, which provides theoretical basis for valuing of safety coefficient of the bearing cable of the cableway.

1 Preface

Cable structures play an important role in many engineering fields, such as civil, bridges, forestry and electrical engineering, and their application is more and more wide with different structural variety[1,2]. Therefore, it is of interest to investigate the dynamics of suspended cables and cable structures. In recent years, numerous journal articles have addressed the nonlinear dynamics of suspended cables. In addition, some studies have focused on the vibration analysis of suspended cables with attached masses, which make the singularities in the deformed cable profile[3].

On the other hand, the moving mass problem of distributed parameter systems is a very important topic in structural dynamics, and the response of structures subjected to a moving mass is of considerable practical importance in different engineering fields[4-6].

The selection of the displacement function is key for calculation method[7-8]. For the one dimensional cable structure, the spline function can be used as efficient interpolation function which has strong continuity and high accuracy[9-11].

In this study, the dynamic model of the suspended cable with attached masses is developed based on the virtual work principle. The stiffness matrix composed by spline function for the cable structure is obtained.

2 Continuum model of suspension cable

[a] Corresponding author: qinjian@epri.sgcc.com.cn

DOI: 10.1051/matecconf/20168201007

2.1 The strain of cable

There is a suspension element under fixed coordinate system, the starting point is X, the endpoint is $X+dX$, the length of the element is ds. Here $ds^2=(X+dX-X)^2=dX^2$. Under the action of external force, the infinitesimal element moves to the new position, the starting point becomes $x=X+dU$, the endpoint becomes $x+dX+dU$, and the length of element becomes ds^*, $ds^{*2}=(x+dX+dU-x)^2=(dX+dU)^2$. The process is shown as Figure 1.

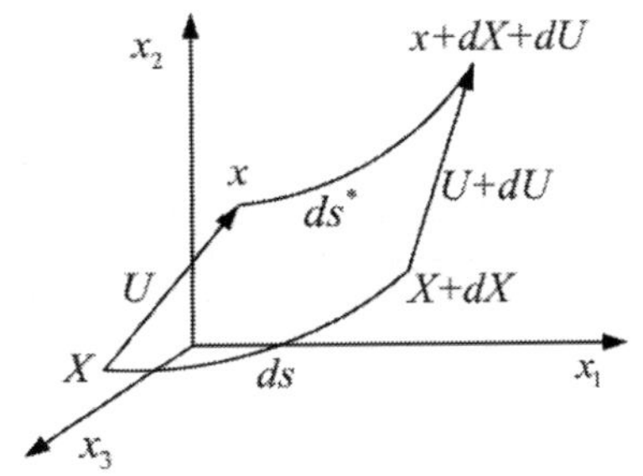

Figure 1. The Infinitesimal Element of Cable.

The transformation from ds to dx can be expressed as

$$dx = \frac{dx}{ds}ds = \frac{d(X+U)}{ds}ds = (\frac{dX}{ds}+\frac{dU}{ds})ds \tag{1}$$

The length changes of infinitesimal element is expressed using the initial coordinates s, i.e., Lagrange coordinates as

$$ds^{*2} - ds^2 = (dX+dU)^2 - dX^2 = 2(dXdU + \frac{1}{2}dUdU) = 2(\frac{dX}{ds}\frac{dU}{ds}+\frac{1}{2}\frac{dU}{ds}\frac{dU}{ds})dsds \tag{2}$$

So the Green strain of the suspension cable is

$$\varepsilon = \frac{1}{2}\frac{ds^{*2}-ds^2}{ds^2} = \frac{dX}{ds}\frac{dU}{ds}+\frac{1}{2}\frac{dU}{ds}\frac{dU}{ds} \tag{3}$$

2.2 Kirchhoff stress of suspension cable

The tension T^* in the current configuration (ds^*) is the true force, which can be seen as the Euler stress tensor (also known as the Cauchy stress tensor).

The Green strain is in the initial configuration (ds), so the stress tensor should be consistent. We move T^* to the endpoint of initial configuration, and transform the direction of T^* according to the coordinates of the endpoint. Then the Kirchhoff stress T in the initial configuration is obtained.

$$T = (\frac{dX}{ds}+\frac{dU}{ds})T^* \tag{4}$$

2.3 The equilibrium equation of suspension cable

The equilibrium equation of the suspension structure is established in the initial configuration.

$$\frac{dT}{ds}+\mathbf{f} = \frac{d}{ds}\left[(\frac{dX}{ds}+\frac{dU}{ds})T^*\right]+\mathbf{f} = 0 \tag{5}$$

Here **f** is volume force.

3 Nonlinear dynamic analysis of suspension cable

3.1 The dynamic equations

For dynamic issue of the suspension cable, its equilibrium equation is to add inertial force and damping force on basis of the static equation, so the dynamic equilibrium equation of the suspension cable in the initial configuration is

$$\frac{d}{ds}\left[(\frac{dX}{ds}+\frac{dU}{ds})T^{*}\right]+\mathbf{f}=\rho\ddot{U}+\mu\dot{U} \tag{6}$$

And the boundary condition of displacement is

$$U|_{S_u}=U_u \tag{7}$$

Initial conditions are

$$U(X,0)=U(X)\,,\ \dot{U}(X,0)=\dot{U}(X) \tag{8}$$

In which, **f** is volume force, ρ is line density of the suspension cable, μ is damping coefficient. $\dot{U}$ and $\ddot{U}$ are first order derivative and second order derivative of U to t, i.e., represent speed and acceleration respectively. Inertial force and damping force in the balance equation are one of the basic characteristics for difference between the elastic dynamic and static force.

The displacement U, strain ε and stress σ are all functions of time. Therefore fixed solution condition of the dynamic issue also includes initial conditions.

It must note the attached masses on the suspension cable are also move synchronously with vibration of the suspension cable. Therefore for dynamic issue, iteration force generated by the attached masses shall be considered in vibration analysis of the suspension cable.

When the attached mass is m_p, action force between the suspension cable and the attached masses is

$$\mathbf{f}=\begin{bmatrix}0\\ m_p g\end{bmatrix}-m_p\ddot{U}_p=m_pG-m_p\ddot{U}_p \tag{9}$$

In which, U_p is displacement vector of the suspension cable at location of the attached masses, G is gravity acceleration vector.

The former formula is placed in the dynamic equation of the suspension cable, and then

$$\frac{d}{ds}\left[(\frac{dX}{ds}+\frac{dU}{ds})T^{*}\right]+m_pG=\rho\ddot{U}+m_p\ddot{U}_p+\mu\dot{U} \tag{10}$$

3.2 Spline function method

The cubic B spline function is a class of piecewise continuous function, and the function space is constructed by the B spline functions at difference nodes such as ϕ_{-1}, ϕ_0, …, ϕ_{n+1} (shown in Figure 2).

DOI: 10.1051/matecconf/20168201007

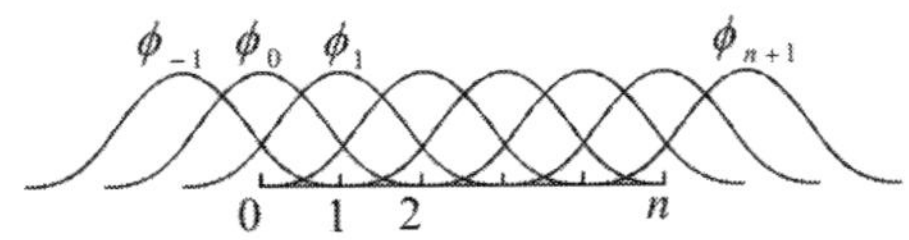

Figure 2. The Base of Function Space.

So the displacement function in the interval [0, s_0] can be interpolated by space base and expressed as the product of spline function and coefficient.

$$U(s,t)=\begin{bmatrix}u_1(s,t)\\u_2(s,t)\\u_3(s,t)\end{bmatrix}=\begin{bmatrix}\Phi(s)B_1(t)\\\Phi(s)B_2(t)\\\Phi(s)B_3(t)\end{bmatrix}=\begin{bmatrix}\Phi & & \\ & \Phi & \\ & & \Phi\end{bmatrix}\begin{bmatrix}B_1\\B_2\\B_3\end{bmatrix}=N(s)\cdot B(t) \tag{11}$$

In the equation, the spline function is $\Phi=[\phi_{-1}(\xi),\phi_0(\xi),\cdots,\phi_n(\xi),\phi_{n+1}(\xi)]$, A is the position coefficient vector of every point of the initial suspension cable, and B is the displacement vector.

Let $C=\left[\frac{dN^T}{ds}\cdot\frac{dN}{ds}\right]$, and then

$$\varepsilon(s,t)=A^TCB+\frac{1}{2}B^TCB=(A+\frac{1}{2}B)^TCB \tag{12}$$

Considering the suspension cable is elastic and follow Hooke's law, tangential force of the suspension cable can be expressed as

$$T=A_0\sigma=EA_0\varepsilon=EA_0(A+\frac{1}{2}B)^TCB \tag{13}$$

Here E and A_0 denote the modulus of elasticity and the cross-sectional area of the cable, respectively.

3.3 Vibration analysis of suspension cable

If displacement U of the elastic cable meets with the geometrical equation and the displacement boundary condition, the integral equation is then obtained according to the virtual work principle in the initial configuration s_0

$$\int_{s_0}(\delta\varepsilon\cdot T^*+\rho\ddot{U}\delta U+m_p\ddot{U}_p\delta U+\mu\dot{U}\delta U-m_pG\delta U)ds=0 \tag{14}$$

The δU is arbitrary, so the motion equation is shown as following:

$$M\ddot{B}(t)+C_t\dot{B}(t)+KB(t)=Q(t) \tag{15}$$

In which M, C_t, K and Q are mass matrix, damp matrix, rigidity matrix and load vector of the system respectively.

$$M=\rho\int_{s_0}N^TNds_0+\sum_k m_kN^T(s_k)N(s_k) \tag{16}$$

$$C_t=\mu\int_{s_0}N^TNds_0 \tag{17}$$

$$K=EA_0\int_{s_0}[C(A+B)(A+\frac{1}{2}B)^TC]ds_0 \tag{18}$$

$$Q = \mathbf{q}\int_{s_0} N^T ds_0 + \sum_k m_k g N^T(s_k) \tag{19}$$

Where **q** is the gravity vector of unit length.

Considering vibration occurs on foundation of the static force balance configuration, increment equation of static force balance shall be analysed when vibration issue of the system is analysed.

So the variation of equation become

$$M\delta\ddot{B}(t) + C_t\delta\dot{B}(t) + (K + K_\sigma)\cdot\delta B(t) = 0 \tag{20}$$

Here

$$K_\sigma = EA_0\int_{s_0}[(A + \frac{1}{2}B)^T CB\cdot C + \frac{1}{2}C(A+B)B^T C]ds_0 \tag{21}$$

Therefore the free vibration equation of the suspension cable without damping is

$$M\delta\ddot{B}(t) + (K + K_\sigma)\cdot\delta B(t) = 0 \tag{22}$$

Set solution of the vibration equation as $B(t) = \phi\sin\omega t$, and substitute it into the above equation, the generalized eigenvalue problem is obtained:

$$[(K + K_\sigma) - \omega^2 M]\phi = 0 \tag{23}$$

Solve the above equation and obtain eigenvalues and eigenvectors of the suspension cable with attached masses, and then obtain the inherent vibration model.

4 Example

The span of one suspension cable is 400m, height difference between the starting point and the end point is -50m. Initial length of the suspension cable is 412m, unit weight is 30N/m, elastic module E=200e9Pa, radius of cable is 12mm. The suspension cable is evenly divided into 40 units.

The nonlinear spline element method is applied to calculate inherent vibration model of the suspension cable, and calculation result is shown as Figure 3. The two kinds of mass conditions are 1000kg and (1000 kg, 1000kg).

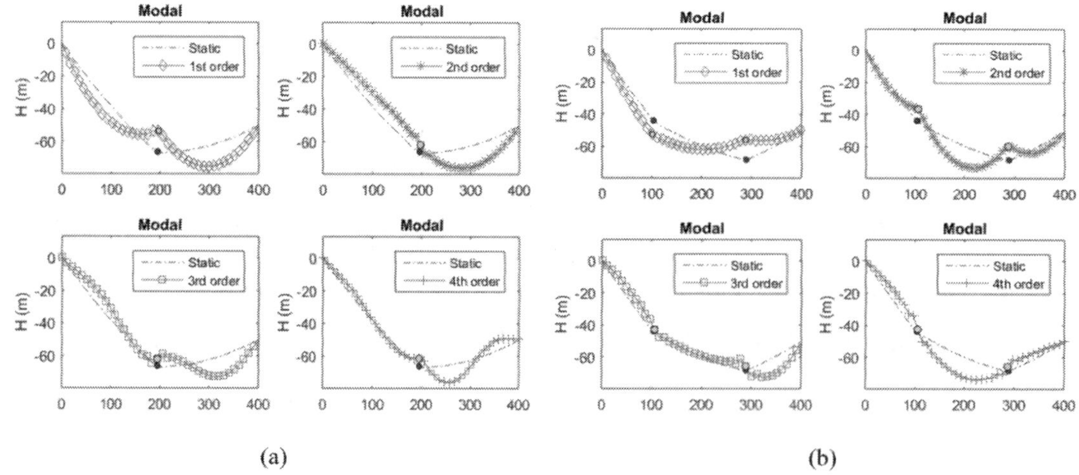

Figure 3. The 1-4th Order Modes with Different Attached Masses

5 Summary

In this paper, the spline function is applied to the vibration analysis of the suspension cable. Based on the principle of virtual work, the formula of the dynamic calculation of the suspension cable with attached masses is proposed.

It can be seen from the calculation results that the attached masses has a great influence on the mode. In the positions of the attached masses, the suspension cable has obvious turning points, and the mass at the lowest end has the largest amplitude and the maximum frequency change. Through the example, the effect of the attached masses to the cable can be used to guide the transportation of cable structures in engineering.

References

1. Okada A. Advanced Structural Design for Cable Structure[J]. Journal of Architecture & Building Science, 2014, **129**
2. Li S, Zhou M, Yang Y. Broken wires diagnosis method numerical simulation based on smart cable structure[J]. *Photonic Sensors*, 2014, **4(4)**:366-372
3. Gavrilov, S.. Non-stationary problems in dynamics of a string on an elastic foundation subjected to a moving load[J]. Journal of Sound and Vibration, 1999, **222(1)**, 345–361
4. Al-Qassab, M., Nair, S.. Wavelet-Galerkin method for the free vibrations of an elastic cable carrying an attached mass[J]. Journal of Sound and Vibration, 2004, **270 (1–2)**, 191–206
5. Cheng, S.P., Perkins, N.C.. Theoretical and experimental analysis of the forced response of sagged cable/mass suspensions[J]. Journal of Applied Mechanics, 1994, **61**, 944–948
6. Di Tomasso, F., Di Egidio, A., Luongo, A., 2003. Modelling and dynamic analysis of strings carrying a single mass. In: Proc. of 16th AIMETA Congress, CD-Rom, Ferrara, Italy (in Italian).
7. Zhang Le-le, Tan Nan-lin, Liu Cai. Simulation and Application of Spline Finite Strip Method in Process of Cold-roll Forming[J]. Journal of System Simulation, 2008, **20(7)**:1703-1706
8. G. Eccher, K.J.R. Rasmussena, R. Zandonini. Geometric nonlinear isoparametric spline finite strip analysis of perforated thin-walled structures[J]. Thin-Walled Structures, 2009, **47**: 219– 232
9. Shen Pengcheng, He Peixiang. The Developments of Spline Finite Element Method in Computational Mechanics[J]. *ADVANCES IN MECHANICS*, 2000, **30(2)**:191-199
10. Qin Rong, Spline function method and program for structural mechanics[M]. Beijing: Science Press, 2015
11. QIN Jian, HUANG Ke-fu. B-Spline Finite Element Method for the Plane Problem of Arbitrary Region[J]. *Engineering Mechanics*, 2010(6):29-34

MATEC Web of Conferences **82**, 01008 (2016) DOI: 10.1051/matecconf/20168201008
D2ME 2016

Catenary Analysis and Calculation Method of Track Rope of Cargo Cableway with Multiple Loads

Jian Qin[a], Jun Chen, Liang Qiao, Jiancheng Wan and Yongjun Xia

China Electric Power Research Institute, 100055, Beijing, China

Abstract. According to actual working condition, the catenary equations of elastic track tope for setting-up is proposed based on the cableway erection requirement, such as the erection angle of end, the initial cable length and midpoint position of cable. The mechanics equilibrium equations, the loads span equations and the consistent equations are presented by analysis of track rope stress state under loads. The nonlinear equations are constructed for elastic track tope with multiple loads and the initial values of newton iteration method are obtained to solve the nonlinear equations. The results of this method are compared with the testing results and numerical results in other literatures and the contrast verifies the reliability of this method. The method is more concise and has smaller amount of calculations with a unified form. It can provide effective means to design the cargo cableway and to check the engineering safety during the erection stage and running stage of cableway.

1 Preface

The overhead cargo cableway has many characteristics, easy line selection, high transportation capability and strong adaptability, etc. The problems of the cableway about design, security assessment, and component selection are more and more important.

Calculation methods of the track rope of the cableway are mainly the analytic method and the finite element method. The analytic method mainly consists of the catenary method [1-3] and the parabola method[4-5]. The catenary method can really reflect the cable shape of the actual suspension rope, the results obtained through theoretical calculation of the catenary method are considered as true solution. Dead weight of the rope structure is assumed to distribute evenly in the parabola theory, it is an approximate calculation method which takes the former two items in the catenary theory. The finite element method mainly includes the two node rod element [6-8] and the multiple nodes curve element [9,10], based on the complete Lagrange description or Euler description, calculation method approximately consists of the increment method and the superposition method. Because it is necessary to carry out simplification or approximate during establishing the finite element format of the rope structure, the obtained calculation models are also different.

For the cargo freight cableway, it is a common acknowledge that track rope without elastic elongation is the catenary. Many references have given out the theoretical reduction process and the relevant calculation equation of the balance equation when the rope structure only bears dead weight[11,12]. For the cargo cableway with the great span, the track rope will generate elastic

[a] Corresponding author: qinjian@epri.sgcc.com.cn

elongation under dead weight and centralized load, the corresponding tension in the rope and the drawing force of the support point will also change. When the track rope has several concentrated loads, these centralized loads are converted to the evenly distributed load in some approximate method, in which tension of track rope is calculated as rope without load. But there is big error with accurate result.

In this paper studies the calculation method of track rope under multiple centralized loads is established to provide practicable theory instruction on calculation and selection of the cargo cableway track rope based on catenary equation of the elastic cable.

2 Catenary equation of elastic cable without centralized load

The following assumptions are introduced in basic theory for design and calculation of the cableway track rope:

(1) The track rope is absolutely flexible, i.e., any section only bears drawing force, and it can't bear bending moment;

(2) The material of the track rope material follows Hooke's law, i.e., the stress and strain relationship of the track rope material is linear elasticity;

(3) The dead load of the track rope is distributed evenly along the arc.

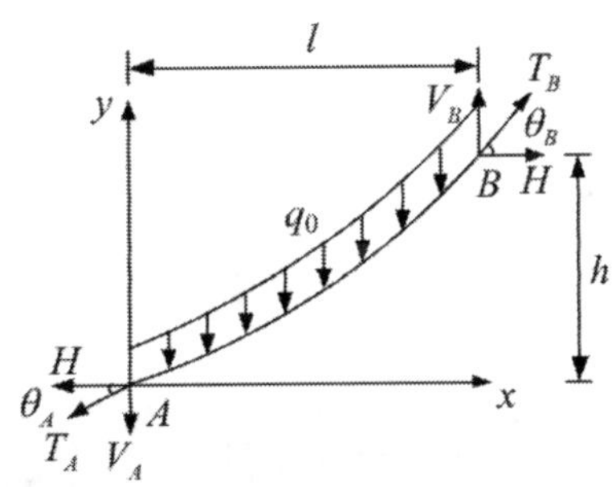

Figure 1. Forcing Schematic Figure of Track Rope without Elasticity

The track rope section shown as figure 1 only bears dead weight load of the rope without elasticity elongation. In which T_A, T_B, V_A, V_B, θ_A and θ_B are tangential tension at end point A and end point B, vertical components of tangential tension, angles between tangential tension and horizontal direction respectively, H is horizontal component of tangential tension, l and h are horizontal span and height different between the end points respectively, q_0 is dead weight intensity of the track rope which is distributed along the arc.

Assume the initial section area of the track rope is A_0, the section area is changed to A with elasticity elongation, intensity of the dead weight distributed along the arc becomes q. Shown as figure 1, the micro-section length of the elastic track rope is ds, the initial arc length is ds_0. Set E as elastic module of the track rope material, the following equation is obtained from Hooke's law: $\mathrm{d}s=[1+T/(EA_0)]\mathrm{d}s_0$. According to the mass conservation of the track rope micro-section after and before deformation: $q_0\mathrm{d}s_0=q\mathrm{d}s$. And it is obtained:

$$q=q_0/[1+T/(EA_0)] \tag{1}$$

Considering forcing condition of the track rope micro-body, the force balance equations at x direction and y direction are obtained:

$$\sum X=0 \;:\; H+\mathrm{d}H-H=0 \tag{2}$$

$$\sum Y=0: H\frac{\mathrm{d}y}{\mathrm{d}x}+\mathrm{d}(H\frac{\mathrm{d}y}{\mathrm{d}x})-H\frac{\mathrm{d}y}{\mathrm{d}x}-q\mathrm{d}s=0 \tag{3}$$

From equation (2), H is constant which means horizontal component of tension of the track rope is equal at any location. According to equation (3), there is

$$\frac{d^2 y}{dx^2} = \frac{q}{H}\frac{ds}{dx} \tag{4}$$

Substituting equation (1) into equation (4), the basic catenary equation of the elastic track rope without centralized load is obtained by $ds/dx = \sqrt{1+(dy/dx)^2}$ and $T = H\sqrt{1+(dy/dx)^2}$ according to deduction process in reference [11]:

$$x = t(a-b) + Kt^2[\sinh(a) - \sinh(b)] \tag{5}$$

$$y = t[\cosh(a) - \cosh(b)] + \frac{Kt^2}{2}[\cosh^2(a) - \cosh^2(b)] \tag{6}$$

$$s_0 = t[\sinh(a) - \sinh(b)] \tag{7}$$

$$s = t[\sinh(a) - \sinh(b)] + \frac{Kt^2}{4}[\sinh(2a) - \sinh(2a)] + \frac{Kt^2}{2}(a-b) \tag{8}$$

In which, $K = q_0/(EA_0)$ (constant), $t = H/q_0$, $a = \sinh^{-1}(V_A/H)$, $b = \sinh^{-1}(V/H)$. V is vertical component of tangential tension at any point. s_0 is initial rope length, s is length of the rope after elastic elongation.

3 Balance equation of elastic track rope under multiple centralized loads

After the setting-up of cableway, the height difference h, the horizontal span l and the initial length s_0 of the whole track rope are determined. The track rope bearing multiple centralized loads is shown as figure 2.

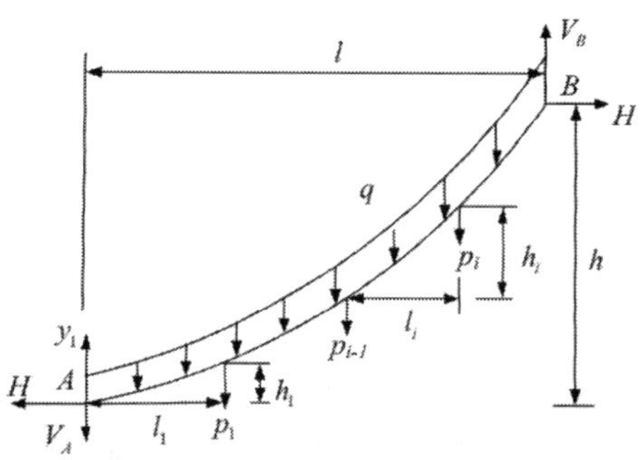

Figure 2. The Track Rope under Multiple Centralized Loads

The rope section splited by centralized load only bears the dead weight of the rope, so the whole rope can be considered as combination of the catenaries. Assuming the track rope has n-1 centralized loads, which are p_1, p_2, …, p_{n-1} respectively, the track rope is divided into n sections, the end points of every section are A_i and B_i, initial arc length is s_i, horizontal span is l_i and height difference is h_i(i=1, 2, …, n).

Established the local coordination system for every section of the track rope, the catenary equation for the every rope section without centralized load is obtained, shown as equations (5)-(8). The nonlinear equations of the whole track rope under multiple centralized loads are established in the following.

3.1 Centralized load

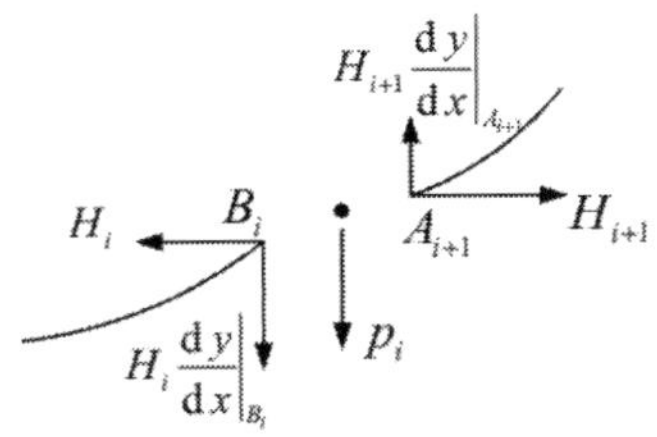

Figure 3. Forcing Analysis at Action Point of Centralized Load

Shown as figure 3, from the force on the action point between every rope section, the force balance conditions at every load point is obtained:

$$H_i - H_{i+1} = 0 \tag{9}$$

$$H_i \left.\frac{\mathrm{d}y}{\mathrm{d}x}\right|_{B_i} - H_{i+1} \left.\frac{\mathrm{d}y}{\mathrm{d}x}\right|_{A_{i+1}} = p_i \ (i=1, 2, \ldots, n\text{-}1) \tag{10}$$

It is known from the balance conditions that the horizontal force of every rope section H_i is same, which is marked as H. And equation (10) can be written as

$$t[\sinh(b_i) - \sinh(a_{i+1})] = \frac{p_i}{q_0} \ (i=1, 2, \ldots, n\text{-}1) \tag{11}$$

Here $t=H/q_0$, $a_i=\sinh^{-1}(V_{Ai}/H)$, $b_i=\sinh^{-1}(V_{Bi}/H)$ ($i=1, 2, \ldots, n$).

3.2 Length of suspension rope

During operation of the cableway, the rope length of the track rope between the centralized load (heavy objects) is fixed and unchanged, initial length s_i of every rope section appears in the equation as the known quantity, the horizontal span l_i and height difference h_i of the rope section will change following different loads and different positions.

Therefore, when initial rope length of every rope section s_i($i=1, 2, \ldots, n$) is known, the rope section length equation is obtained according to equation (7):

$$t[\sinh(a_i) - \sinh(b_i)] = s_i \qquad (i=1, 2, \ldots, n) \tag{12}$$

3.3 Compatibility condition

1) Every section of the track rope shall meet with the compatibility equation of the whole horizontal span $\sum_{i=1}^{n} l_i = l$, according to equation (5), it can be obtained:

$$t\sum_{i=1}^{n}(a_i - b_i) + Kt^2\sum_{i=1}^{n}[\sinh(a_i) - \sinh(b_i)] = l \tag{13}$$

2) According to deformed compatibility equation in height $\sum_{i=1}^{n} h_i = h$ and equation (6)

$$t\sum_{i=1}^{n}[\cosh(a_i)-\cosh(b_i)]+\frac{Kt^2}{2}\sum_{i=1}^{n}[\cosh^2(a_i)-\cosh^2(b_i)]=h \quad (14)$$

Here $K=q_0/(EA_0)$.

3.4 Nonlinear equation set

Summarizing the above equations, the nonlinear equation set of the elastic track rope under multiple centralized loads is obtained:

$$\begin{cases} t[\sinh(b_i)-\sinh(a_{i+1})]-\dfrac{p_i}{q_0}=0 & (i=1,2,\cdots,n-1) \\ t[\sinh(a_i)-\sinh(b_i)]-s_i=0 & (i=1,2,\cdots,n) \\ t\sum_{i=1}^{n}(a_i-b_i)+Kt^2\sum_{i=1}^{n}[\sinh(a_i)-\sinh(b_i)]-l=0 & \\ t\sum_{i=1}^{n}[\cosh(a_i)-\cosh(b_i)]+\dfrac{Kt^2}{2}\sum_{i=1}^{n}[\cosh^2(a_i)-\cosh^2(b_i)]-h=0 & \end{cases} \quad (15)$$

The equation set includes 2n+1 nonlinear equations, 2n+1 unknown number ai, bi(i=1, 2 ,…, n) and t, therefore the nonlinear equation set is closed, which can be expressed as:

$$F(X)=0 \quad (16)$$

In which $X=[a_1, a_2, \ldots, a_n, b_1, b_2, \ldots, b_n, t]^T$.

After X is calculated, the vertical component V_x and tangential tension T_x at any rope length s_x point of the track rope are obtained respectively:

$$V_x = tq_0\sinh(a_m)+q_0\left(s_x-\sum_{i=1}^{m}s_i\right) \quad (17)$$

$$T_x=\sqrt{(tq_0)^2+V_x^2} \quad (18)$$

In which, m is number of centralized load in the $[0,s_x]$.

4 Examples

In order to validate calculation accuracy of the cargo cableway track rope, testing data in the reference document [1] and calculation data in reference document [8] are applied for comparison.

Load weight in the reference[1] is P=6130N. The cross area of the track rope is A_0=289.95mm^2, weight of unit length is q_0=27.0N/m, elastic module E=90GPa. The cableway set up for testing is two span with single track rope. Testing data are complete in the first span of the cableway. The horizontal span of first span is l=70.9m, height difference h=15.2m. This paper selects the testing data with deflection coefficient $S_f=f/l$=0.05 for comparison and validation.

Table 1. Testing Comparison and Validation

Comparison items	Test value	Calcula-tion value	Error %

No-load	Drawing force of lower support point /N		5000	4911	-1.780
	Torsion at 0.3l location /m		3.077	3.060	-0.552
	Torsion at 0.5l location /m		3.475	3.544	1.986
	Torsion at 0.8l location /m		2.105	2.292	8.884
On load	Load at 0.3l location	Drawing force of lower support point /N	23500	27002	14.902
		Torsion of loading point /m	4.097	3.881	-5.272
	Load at 0.5l location	Drawing force of lower support point /N	25300	28812	13.881
		Torsion of loading point /m	4.555	4.292	-5.774
	Load at 0.8l location	Drawing force of lower support point /N	20500	22625	10.366
		Torsion of loading point /m	3.570	3.380	-5.322

Seen from comparison and validation data in Table 1, calculated drawing force of the lower support point of the track rope and deflection value at different position are almost same as testing value. When there is centralized load, drawing force of the lower support point of the track rope obtained through calculation is greater than testing value, deflection value at different position is smaller than testing value. Friction force generated between the saddle and the track rope at no-load belongs to static friction force, no displacement occurs in the track rope. When the cableway is loaded, friction force generated between the saddle and the track rope is insufficient to overcome tension difference between two sides of the saddle. The track rope moves, the sag increases, and tension reduces, corresponding deflection at different position increases.

The reference [8] calculates the large sag single span cableway with the nonlinear finite element method in the two node ropes unit with Euler description. The horizontal span of the cableway l=305m, height difference is 0, and the middle deflection coefficient S_f=0.1. The elastic module of the rope is E=130GPa, the section area is A_0=550mm^2, centralized load acting on C point 122m away from the end point is P=35.6kN.

Table 2. Value Calculation Comparison and Validation

Comparison item		Value in reference document	Value in text	Error %
No load	Horizontal tension /kN	17.728	17.955	1.280
	Sag at C point /m	29.923	29.296	-2.095
On load	Displacement of C point after load acts /m	5.64	5.56	1.418

Seen from Table 2, the calculation value in this paper after and before acting of centralized load are basically same as the results in the reference documents, it means the calculation method in this paper has extensive adaptability.

5 Conclusions

The track rope is an important weight bearing member in the cargo cableway, its shape and tension calculation relationship is directly related to speciation and model selection of the track rope, etc. This paper establishes the balance equations of the elastic track rope under the multiple centralized loads based on the catenary equation of the elastic track rope. The formation of equations is brief and the structure is clear. The Newton iteration method is applied to calculate the equations, in which calculation quantity is small, and accuracy of result is high. Through comparison with result data in other reference documents, reliability of the calculation method in this paper is validated, which can meet with requirement of design and engineering safety in the cargo cableway.

References

1. Zheng Lifeng. Theory study of cableway on the catenary method [J]. Fujian: Fujian Agriculture and Forestry University. 2002.
2. CHEN Changsong, CHEN Zhengqing, YAN Donghuang. Accurate iteration method to calculate the initial [J]. Engineering Mechanics, 2006, **23(8)**: 62~68.
3. QIN Jian, XIA Yongjun. The matrix iteration method for analysis of suspension cable based on segmental catenary theory [J]. Chinese Journal of Engineering Design, 2013, **20(5)**: 404~408.
4. Qian Changzhao, Chen Changping, Chen Zili. Direct solution procedure obtaining the line shape of cable with segmentation parabolic function [J]. Journal of Xiamen University of Technology, 2012, **20(3)**: 71~74.
5. Zhou Xinnian, Zhan Zhengyi. The system of multi-span cableway design based on parabola method (JIA SHI) [J]. Journal of Zhejiang Forestry College, 2000, **17(1)**:50~55.
6. Ahmadi-Kashani, K. Representation of cables in space to uniformly distributed loads [J]. International Journal of Space Structures, 1988, **3(4)**: 221~230.
7. Zhang Qilin, Zhang Li, Zhou Dai, U.Peil. Fem with Spline Function for Nonlinear Analysis of Continuous Long Cables [J]. Engineering Mechanics, 1999, **16(1)**: 115~122.
8. Tang Jianmin, Zhao Yin. A Euler Geometrically Non-linear Finite Element Method with Two-node Cable Element for the Analysis of Cable Structure [J]. Shanghai Journal of Mechanics, 1999, **20(1)**: 89~94.
9. Yuan Xingfei, Dong Shilin. A two-node curved cable element for nonlinear analysis [J]. Engineering Mechanics, 1999, **16(4)**: 59~64.
10. Yang Menggang, Chen Zhengqing. Nonlinear analysis of cable structure using a two-node curved cable element of high precision [J]. Engineering Mechanics, 2003, **20(1)**:42~47.
11. Zhang Zhiguo, Jin Mingjun, Zou Zhenzhu. Static solution of suspension cables under tare load [J]. China Railway Science, 2004, **25(3)**: 67~70.
12. Bai Xuesong, Miao Qian. Research on calculation model based on suspension cable theories in construction cargo cableway [J]. Power System Technology, 2008, **32(20)**: 90~94.

MATEC Web of Conferences **82**, 01009 (2016) DOI: 10.1051/matecconf/20168201009
D2ME 2016

Kinetic Modelling and Model based Control of SCR Deposit

Feng Qian[1,a] and Lin Lv[1]

[1]*Key Laboratory of High Performance Ship of Ministry of Education, Wuhan University of Technology, Wuhan, China*

Abstract. Selective catalytic reduction (SCR) is widely used for on-highway as well as off-road applications to fulfil stringent nitrogen oxide (NO_X) emission standards. However, in the practical operations, urea deposits occur under unfavourable conditions, which bring about urea dosing strategy failures, affect SCR system de-NO_X performance and decrease fuel economy by increase engine back pressure. This paper presents a 1D kinetic model of SCR system, which can predict SCR reaction and deposit formation process. Furthermore, a model-based control strategy has been designed for real engine applications. The ETC test shows the deposits quantity can be reduced by 41.8% with this model.

1 Introduction

Nowadays, the selective catalytic reduction is a promising solution to fulfill stringent nitrogen oxide (NO_X) emission standards enforced by worldwide regulation bodies for lean burning engines. However, in the practical operations, the rapid change of engine working conditions leads to a rapid change of exhaust temperature, space velocity (SV) and urea dosage rate, which makes it difficult to obtain a complete evaporation and decomposition of urea before the catalyst. Consequently, urea deposits occur under unfavorable conditions, which bring about a failure of dosing strategy. Moreover, the formed deposits may increase the back pressure of the engine and affect engine system performance.

Theoretically, it would be no deposits formation in the SCR system if there is no contact between urea droplets and pipe/mixer wall, which is impossible due to the restricted packaging space of the practical applications. Therefore, the SCR system performance which include the uniformity of the reducing agent, the pressure drop, the de-NO_X efficiency and the risk of deposits of the system, should be evaluated in the product development stage. Owing to the variety of different exhaust gas configurations, the evaluation requires appropriate support from numerical simulation in order to obtain the required objectives. Nowadays, the evaluation of the spray preparation and de-NOX efficiency has become state of the art [1-4]. In contrast, the studies of the reliable prediction methods for deposit risk through numerical simulation are relatively scarce.

This paper presents a kinetic model for SCR system, which can predict SCR reaction and deposit formation process. Furthermore, could be used for urea dosing control strategy optimization.

2 Experimental Setup

[a] Corresponding author: frank.qian1@gmail.com

The experimental setup consists of an 11.12 L diesel engine as Table 1, a 21.75 L V_2O_5-WO_3/TiO_2 catalyst. The aftertreatment system is equipped with urea dosing pump, NO_X sensors, inlet and outlet catalyst temperature sensors, an AVL AMA i60 emission analyzer and a Siemens ammonia analyser.

Table 1. Engine Information.

Engine model	Parameters
Features	Inline 6-cylinder, CRDI, TCI
Displacement	11.12 L
Cylinder bore × travel	123mm×156mm
Compression ratio	16.4
Rated power	303 kW at 1900 rev/min
Maximum torque	1870 Nm at 1300 rev/min

3 1D Kinetic Model Description

3.1 Structure

The injection, wallfilm and deposits formation process are shown in Figure 1. The liquid that injected into the exhaust pipe impact on the solid surface and a part of the liquid become wallfilm. The deposits would appear due to urea precipitation and the by-products generated in the urea thermolysis.

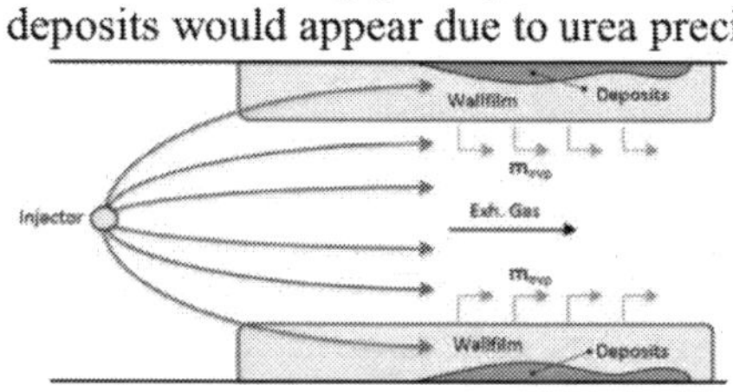

Figure 1. Injection Process.

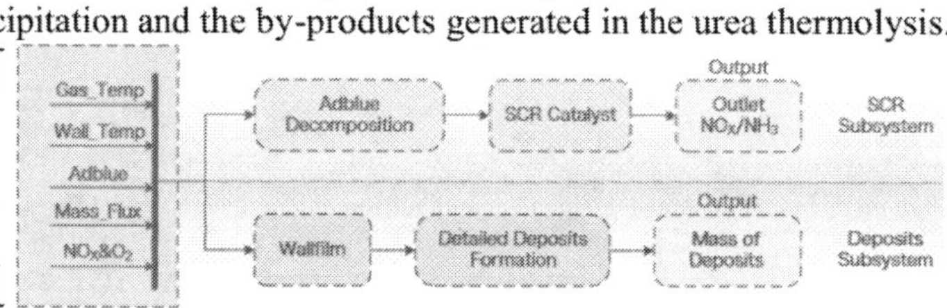

Figure 2. The structure of the 1D kinetic model.

Many researches have shown that the by-products formed in the urea thermolysis can be ignored in the simulation of system-out NO_X and NH_3. Therefore, the 1D Model presented in this paper calculates the NO_X/NH_3 concentration and deposits respectively Figure 2, which would obtain a good balance between the computation precision and speed.

3.2 SCR Subsystem

The SCR subsystem includes tow pars, one is the urea decomposition block which is used to simulate the water evaporation, urea thermolysis and HNCO hydrolysis, the other is the SCR catalyst block which is used to simulate the reactions occurring inside the porous catalyst. In the urea decomposition block, instantaneous water evaporation in the gas phase is assumed and heat of evaporation is taken from the gas phase. The urea thermolysis and HNCO hydrolysis was simulated in CSTR method and the reaction rates are calculated by equation (1) and (2), respectively.

$$\dot{m}_{urea} = -m_{urea} \cdot K_{urea} \cdot \exp\left(-\frac{E_{urea}}{R \cdot T_g}\right) \tag{1}$$

$$\dot{m}_{HNCO} = -m_{HNCO} \cdot K_{HNCO} \cdot \exp\left(-\frac{E_{HNCO}}{R \cdot T_g}\right) \tag{2}$$

Where Tg is the exhaust gas temperature.

SCR Catalyst block is based on the geometry of a honeycomb monolith that consists of numerous parallel open channels. As the geometrical properties of all channels, their catalyst distribution, and the inlet conditions are assumed identical, the flow through the monolith is modeled by solving the corresponding mass and energy balances for a one-dimensional single channel. Seven reactions are considered as Table 2, including urea thermolysis, HNCO hydrolysis, NH_3 ad/desorption, standard SCR, fast SCR, slow SCR, NH_3 oxidation, NO/NO_2 equilibrium, N_2O formation.

Table 2. Reaction Mechanism for SCR Catalyst Block.

Reaction	Rate equation
$(NH_2)_2CO \rightarrow NH_3 + HCNO$	$r_1 = K_1 \cdot \exp\left(-\frac{E_1}{RT_s}\right) \cdot c_{urea}$
$HCNO + H_2O \rightarrow NH_3 + CO_2$	$r_2 = K_2 \cdot \exp\left(-\frac{E_2}{RT_s}\right) \cdot c_{HCNO} \cdot c_{H_2O}$
$NH_3 + * \rightarrow NH_3^*$	$r_3 = K_3 \cdot \exp\left(-\frac{E_3}{RT_s}\right) \cdot c_{NH_3} \cdot (1-\theta_{NH_3})$
$NH_3^* \rightarrow NH_3 + *$	$r_4 = K_4 \cdot \exp\left(-\frac{E_4 \cdot (1-\varepsilon \cdot \theta_{NH_3})}{RT_s}\right) \cdot \theta_{NH_3}$
$4NH_3^* + 4NO + O_2 \rightarrow 4N_2 + 6H_2O$	$r_5 = K_5 \cdot \exp\left(\frac{-E_5}{RT_s}\right) \cdot c_{NO} \cdot \theta_{NH_3}^* \cdot \left(1-\exp(-\frac{\theta_{NH_3}}{\theta_{NH_3}^*})\right)$
$4NH_3^* + 2NO + 2NO_2 \rightarrow 4N_2 + 6H_2O$	$r_6 = K_6 \cdot \exp\left(\frac{-E_6}{RT_s}\right) \cdot c_{NO} \cdot c_{NO_2} \cdot \theta_{NH_3}^* \cdot \left(1-\exp(-\frac{\theta_{NH_3}}{\theta_{NH_3}^*})\right)$
$8NH_3^* + 6NO_2 \rightarrow 7N_2 + 12H_2O$	$r_7 = K_7 \cdot \exp\left(\frac{-E_7}{RT_s}\right) \cdot c_{NO_2} \cdot \theta_{NH_3}^* \cdot \left(1-\exp(-\frac{\theta_{NH_3}}{\theta_{NH_3}^*})\right)$
$4NH_3^* + 3O_2 \rightarrow 2N_2 + 6H_2O$	$r_8 = K_8 \cdot \exp\left(-\frac{E_8}{RT_s}\right) \cdot \theta_{NH_3}$
$NO + 0.5O_2 \leftrightarrow NO_2$	$r_9 = K_9 \cdot T_s^4 \exp\left(-\frac{E_9}{RT_s}\right) \cdot \left(c_{NO}^2 \cdot c_{O_2} - \frac{c_{NO_2}^2}{K_{Eqv}}\right) \cdot (1-\theta_{NH_3})$
$2NH_3^* + 2NO_2 \rightarrow N_2 + N_2O + 3H_2O$	$r_{10} = K_{10} \cdot \exp\left(\frac{-E_{10}}{RT_s}\right) \cdot c_{NO_2} \cdot \theta_{NH_3}$

3.3 Deposits Subsystem

In the wallfilm block, the fraction xstored which is stored in the wallfilm during the injection process at different temperature, mass flux and dosage. It worth noting that, in this case, the fraction is significantly increased with the increasing of deposits mass due to the enclosed area created by the deposits Figure 3, which is modeled by a piecewise function.

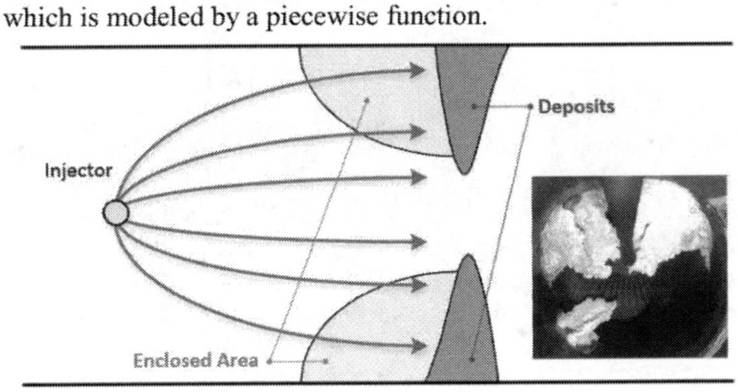

Figure 3. The enclosed area created by the deposits.

The mass and temperature of wallfilm can be calculated by equation (3) and (4), respectively.

$$\frac{\partial m_i}{\partial t} = x_{stored} \cdot \dot{m}_{Inj,i} + \dot{m}_{evap,i} + \dot{m}_{de,i} \quad (3)$$

$$\rho_f \cdot V_f \cdot \frac{\partial (c_{p,f} \cdot T_f)}{\partial t} = \dot{m}_{In} \cdot c_{p,f} \cdot (T_{inj} - T_f) + A_f \cdot k_{fs} \cdot (T_s - T_f) + A_f \cdot k_{fg} \cdot (T_g - T_f) \quad (4)$$

Where m_{Inj} is mass of the injected urea, m_{evap} is mass of evaporated water, m_{de} is urea mass which involved in the themolysis reaction, ρ_f is wallfilm density, $c_{p,f}$ is wallfilm specific heat, V_f is wallfilm volume, T is temperature, A_f is wallfilm area, k is heat transfer coefficient. The wallfilm exchanges heat with the gas phase and with the solid wall are calculated by equation (5) and (6), respectively.

$$k_{fg} = 0.0234 \cdot \Pr \cdot \mathrm{Re} \cdot \frac{\lambda_g}{d_{hyd}} \cdot A_f \quad (5)$$

$$k_{fs} = \frac{2}{\frac{d_f}{\lambda_{film}} + \frac{d_s}{\lambda_{solid}}} \cdot A_f \quad (6)$$

Where λ is thermal conductivity, dhyd is pipe diameter, df and ds are thickness of the wallfilm and pipe wall respectively.

Kontin showed that the impact of oversaturation in the urea water solutions leading to urea precipitation can be significant at low temperatures [5]. As a consequence, urea can be present in different forms (solid, dissolved, molten) depending on the operating conditions. Urea stat to decompose before melting, so that the existence of the molten phase is restricted to a quite narrow temperature range [6-7]. Therefore, the reactions are supposed to take place in solid phase and the mechanism presented by Ebrahimian [8] is used in this paper.

4 Model-based Optimization for Urea Dosing Strategy

4.1 Urea Dosing Strategy

The 1D Kinetic model can be applied to optimize the urea dosing strategy. This paper takes the strategy shown in the Figure 4 as an example to illustrates the optimization process under the objective of minimizing system-out NO_X, NH_3 slip and deposits.

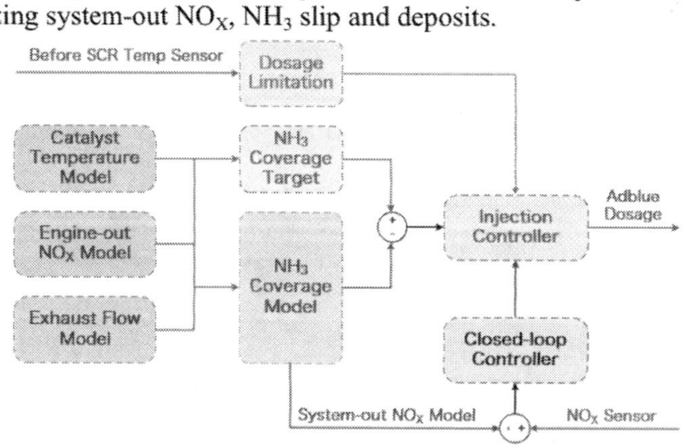

Figure 4. Urea Dosing Strategy.

As shown in Figure 4, the core of the open-loop part is an embedded chemical model which accounting for the main parameters that influence NO_X conversion and resulting NH3 coverage. The amount of ammonia added to the SCR catalyst at every time instance is dictated by an optimized look-up table (Coverage Table), which relates average catalyst temperatures to optimal target average ammonia surface coverage. Moreover, the urea dosage at different temperature is limited by an optimized look-up table (limitation Table), which relates inlet gas temperature to optimal dosage limitation. The urea dosage rate is calculated by the equation (7)

$$V_{Adblue,open} = \begin{cases} (\theta_{desired} - \theta_{actual}) \cdot \sigma \cdot A_{trans} \cdot V_{cat} \cdot 0.5 \cdot 60/0.325/1.085/0.2 \cdot 3600 & < limit \\ limitation & > limit \end{cases} \quad (7)$$

Where V_{Adblue} is urea dosing rate (ml/h), σ is site density, Atrans is geometrical reaction surface.

In the closed-loop part, a two-stage integral prediction method (TIP method) is used to provide robustness of the system in the real world. Research shows that the NH_3 slip mainly occurs under the transient conditions with an immediate increase of exhaust temperature or a sudden reduction of engine load [9]. Therefore, a pre-integrator in the closed-loop controller is used in the transient conditions and a main integrator is used in the steady conditions. Then, if there is a significantly increase of temperature, or the torque keeps reducing in 1 sec or the engine gets into motored condition, the data in the pre-integrator would be removed and the closed-loop correction factor is calculated only with the data in the main integrator. However, the data in pre-integrator and in main integrator would both be used. What's more, in order to minimize the over-shooting risk, a moving average numeric filter is used in the calculation. The correction factor and the final urea dosage rate is calculated by equation (8) and (9).

$$CF_{closed} = \frac{1}{N} \cdot \sum_{i=0}^{N-1} \left(\frac{InNO_X - ObNO_X}{InNO_X - RfNO_X} \right) \tag{8}$$

$$V_{Adblue,final} = V_{Adblue,open} \cdot CF_{closed} \tag{9}$$

Where In_{NOX}, Ob_{NOX} and Rf_{NOX} is the integral of the engine-out NO_X mass flux, observed NO_X mass flux (NO_X sensor) and referenced NO_X mass flux (System-out NO_X model). N is filter order (N=3). In order to prevent the failure of closed-loop control, the correction factor CF_{closed} would be reset to 1 if the CF_{closed} and the observed NOx mass flux are continuously increased at the same time.

4.2 Model-based Optimization

In this work, the optimization of urea dosing strategy was based on ETC test. The baseline engine-out NO_X of ETC test is 11 g/kWh and system-out NO_X target is 2 g/kWh due to NO_X reduction capability of catalyst. Therefore, beside meet NO_X emission target, also need to follow below constraints:

$$\max\left[NH_{3,slip}\right] \leq 25\,ppm \tag{10}$$

$$Mass_{Deposits} \leq 1g \tag{11}$$

Where NH_3 slip is the mole fraction of system-out NH_3, $Mass_{Deposits}$ is the mass of deposits.

The optimization of the coverage table and limitation table is completed using a trust-region algorithm for a constrained, non-linear optimization problem. The values for the two tables before and after the optimization are shown in the Figure 5.

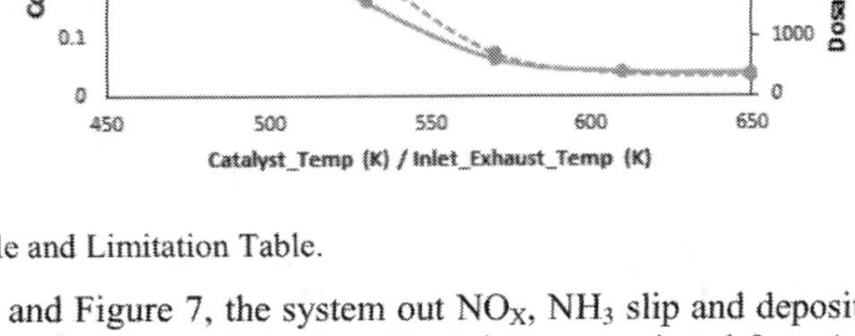

Figure 5. Coverage Table and Limitation Table.

As shown in Figure 6 and Figure 7, the system out NO_X, NH_3 slip and deposits mass of the ETC test are all satisfy the target after the optimization. Deposits mass reduced from 1.196g to 0.696g and the reduction ratio is 41.8%.

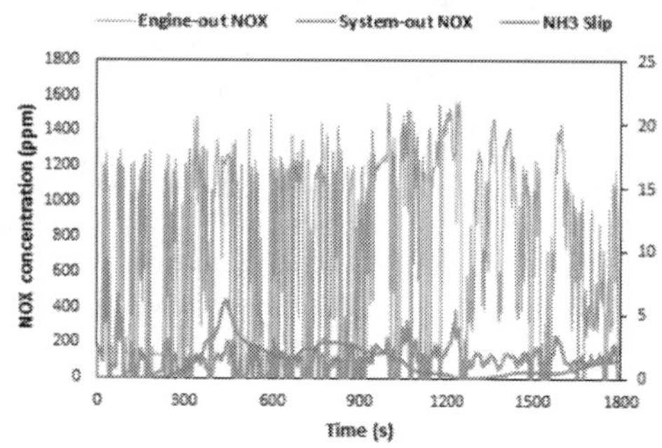

Figure 6. System-out NO_X and NH_3 slip of ETC

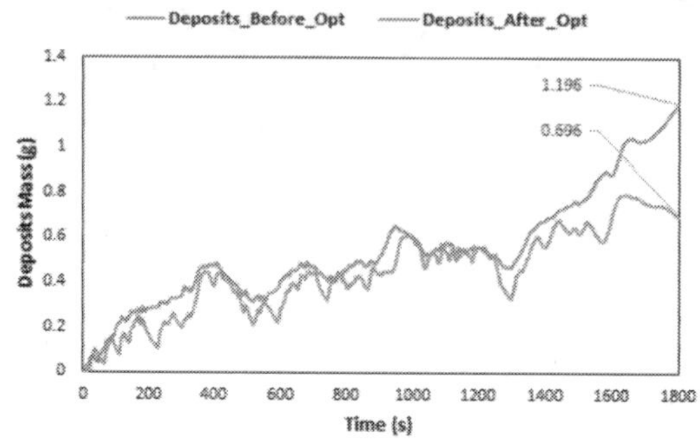

Figure 7. Deposits mass before and after the optimization.

5 Conclusion

This paper gives a comprehensive analysis of the 1D Kinetic model for SCR deposits. From the results, the following conclusions can be drawn:

1. The model can predict the SCR reaction and deposit formation process at the same time, the output contains the NO_X specific emission, NH_3 slip and deposits mass.
2. The model can predict deposit component including urea, biuret, CYA, ammelide and has good agreement with the experiment data.
3. A 1D Kinetic model is presented which can be applied to optimize the urea dosing strategy. The deposits mass of ETC test reduced by 41.8% after the optimization.

Acknowledgements

This work was supported by the National Natural Science Foundation of China (No. 51379165). The author appreciates the support of Key Laboratory of High Performance Ship Technology of Ministry of Education (Wuhan University of Technology).

References

1. Lee, Jung-Gil; Kim, Young-Deuk; Kim, Woo-Seung. Design and Implementation of Mixing Chambers to Improve Thermal Decomposition of Urea for NOX Abatement. Environmental Engineering Science. Vol.**29**, NO.10, Oct, 2012
2. Tan Feng, Lin Lü. Development of a variable resistance-capacitance model with time delay for urea-SCR system. Environ. Eng. Res. 2015; **20(2)**: 155-161
3. Tan Feng, Lin Lu. The characteristics of ammonia storage and the development of model-based control for diesel engine urea-SCR system. Journal of Industrial and Engineering Chemistry. Volume **28**, 25 August 2015, Pages 97–109
4. FENG Tan, LU Lin. Effect of Washcoat Pore Structure on NOX conversion Efficiency of SCR System. Chinese Internal Combustion Engine Engineering. Vol.**36**, NO.1, February. 2015
5. KONTIN S., HÖFLER A., KOCH R. and BAUER H.-J., Heat and Mass Transfer accompanied by Crystallisation of single Particles containing Urea-water solution, ILASSEurope 2010, 23rd Annual Conference on Liquid Atomization and Spray Systems, Brno, Czech Republic, 2010
6. WANG X.Z., MA C.Y., Morphological population balance model in principal component space, AIChE J., 55,9, pp.2370-2381, 2009
7. SCHMIDT A., Verfahrenstechnische Probleme bei der Herstellung von Melamin aus Harnstoff bei Atmospharendruck, Österr Chemiker-Ztg, **68**, pp.175-179, 1967
8. Vahid Ebrahimian, Andre Nicolle, and Chawki Habchi. Detailed Modeling of the Evaporation and Thermal Decomposition of Urea-Water Solution in SCR Systems. AIChE Journal. July 2012 Vol. **58**, No. 7
9. FENG Tan, LU Lin. Development of Low Temperature Control Strategy for Diesel Engine Urea-SCR system. Chinese Internal Combustion Engine Engineering. Vol.**35**, NO.6, December

MATEC Web of Conferences **82**, 01010 (2016) DOI: 10.1051/matecconf/20168201010
D2ME 2016

Topology Optimisation of Wireless Sensor Networks

Aye Min Thike[1], Sergey Lupin[1], Roustiam Chakirov[2] and Yuriy Vagapov[3,a]

[1]*National Research University of Electronic Technology, Zelenograd, Moscow, 124498, Russia*
[2]*Bonn-Rhein-Sieg University of Applied Science, 20 Grantham-Allee, Sankt Augustin, D-53757, Germany*
[3]*Glyndwr University, Plas Coch, Mold Road, Wrexham, LL11 2AW, UK*

Abstract. Wireless sensor networks are widely used in a variety of fields including industrial environments. In case of a clustered network the location of cluster head affects the reliability of the network operation. Finding of the optimum location of the cluster head, therefore, is critical for the design of a network. This paper discusses the optimisation approach, based on the brute force algorithm, in the context of topology optimisation of a cluster structure centralised wireless sensor network. Two examples are given to verify the approach that demonstrate the implementation of the brute force algorithm to find an optimum location of the cluster head.

1 Introduction

Wireless sensor networks (WSN) have received much attention in the past decade due to advancements in the area of wireless telecommunication and computer engineering. WSNs are widely used in various applications to gather and collect information where monitoring, observation and surveillance are crucial [1]. A wide variety of communications solutions for industrial automation, mobile object tracking, medical monitoring, environmental monitoring etc. are based on WSNs [2].

The main component of WSNs is a sensor node comprising a sensor element, micro-controller, power supply and radio-communication circuit. The sensor nodes designed for industrial applications are embedded into industrial environment in order to provide monitoring of temperature, pressure, vibration, sound and radiation etc. They can also operate as components of industrial automated systems generating information for control algorithms. The size of a modern sensor node is less than a few cubic centimetres [3].

A sensor node is usually a low-power device where data processing through the sensor and microcontroller requires much less energy than radio transmission. Because of this the range of its reliable telecommunication is quite limited. In order to provide a reliable operation of a network WSN should be designed to guarantee stable communication of the sensor nodes at low power consumption. Topology of WSN is a key parameter affecting on provision of reliable communication and, therefore, a crucial part of the network design [4, 5].

In an industrial environment the position of a sensor node is often fixed and determined by the particular equipment needs. The WSN topology generated under this circumstance could not support a reliable radio communication between the nodes. To improve commutation reliably the sensor nodes are divided into clusters where the cluster head can provide radio communication with the nodes at a

[a] Corresponding author : y.vagapov@glyndwr.ac.uk

longer distance. As an example, Figure 1 demonstrates a concept of a 7-node cluster of centralised WSN.

Finding of the optimum location of a cluster head is the main problem of WSN topology design particularly for an industrial environment. The paper discusses optimisation approach based on the brute force algorithm in context of topology optimisation of a centralised WSN having a cluster structure.

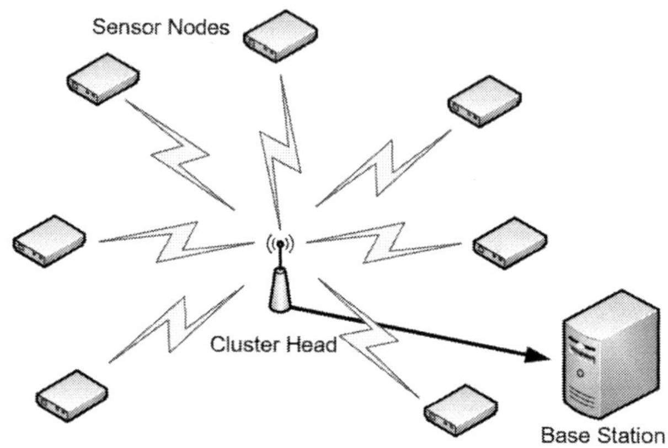

Figure 1. Concept of a cluster of centralised sensor wireless network.

2 Topology optimisation

Optimisation problems (including topology optimisation) are often solved using genetic algorithms. These algorithms rely on heuristics based on the principle of natural selection. Genetic algorithms can be successfully applied for a variety of optimisation tasks inappropriate for standard optimisation algorithms, particularly, if an objective function is discontinuous, not differentiable, stochastic or substantially nonlinear. [2, 6, 7] demonstrate examples of successful implementation of the genetic algorithms for various WSN designs. In particular, [7] proposed a genetic algorithm to determine an optimal location of sensor nodes in the network in order to minimise energy consumed by nodes to transmit data. The work also investigated various approaches to the solution including the combining of sensor nodes into clusters to reduce computational complexity of the algorithm. A major disadvantage of genetic algorithms however is related to difficulties of their adaptation to a task formulation.

An alternative approach to optimisation is the introduction of a brute force algorithm. This algorithm is more universal and can be easily implemented with great ease whilst comparing genetic algorithms. However, brute force algorithm is not widely used in optimisation tasks due to high demand in computational resources. Conversely the recent advancements in the area of computing such as multi-core processors and accelerators significantly increased the computing capability of modern computers making brute force algorithm applicable for optimisation task solutions.

Application of the brute force algorithm for WSN topology optimisation can be formulated as following:

1. Let $\{\mathbf{A}_N\}$ be a finite set of network elements which can occupy possible positions $\{\mathbf{P}_M\}$. In practice, the number of positions is much larger than the number of elements: $M >> N$.

2. Find the distribution of elements $\{\mathbf{A}_N\}$ at the positions $\{\mathbf{P}_M\}$ corresponding to an extrema of some functional. In general, the task can be multi-objective.

This formulation of the problem allows for implementation of a brute force algorithm as a multi-threaded application where computational complexity of the algorithm is defined as $O(M^N)$. The examples below demonstrate of implementation of a brute force algorithm to find the optimum topology for a centralised WSN having one (example 1) and two (example 2) cluster heads.

3 Topology optimisation criteria

The optimum criteria for the cluster head allocation can be developed using the following conditions.

1. Power of the communication signal $P_i(R)$ is inversely proportional to square of the distance between the cluster head and a sensor node R.

$$P_i(R) = P_i^0 / R^2 \qquad (1)$$

where P_i^0 is the rated power of communication signal of i-th sensor node; R is the distance between the cluster head and a sensor node.

2. The communication between the cluster head and a sensor node is possible if a sensor node is located in the area of reliable wireless communication.

$$R \leq R_{max} \qquad (2)$$

where R_{max} is max distance of reliable communication

3. The network consists of a number of sensor nodes $\{\mathbf{S}_k\}$ where each sensor node S is characterised by location coordinates and number transferred parameters into the network D.

$$S_i \leq (X_i, Y_i, D_i) \qquad (3)$$

Therefore the optimisation criteria can be formulated as following:

1. Maximising number of sensor nodes communicating to the cluster head.

$$F_1 = \sum_{1}^{K} C(S_i) \rightarrow \max \qquad (4)$$

where $C(S_i) = 1$ if the distance between a sensor and any cluster head is less than R_{max}.

2. Maximising the number of parameters transferred by sensors to the network.

$$F_2 = \sum_{1}^{K} D(S_i) * C(S_i) \rightarrow \max \qquad (5)$$

where $D(S_i)$ is a number of the object parameters transferring by a sensor node S_i to a cluster head.

3. Maximising intensity of communication signal of the cluster head for all sensor nodes.

$$F_3 = P_i^0 / R^2 \rightarrow \max \qquad (6)$$

Computational complexity of the optimisation under above criteria is approximately the same. The given below example of the optimisation utilises the criterion function F_3.

4 Examples of cluster head topology optimisation

In order to verify the proposed method of topology optimisation two examples of a WSN are considered. The model of WSN used in each example consists of 9 identical sensor nodes allocated in a flat industrial environment. The coordinates of the sensor node locations are chosen arbitrary; their details are given in Table 1. The sensor nodes are considered as low-power devices having a limited range of radio communication. The nodes communicate with the cluster head which transfers the information to the base station for further processing. It is assumed that all nodes are operating under a stable clock synchronisation.

The optimisation problem in the first example is formulated as a determination of the optimum location of the cluster head in order to provide a power saving operation of the nodes. The optimisation criterion is the intensity of the wireless communication signal at the network nodes' locations. It is assumed that the intensity of wireless communication signal is inversely proportional to square of the distance between the cluster head and a sensor node (criterion F_3). The criterion function for each variant of the cluster head allocation is based on sensor having minimum value of the criterion F_3. The optimal solution should provide the best value of the criterion. It is obvious that solution accuracy depends on the accuracy of the cluster head coordinates representing grid cell size.

Table 1. Coordinates of the sensor nodes.

Number of sensor node	Axis X	Axis Y
1	13.5	8.7
2	13.1	9.3
3	13.9	10.6
4	4.8	8.9
5	5.7	10.6
6	6.4	9.5
7	9.2	13.3
8	10.1	15.3
9	11.2	14.1

The solution for the first example represents the coordinates of the cluster head optimum location. According the optimisation algorithm, the area of flat industrial environment has been represented using a grid layout where the solution coordinates are the centre of a gridded region. The optimisation has been conducted using three grid cell sizes: 2.0, 1.0, and 0.5. The results of the optimisation are given in Table 2. Figure 2 shows the solution coordinates for three grid cell sizes. It has been found that the accuracy of the solution depends on the grid size. The best (highest) value of the criterion has been obtained for the smallest grid cell size. The reduction in the grid cell sizes however will significantly increase the employment of computational resources and calculation time whereas the grid cell size of 0.5 will provide a satisfactory accuracy.

Table 2. Solution coordinates for the cluster head at various grid cell size.

Grid cell size	Solution coordinates for the cluster head		Criterion function
	X	Y	
2	11	9	0.0260078
1	10.5	8.5	0.0306279
0.5	10.25	9.25	0.0335289

Another example divides the sensor nodes area onto two clusters having two cluster heads. The optimisation problem for the second example is formulated as the coordinates of the optimum location of two cluster heads which form two clusters. The same conditions in terms of the grid cell size have been applied to the second example. The results of the optimisation are given in Table 3. Figure 3 shows the solution coordinates for three grid cell sizes where the sensor nodes are divided onto clusters. As in the previous example, the best (highest) value of the criterion has been obtained for the smallest grid cell size. It should be noted that the algorithm generates two solutions for each pair of the cluster heads. In fact two solutions are identical representing two symmetrical coordinates for the cluster heads.

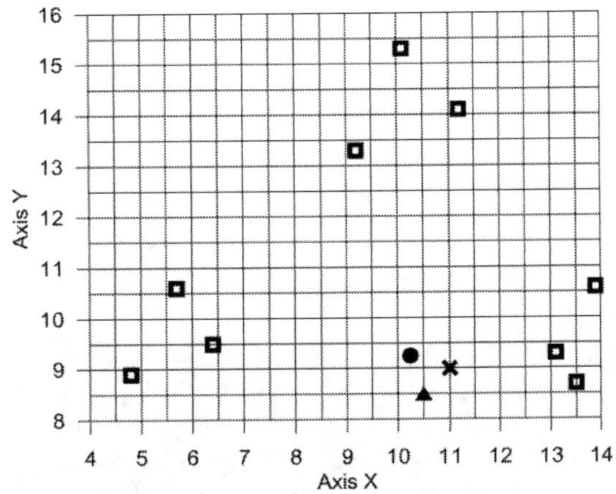

Figure 2. Solution coordinates of one cluster head, where □ – sensor nodes, **x** – the solution coordinates of the cluster head at grid cell size of 2.0, ▲ – the solution coordinates of the cluster head at grid cell size of 1.0, • – the solution coordinates of the cluster head at grid cell size of 0.5.

Table 3. Solution coordinates for two cluster heads at various grid cell size.

Grid cell size	Solution coordinates for two cluster heads				Criterion function
	X_1	Y_1	X_2	Y_2	
2	7	11	13	9	0.0729927
2	13	9	7	11	0.0729927
1	6.5	11.5	12.5	9.5	0.0735294
1	12.5	9.5	6.5	11.5	0.0735294
0.5	6.75	11.75	12.75	8.75	0.079968
0.5	12.75	8.75	6.75	11.75	0.079968
0.25	7.125	11.375	12.625	9.125	0.0867209
0.25	12.625	9.125	7.125	11.375	0.0867209

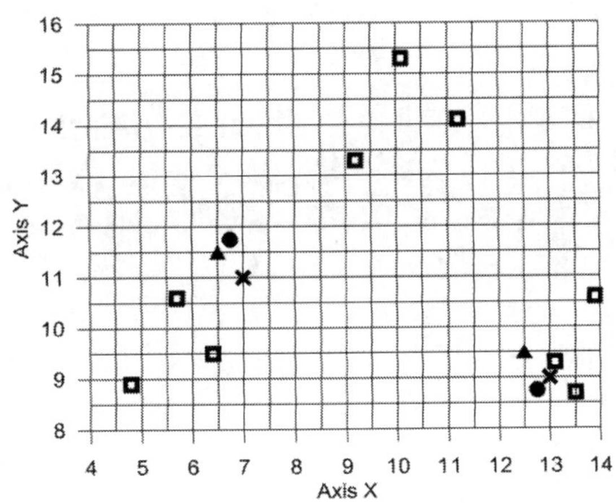

Figure 3. Solution coordinates of two cluster heads, where □ – sensor nodes, **x** – the solution coordinates of the cluster heads at grid cell size of 2.0, ▲ – the solution coordinates of the cluster heads at grid cell size of 1.0, • – the solution coordinates of the cluster heads at grid cell size of 0.5.

5 Conclusion

The examples demonstrate an effective implementation of the brute force algorithm to determine the optimum location of the cluster heads in WSNs where the optimisation criterion is intensity of radio signal at the network nodes' locations. The algorithm is invariant with respect to the criterion function and can be used for optimisation under another criterion without substantial modifications. The proposed algorithm was used for the location optimisation on a flat area. The further work is related to implementation of the algorithm for topology optimisation in a three dimensional space area.

A significant drawback of the proposed method is a high computational complexity. For example, the solution for two cluster heads under the grid cell size of 0.5 required a search time of 150 seconds. For relatively small numbers of cluster heads and sensor nodes the solution can be calculated using a powerful workstation. However for large values the solution computational algorithm should be executed under a multi-threaded approach using, for example, Intel Xeon Phi accelerators. Another approach to reduce computational complexity is based on a topology fragmentation. Under this approach the solution can be found as a set of optimum solutions obtained for a number of local topologies. The further work aims to investigate implementation of topology fragmentation algorithms for topology optimisation of WSNs.

The paper has been supported by RFBR No 16-07-01055\165 "Adaptation of resource demanding algorithms to distributed computational environment".

References

1. P.-J. Chuang, Y.-J. Jiang, Effective neural network-based node localisation scheme for wireless sensor networks, IET Wireless Sensor Systems, **4**(2), 97-103 (2014)
2. N. Panigrahi, P.M. Khumar, Optimal topological balancing strategy for performance optimisation of consensus-based clock synchronisation protocols in wireless sensor networks: A genetic algorithm-based approach, IET Wireless Sensor Systems, **4**(4), 213-222 (2014)
3. X. Shen , Z. Wang, Y. Sun, Wireless sensor networks for industrial applications, 5th World Congress on Intelligent Control and Automation, 3636-3640 (2004)
4. L F. Ge, Y. Wang, An adaptive energy efficient topology for wireless sensor networks, 2nd Int. Conf. on Future Generation Communication and Networking, 178-192 (2008)
5. H. Chen, C.K. Tse, J. Feng, Impact of topology on performance and energy efficiency in wireless sensor networks for source extraction, IEEE Trans. on Parallel and Distributed Systems, **20**(6), 886-897 (2009)
6. C. Charalambous, S. Cui, A biologically inspired networking model for wireless sensor networks, IEEE Network, **24**(3), 6-13 (2010)
7. A.P. Bhondekar, V. Renu, M. Singla, C. Ghanshyam, P. Kapur. Genetic algorithm based node placement methodology for wireless sensor networks, Int. Multi Conf. of Engineers and Computer Scientists, **1**, 106-112 (2009)

A Hug Behavior Generation Model Based on Analyses of Human Behaviors for Hug Robot System

Mitsuru Jindai[1,a], Shunsuke Ota[1] and Tohru Sasaki[1]

[1]*Graduate School of Science and Engineering, University of Toyama, 3190 Gofuku, Toyama, 930-8555, Japan*

Abstract. In human face-to-face communication, embodied sharing using the synchrony of embodied rhythms is promoted by embodied interactions. In particular, embodied interactions with direct contact are considered to synchronize embodied rhythms effectively. Hug behavior is one of the types of embodied interactions that involve direct contact. In hug behaviors, humans whole-body contact with each other. In the case of interaction between a human and a robot, the robot synchronizes embodied rhythms effectively using hug behaviors. Thus, a behavior model that can generate a hug behavior with humans is required for robots. Therefore, in this paper, we analyzed hug behaviors between humans and proposed a hug behavior generation model for a hug robot system based in the results of the analysis. This model generates a hug behavior in response to human behavior when a human requests a hug from the robot. Next, we developed a hug robot system that uses the proposed hug behavior generation model. Using the robot system, we demonstrated the effectiveness of the proposed hug behavior generation model by means of a sensory evaluation.

1 Introduction

In human face-to-face communication, embodied sharing using the synchrony of embodied rhythms is promoted by embodied interactions. Therefore, embodied interactions are important to the smooth initiation of coexistence and communication. In the case of interaction between a human and a robot, the robot can begin to communicate and coexist smoothly with the human without producing feelings of aversion in the human if the robot generates an embodied interaction that is emotionally acceptable to the human. A robot that can generate such an interaction is emotionally acceptable to humans because it produces feelings of security. Embodied interactions that involve direct contact are considered to synchronize embodied rhythms particularly effectively.

In hug behaviors, humans make whole-body contact with each other. Therefore, hug behaviors are expected to synchronize embodied rhythms effectively. In the case of an interaction between a human and a robot, the robot can effectively synchronizes embodied rhythms by hug behaviors. Hug behaviors between robots and humans have been examined in previous studies [1]. These studies examined remote hug communications with partners or hugging robots. However, no previous studies have examined the hand behaviors of hugs or the timing of their initiation. Thus, a behavior model that can generate hug behavior with humans is required.

[a] Corresponding author : jindai@eng.u-toyama.ac.jp

Therefore, in this paper, we propose a hug behavior generation model based on an analysis of human behaviors. We first analyze hug behaviors between humans. Based on the results of this analysis, we propose a hug behavior generation model for a hug robot system. This model generates a hug behavior in response to human behavior when a human requests a hug from the robot. Furthermore, a hug robot system that uses the proposed hug behavior generation model is developed. Using the developed robot system, the effectiveness of the proposed hug behavior generation model is demonstrated by sensory evaluation.

2 Analysis of Hug Behavior between Humans

2.1 Hug Experiment

Hug behaviors between humans were measured experimentally to provide a basis for modelling similar behaviors between a human and a robot. In the experiment, human behaviors were measured using a three-dimensional motion capture system (VICON). The request side and response side were determined beforehand. Eleven reflection markers (head, both shoulders, both elbows, both wrists, both hands, and both ankles) were attached to the request-side and response-side subjects. Furthermore, microphones were used to input voice utterances of the request-side and response-side subjects.

Figure 1 shows a typical experimental scene. The subjects initially stand face to face at a distance of 3,000 mm, with their arms extended downward. The request-side subject approaches the response-side subject and starts to move his or her hands at an arbitrary time to hug the response-side subject. The response-side subject then moves his or her hands in response to this behavior by the request-side subject. The subjects were 30 healthy right-handed students (15 pairs) aged between 20 and 24, and each pair hugged 10 times.

Figure 1. Experimental scene of hug behavior between humans

2.2 Analysis Results

The rotational behaviors of the subjects' arm joints during a hug were analyzed to determine the hand trajectories. Furthermore, the timing of the initiation of hand behavior was analyzed.

2.2.1 Angular velocities of joints

The angular velocities of the joints of the response-side subjects were measured to analyze the hand behaviors of the response-sides subjects during hug behaviors between humans. Examples of measured velocity profiles are shown in Figures 2 and 3. Figure 2 shows angular velocity profiles of the elbow and shoulder, and Figure 3 shows an angular velocity profile of the external and internal rotation of the upper arm. Based on these figures, the velocity profiles were classified as belonging to either of two types. The first type is a bell-shaped velocity profile, as shown in Figure 2. In this velocity profile, there is a positive or a negative peak. The second type is a wave-shaped velocity profile, as shown in Figure 3. In this velocity profile, there is both a positive and a negative peak. The velocity profiles of the shoulder, elbow, forearm, and wrist were classified as being bell-shaped. The

DOI: 10.1051/matecconf/20168201011

velocity profile of the external and internal rotation of the upper arm was classified as being wave-shaped. These figures illustrate that the angular velocity profiles have smooth transitions. This indicates that the response-side subjects moved their right and left hands along smooth trajectories.

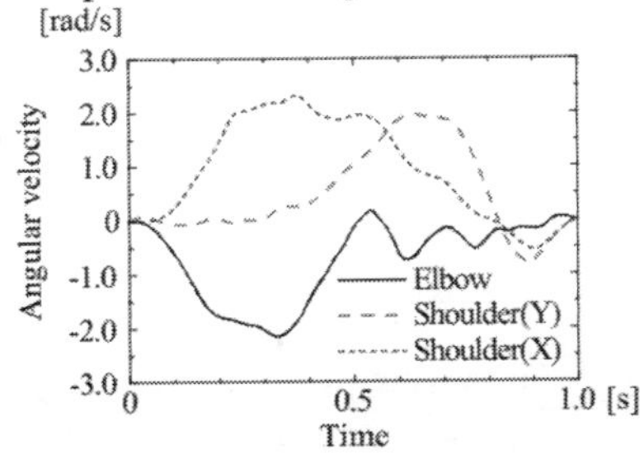

Figure 2. Velocity profiles of elbow and shoulder

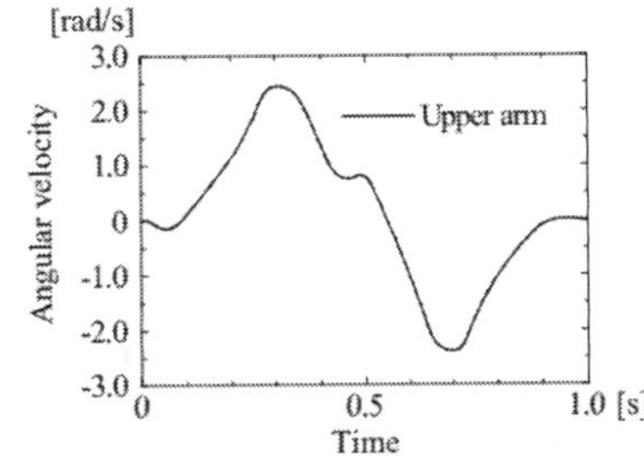

Figure 3. Velocity profiles of upper arm

2.2.2 Initiation of hand behaviors

In hug behaviors between humans, the response-side humans move their hands according to the behaviors of the request-side humans. Therefore, the timing of the initiation of hand behaviors is important for smooth hug behaviors. In this study, the time gaps between hand behaviors of the response-side subjects and various behaviors of the request-side subjects were measured. Hand behaviors, utterances of voice greetings, and the termination of approaching were the focuses of the observed hug behavior of the request-side subjects.

(1) Time gap of initiation of hand behaviors

The time gaps between the initiation of hand behaviors by the response side and the request side during the hug behaviors were measured. Figure 4 shows a histogram of the time gaps. As this figure shows, the time gaps that were measured for most of the subjects were widely distributed between -0.1 s and 1.2 s, with a peak of 0.3 s. A negative time gap indicates that the hand behavior of the response side began before the hand behavior of the request side, whereas a positive time gap indicates that the hand behavior of the response side began after the initiation of the hand behavior of the request side. These results suggest that most of the response-side subjects began their hug behavior 0.3 s after the initiation of the hug behavior of the request-side subjects.

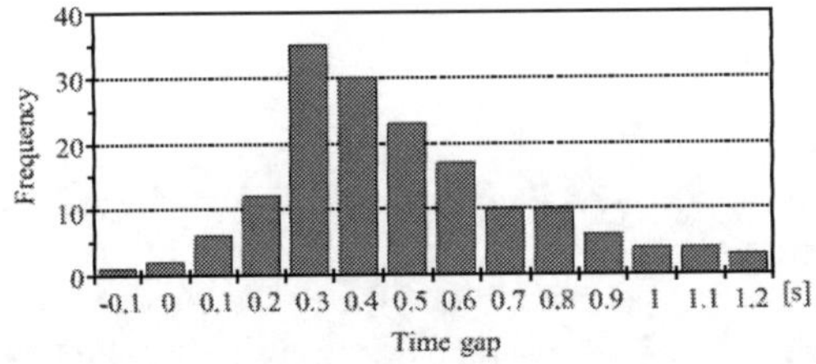

Figure 4. Time gap of hand behaviors

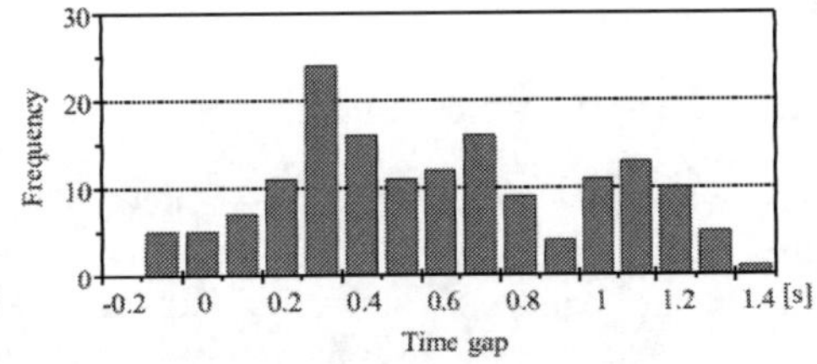

Figure 5. Time gap between hand behavior and voice

(2) Time gap between initiation of hand behavior and voice utterance

The time gaps between the initiation of hand behaviors of the response-side subjects and the initiation of voice utterances of the request-side subjects during the hug behaviors were measured. Figure 5 shows a histogram of the time gaps. As the figure shows, the time gaps measured for most of the subjects were widely distributed between -0.1 s and 1.4 s, with a peak of 0.3 s. A positive time gap indicates that the hand behavior began after the voice greeting. These results suggest that most of the response-side subjects begin their hug response behavior 0.3 s after the voice utterance of the request-side subjects.

(3) Time gap between initiation of hand behavior and termination of approaching

The time gaps between the initiation of hand behaviors of the response-side subjects and the termination of approaching of the request-side subjects were also measured. Figure 6 shows a histogram of the time gaps. As this figure shows, the time gap measured for most of the subjects were widely distributed between -0.4 s and 1.2 s, with a peak at 0.4 s. A positive time gap indicates that the hand behavior began after the termination of approaching. These results suggest that most of the response-side subjects began their hug response behavior 0.4 s after the termination of approaching behavior by the request-side subject.

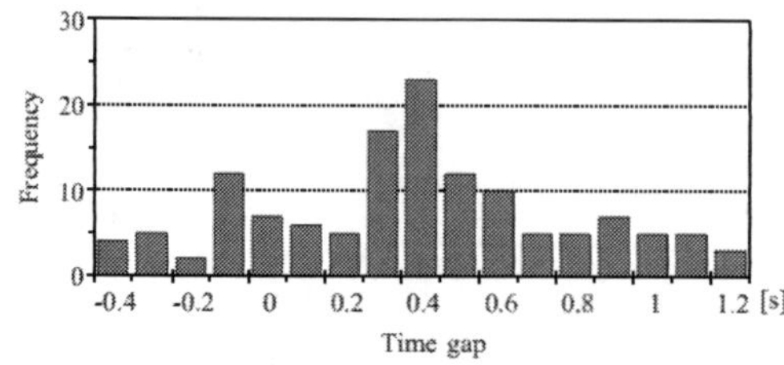

Figure 6. Time gap between hand behavior and termination

3 Hug Behavior Generation Model

A hug behavior generation model is proposed for generation of hug behavior. Based on the analysis of the hug behaviors between humans, it was determined that the velocity patterns of human arm joint movements during hug behavior are bell-shaped and wave-shaped. Therefore, a hug behavior generation model is proposed that can generate hug behavior described by these two velocity patterns.

3.1 Bell-Shaped Velocity Pattern

A bell-shaped velocity pattern with an adjustable peak time is generated on the basis of the minimum jerk model [2]. The minimum jerk model can accurately reproduce the point-to-point motion of human hands. In this model, the acceleration and velocity patterns exhibit smooth transitions. Behaviors that are generated on the basis of the minimum jerk model are expected to be preferred by humans. Therefore, a bell-shaped velocity pattern is generated from two velocity patterns that are generated by the minimum jerk model [3].

3.2 Wave-Shaped Velocity Pattern

The wave-shaped velocity pattern is generated using a fifth-order curve. The wave-shaped velocity pattern is shown in Figure 7. In the wave-shaped velocity pattern, its shape is determined by a movement time, maximum velocity, and initial and target angles. The fifth-order curve is given by Equation (1). The movement time and peak time are denoted by T_f and T_p, respectively. The time T_c is the time at which the velocity becomes 0. The value of the constant K is calculated from the initial angle R_0 and the target angle R_f, using Equation (2). The peak time T_p is calculated using Equation (3). Using these equations, the time T_c is determined as the velocity at time T_p reaches the maximum velocity V_{max}.

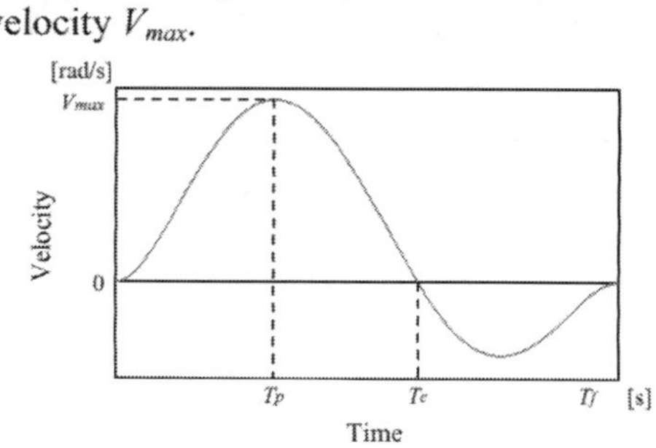

Figure 7. Velocity profiles of elbow and shoulder

$$V(t) = K(T_f - t)^2(T_c - t)t^2 \qquad (1)$$

$$K = \frac{60(R_f - R_0)}{2T_f^{\,5}T_c - T_f^{\,6}} \qquad (2)$$

$$T_p = \frac{(4T_c + 3T_f) - \sqrt{16T_c^{\,2} - 16T_fT_c + 9T_f^{\,2}}}{10} \qquad (3)$$

 DOI: 10.1051/matecconf/20168201011

3.3 Generated Hug behavior

In the proposed hug behavior generation model, the rotational behaviors of the shoulder, elbow, forearm, and wrist joints are generated using the bell-shaped velocity pattern, and the rotational behavior of the upper arm is generated using the wave-shaped velocity pattern. To produce a hand behavior similar to that of a human, the movement time, maximum velocity, peak time, and initial and target angles are matched to those of humans, as determined from the analysis of hug behaviors between humans.

4 Experiment Using Hug Robot System

A hug robot system that uses the proposed hug behavior generation model was developed as shown in figure 8. Using the hug robot system, an experiment was performed. Figure 9 shows a typical experimental scene. Three modes were used in this experiment. In mode (a), the robot began hug behavior 0.3 s after the initiation of hug behavior by a human. In mode (b), the robot began hug behavior 0.3 s after a voice utterance by a human. In mode (c), the robot began hug behavior 0.4 s after the termination of approaching behavior by a human. These modes were selected based on the results of the analysis of hug behaviors between humans, as discussed in the previous section.

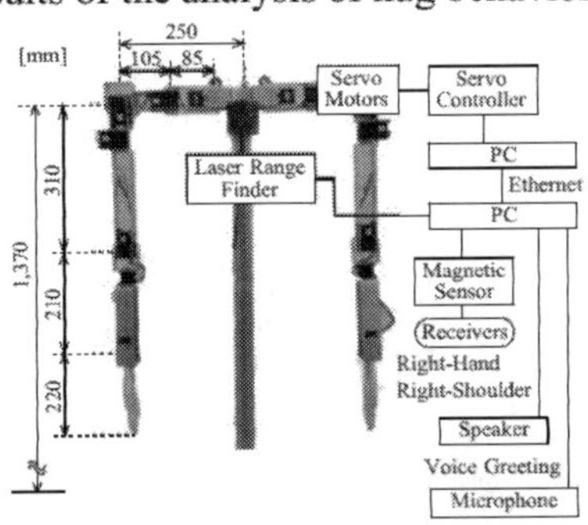

Figure 8. Hug Robot System

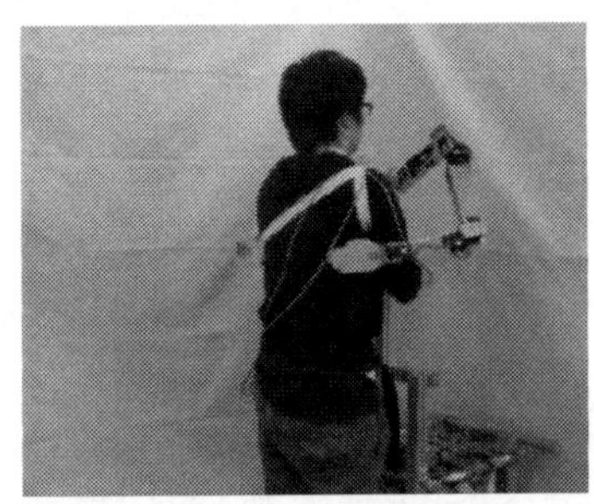

Figure 9. Experimental scene

4.1 Experimental Method

In the experiment, a human and the robot initially stood face to face at a distance of 3,000 mm. Next, the human approached the robot and started to move his or her hand at an arbitrary time to request a hug. The robot then moved its hand in response to this behavior. The experiment was performed for a hug behavior accompanied by a voice greeting. The robot uttered a voice greeting 0.1 s after the initiation of its hand behavior. This timing was selected based on the analysis of hug behaviors between humans.

In the sensory evaluation, paired comparison was made first. Subsequently, seven-point bipolar rating was made. Paired comparison was performed for all six pairs of mode combinations, and then seven-point bipolar rating on a scale from -3 (not at all) to 3 (extremely) for the following four items: “Ease of hug,” “Velocity (Comfortable),” “Security,” and “Politeness.” The subjects used the modes randomly. The subjects were 30 healthy students aged between 20 and 24 years. All subjects were males

DOI: 10.1051/matecconf/20168201011

4.2 Experimental results

The result of the paired comparison is shown in Table 1. The number of subjects that preferred the column mode to the row mode is shown in each cell of the table. Table 1 indicates that mode (b) was preferred by the largest number of subjects.

The results of the seven-point bipolar rating are shown in Figure 10. Modes (b) and (c) were rated higher than mode (a), with significant differences at the 1 and 5 % significance levels for the "Politeness" item. However, there were no significant differences between modes (b) and (c) for any of the items.

The results obtained using the paired comparison indicate that mode (b) was the preferred mode, and using the seven-point bipolar rating scale, all items related to mode (b) had positive values. Therefore, it can be concluded that the hug behavior described by mode (b) is preferred by humans over that described by the other modes.

Table 1. Result of the paired comparison

	(a)	(b)	(c)	Total
(a)		24	34	58
(b)	36		36	72
(c)	26	24		50

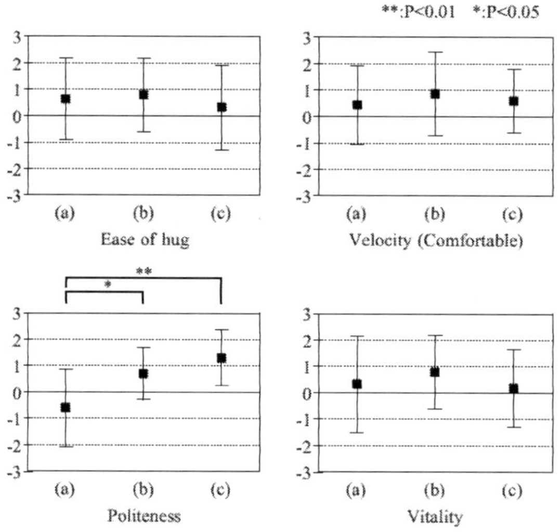

Figure 10. Results of seven-point bipolar rating

4.3 Discussion of Experimental Results

The experimental results suggest that in a hug interaction between a human and a robot, the robot should move its hands in response to a voice utterance by the human when the robot is the responder. The timing of a voice utterance by a human was found to be important to the generation of smooth hug behavior by a robot. Furthermore, a hug robot system can generate a hug behavior that is preferred by many humans. Therefore, the effectiveness of the proposed model has been demonstrated.

5 Conclusion

In this paper, we proposed a hug behavior generation model based on an analysis of human hug behaviors. This model generates a hug behavior in response to human behavior when a human requests a hug from a robot. Furthermore, a hug robot system that uses the proposed hug behavior generation model was developed. From experiment, in a hug interaction between a human and a robot, the timing of a voice utterance by the human was found to be important to the generation of smooth hug behavior by the robot. Furthermore, the effectiveness of the proposed hug behavior generation model has been demonstrated.

Acknowledgment

This work was supported by KAKENHI Grant Number 25330239 of the Japan Society for the Promotion of Science (JSPS), Japan.

References

1. J. Nakanishi, K. Kuwamura, T. Minato, S. Nishio and H. Ishiguro, Evoking Affection for a Communication Partner by a Robotic Communication Medium, Proceedings of the 1st International Conference on Human-Agent Interaction, III-1-4 (2013)
2. T.Flash and N.Hogan, The Coordination of Arm Movements: An Experimentally Confirmed Mathematical Model, Journal of Neurosciences Vol.**5**, No.7, pp.1688-1703 (1985)
3. M. Jindai and T. Watanabe, Development of a Handshake Request Motion Model Based on Analysis of Handshake Motion between Humans, Proceedings of the 2011 IEEE/ASME International Conference on Advanced Intelligent Mechatronics, pp.560-565 (2011)

MATEC Web of Conferences **82**, 01012 (2016) DOI: 10.1051/matecconf/20168201012
D2ME 2016

Study on Circle Center Positioning Error by Fitting of Measurement System for Gear Chamfering

Zhang chi[a], Xu Zengpu, Liu Guohao, Wang Yongqiang and Zhou Congling

Department of Mechanical Engineering,Tianjin University of Science and Technology: No.1038 Dagu Nanlu,Hexi District, Tianjin,China

Abstract. This paper briefly introduced the working principle of non-contact measurement system of gear chamfering based on computer vision technology and laser range-finding technology. Then the circle center positioning method during the process of the first-stage positioning measurement of measurement system was studied, and the influence of circle center positioning error on the precision of starting point for measurement was analyzed. The standard circle with a cross and a radius of 20mm was used for the measurement experiment, as well as the precision error rule of fitting circle center positioning and influence relationship of the starting point error of measurement were studied. Thus a kind of effective method to improve the measurement precision was put forward.

1 Introduction

The machining precision of gear with chamfering in the automobile gearbox would directly affect the technical indexes of automobile transmission system [1-2]. At present, the non-contact measurement method based on computer vision technology and laser range-finding technology has become the main means to detect the machining precision of gear chamfering. In order to obtain the actual location of the gear, the least square method was used to fit the distribution points in the dedendum circle of gear so as to obtain the circle center coordinate of fitting circle during the actual measurement process. Whether the circle center positioning is accurate or not directly affects the subsequent positioning of starting points for measurement in the cross section of gear chamfering as well as determines the accuracy of the final measurement results. Hence, it is necessary to analyze and study the positioning precision of fitting circle center in the first-stage vision measurement system.

2 Working Principle of Measurement System

2.1 Measurement object and parameters

Figure 1 (a). showed the real picture of gear with chamfering. The main parameters to be measured were the angle M of the modeling section and the distance N from the intersecting line of modeling

[a] Corresponding author : zhangchi1025@163.com

surface to the specified surface, as shown in Figure 1 (b) [3-4].

(a) Real picture of gear with chamfering.

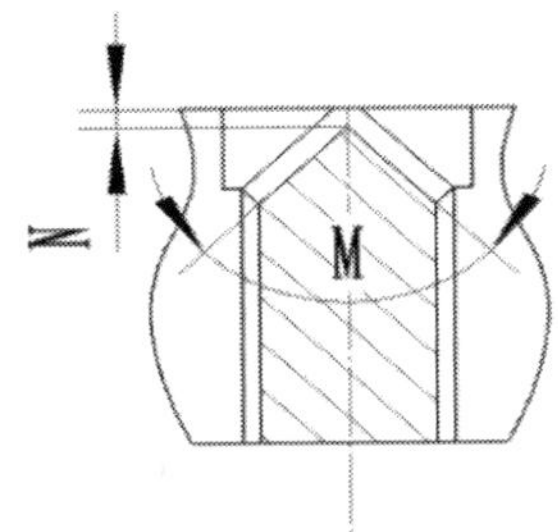

(b) Parameters to be measured.

Figure 1. Structure and measurement parameters of gear chamfering.

2.2 Working principle of measurement system

The non-contact measurement system of gear with chamfering was composed of XY two-dimensional precise mobile platform, camera (CCD array sensor), backward light source, laser displacement sensor, marble base, rotating stage, computer, and so on (shown in Figure 2.) This system used the two-stage collaborative measurement way. The first-stage measurement adopted both the array image sensor and calibration algorithm to complete the preliminary measurement positioning [5]. Likewise, the second-stage measurement used two-dimensional high-precision electric platform with a laser displacement sensor to finish the high-precision measurement for workpiece contour sizes according to the planning path, and then the contour and parameters of gear chamfering were worked out by program.

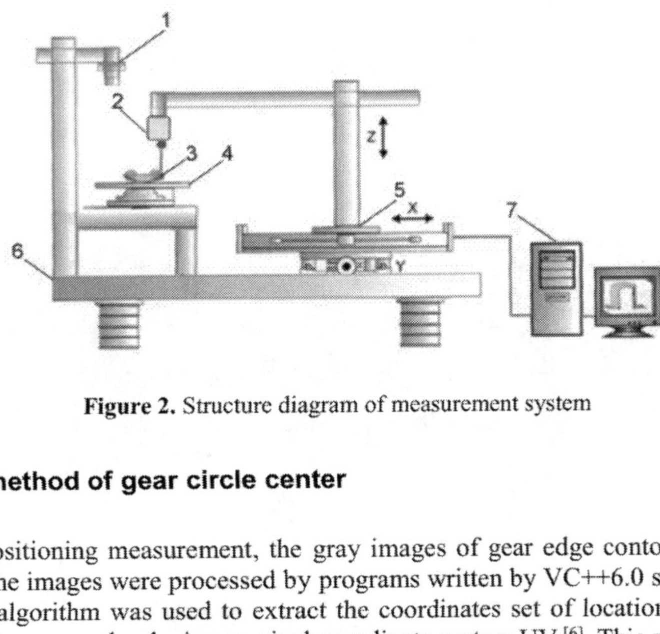

Figure 2. Structure diagram of measurement system

2.3 Positioning method of gear circle center

In the first-stage positioning measurement, the gray images of gear edge contour were collected by cameras, and then the images were processed by programs written by VC++6.0 software. Furthermore, the edge detection algorithm was used to extract the coordinates set of location points for the outer circle edge of gear images under the image pixel coordinate system UV [6]. This set can be divided into three subsets according to different locations of edge points, including the points set of addendum circle, the points set of dedendum circle, and the points set of tooth profile. The points set of dedendum circle was extracted from them, and then least square circle fitting algorithm was used to obtain the coordinate of the gear circle center. The set composed of all the points on the dedendum circle was denoted by D, and the distance from any point D_i (u_i, v_i) to the circle center was R_i. The area of the circle with a radius of R_i was s_i. In addition, the fitting circle center coordinate was (x_0, y_0), the standard circle radius was R, and its area was s. Then the area error of two circles was:

$$\begin{aligned} \varepsilon_i &= s_i - s = \pi R_i^2 - \pi R^2 \\ &= \pi[(u_i - x_0)^2 + (v_i - y_0)^2 - R^2] \end{aligned} \quad (1)$$

The quadratic sum function of error was expressed as:

$$F(x_0, y_0, R) = \pi^2 \sum_{i=1}^{n} [(u_i - x_0)^2 + (v_i - y_0) - R^2]^2 \quad (2)$$

According to the least squares principle, the values of x_0, y_0 and R were achieved as the minimum of F (x_0, y_0, R) was taken. Thus the partial derivative of F (x_0, y_0, R) was worked out and then was equal to zero, namely,

$$\frac{\partial F}{\partial x_0} = \frac{\partial F}{\partial y_0} = \frac{\partial F}{\partial R} = 0 \quad (3)$$

$$\begin{aligned} \frac{\partial F}{\partial x_0} &= \pi^2 \sum_{i=1}^{n} (-4)[(u_i - x_0)^2 + (v_i - y_0)^2 - R^2](u_i - x_0) = 0 \\ \frac{\partial F}{\partial y_0} &= \pi^2 \sum_{i=1}^{n} (-4)[(u_i - x_0)^2 + (v_i - y_0)^2 - R^2](v_i - y_0) = 0 \\ \frac{\partial F}{\partial R} &= \pi^2 \sum_{i=1}^{n} (-4)[(u_i - x_0)^2 + (v_i - y_0)^2 - R^2]R = 0 \end{aligned} \quad (4)$$

Equation (4) was solved, then

$$\begin{cases} x_0 = \dfrac{hq - tc}{2(pq - c^2)} \\ y_0 = \dfrac{tp - hc}{2(pq - c^2)} \end{cases} \quad (5)$$

Where

$$p = m\sum x_j^2 - (\sum x_j)^2$$

$$q = m\sum y_j^2 - (\sum y_j)^2$$

$$h = m\sum x_j^3 + n\sum x_j y_j^2 - \sum (x_j^2 + y_j^2)\sum x_j$$

$$t = m\sum y_j^3 + m\sum x_j^2 y_j - \sum (x_j^2 + y_j^2)\sum y_j$$

$$c = m\sum x_j y_j - \sum (x_j)\sum y_j$$

The precise circle center coordinate O (O_u, O_v) of the gear could be obtained by equation (5).

2.4 Influence of positioning error of circle center on the precision of starting point for measurement

There existed error in the position of circle center coordinate, hence, the error would also appear in the starting point position of gear chamfering.

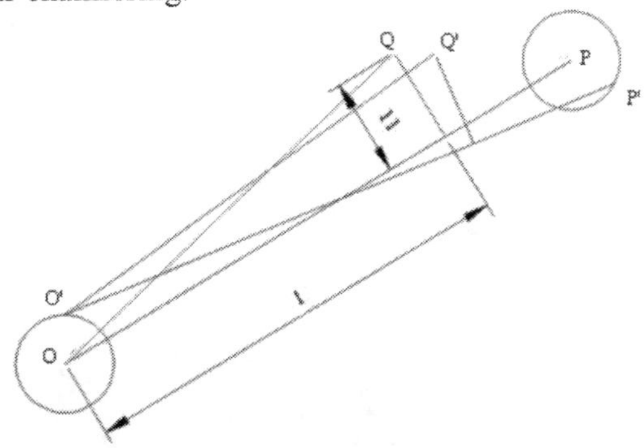

Figure 3. Analysis method of influence of circle center error on the starting points for measurement.

The coordinates of points Q and Q' deduced from Figure 3 were shown in equations (6)-(9).

$$Qx = \sqrt{l^2 + l1^2} \cdot \cos(\arctan(\frac{Py - Oy}{Px - Ox}) + \arctan(\frac{l1}{l})) + Ox \quad (6)$$

$$Qy = \sqrt{l^2 + l1^2} \cdot \sin(\arctan(\frac{Py - Oy}{Px - Ox}) + \arctan(\frac{l1}{l})) + Oy \quad (7)$$

$$Q'x = \sqrt{l^2 + l1^2} \cdot \cos(\arctan(\frac{Py + \delta y - Oy - \Delta y}{Px + \delta x - Ox - \Delta x}) + \arctan(\frac{l1}{l})) + Ox + \Delta x \quad (8)$$

$$Q'y = \sqrt{l^2 + l1^2} \cdot \sin(\arctan(\frac{Py + \delta y - Oy - \Delta y}{Px + \delta x - Ox - \Delta x}) + \arctan(\frac{l1}{l})) + Oy + \Delta y \quad (9)$$

Where Q was the theoretical starting point for measurement. Q' was the measurement starting point deduced by fitting. l was the distance from the measuring cross section to the circle center of gear. $l1$ was half the distance from the measuring cross section to the circle center of gear. O was the theoretical circle center of gear. O' was the circle center of gear by fitting. P was the theoretical central point of addendum arc chord. P' was the central point of addendum arc chord obtained by fitting.

3. Measurement Experiment

3.1 Experiment scheme

In order to verify the detection precision of measurement system, the standard circle with a cross and a radius of 20 mm (shown in Figure 4) was used for field measurement. The experiment method was as follows. Firstly, the camera was used to take checkerboard images, and then the sizes were calibrated. After the calibration, the checkerboard was removed, and the standard circle was brought. Then the standard circle images were taken (standard circle was placed below the camera lens and kept away from the camera lens. Then the standard circle images were taken one by one.) The computer was used to process the images one by one to obtain the points on the outer contour of standard circle. Thus the circle center coordinate was obtained by fitting. Then the cross lines were manually fitted to obtain the intersection, and the intersection was taken as the theoretical circle center. Finally, the circle center coordinate of fitting circle was compared with the hypothetically theoretical circle center coordinate to work out the positioning precision error of the circle center coordinate of fitting circle in the measurement system, and the measurement results were shown in Table 1.

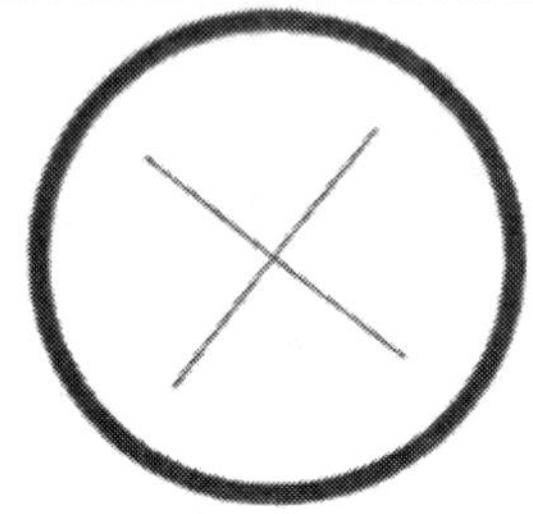

Figure 4. Standard circle with a cross.

Table1. Measurement results on error of center location in measuring

Distance	Error (mm)								
Near	0.03373	0.02236	0.02961	0.07689	0.081437	0.06933	0.06906	0.06594	0.06296
	0.04070	0.03190	0.01431	0.04472	0.037643	0.03088	0.05024	0.05930	0.06184
	0.09701	0.04687	0.03640	0.02707	0.027586	0.01216	0.02690	0.03512	0.06640
	0.10751	0.06726	0.00412	0.07905	0.045099	0.01526	0.02816	0.03417	0.08338
	0.08297	0.04812	0.03101	0.08586	0.087143	0.02213	0.01843	0.04111	0.09861
Distant	0.04904	0.061770	0.03114	0.07975	0.095707	0.05261	0.02047	0.04494	0.03257

As shown in Table 1, when the standard circle moved from the near to the distant, there was no obvious change trend for the error between the fitting circle center and theoretical circle center. Hence, all the measuring data needed the statistical analysis, as shown in Figure 4.

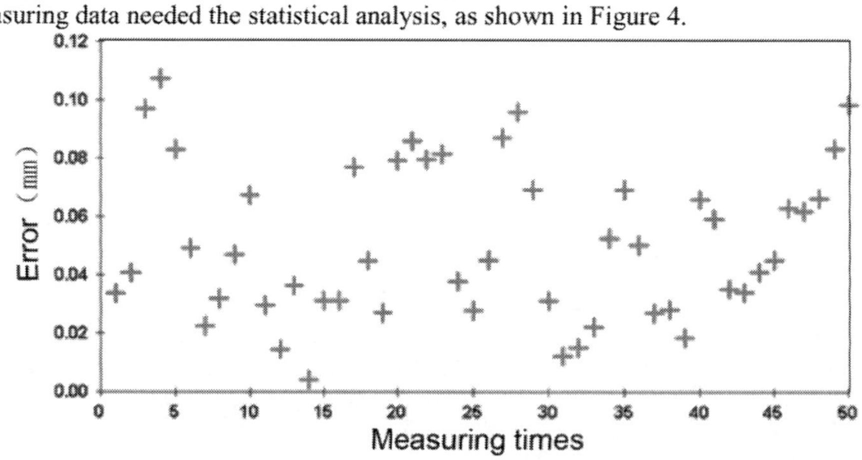

Figure 5. Measurement data error distribution of circle center positioning errors.

As shown in Figure 5, few error data distributed in two sides, the error data mostly distributed between 0.03 mm and 0.08mm, and the mean of errors was 0.0509mm. In order to further verify the influence of circle center error on the starting point position of measurement section, the software program codes written according to the experiment results were as follows.

```
x1=r1*cos(sita);% X coordinate of circle center error by fitting;
y1=r1*sin(sita);% Y coordinate of circle center error by fitting
x2=0;% X coordinate of center error by fitting;
y2=0;% Y coordinate of center error by fitting;
%    x1=0;% X coordinate of circle center error by fitting;
%    y1=0;% Y coordinate of circle center error by fitting;
%    x2=r2*cos(beta);% X coordinate of center error by fitting;
%    y2=r2*sin(beta);% Y coordinate of center error by fitting;
alpha1=atan((p2-o2)/(p1-o1));% Angle between theoretical normal and horizontal line, and this value cannot be pi/2(p1 != o1);
alpha2=atan(l/l1);% Fixed angle;
alpha=alpha1+alpha2;% The sum of theoretical angles;
gamma1=atan((p2+y2-o2-y1)/(p1+x2-o1-x1));% Angle between fitting normal and horizontal line;
gamma=gamma1+alpha2;% The sum of fitting angles;
x11=sqrt(l*l+l1*l1)*cos(alpha)+o1;% X coordinate of theoretical starting points for measurement;
y11=sqrt(l*l+l1*l1)*sin(alpha)+o2;% Y coordinate of theoretical starting points for measurement;
x22=sqrt(l*l+l1*l1)*cos(gamma)+o1+x1;% X coordinate of starting points for measurement by fitting;
y22=sqrt(l*l+l1*l1)*sin(gamma)+o2+y1;% Y coordinate of starting points for measurement by fitting;
d=sqrt(((x11-x22)^2)+((y11-y22)^2));% Distance between theoretical points and fitting points;
```

The programs were run, and then the analysis results were obtained, as shown in Table 2.

Table 2. Effect of central fitting error on the starting points for measurement.

Name	Measurement data					
Circle center error (mm)	0.01	0.02	0.03	0.04	0.05	0.06
Error of starting points for measurement (mm)	0.01013	0.02025	0.03038	0.04050	0.05063	0.06076

It can be seen from the measuring data that the error of circle center fitting was proportional to the error of starting points for measurement. Namely, the error of starting points for measurement was about 1.013 times as large as the error of circle center. The camera pixel was equivalent to the actual size of 0.178mm, and the measuring error was within a pixel. Hence, in order to improve the circle center precision of fitting circle, the camera and lens with high resolution should be chosen so as to reduce pixel minimal size and decrease circle center error of fitting circle, and thus the positioning precision of starting points for measurement was improved.

4. Conclusions

This paper studied the circle center positioning precision of fitting circle in the non-contact measurement system of gear with chamfering and determined that the precision of fitting circle center was 0.0509mm in the measurement system by the experiment. Then the influence of circle center positioning error on the position errors of starting points for measurement on the measuring cross section was analyzed. Finally, it is concluded that the fitting precision of circle center was improved by improving the resolution of camera and decreasing the distinguishable minimal size precision. This conclusion can provide the theoretical basis for perfecting the working performance and measuring precision of measurement system.

References

1. Xu Zengpu, Zhang Chi, Cai Chunming, and et al. Study on cooperative high precision measurement method of gear chamfering profile. Manufacturing Automation [J] (2015)
2. Geometrical modeling and adjustment simulation of the chamfering cutter of gear curved face gear, written by Zhanwen Zhu, Renlong Liu, Wei Guo, Tianjin University Paper (2006)
3. XU Z P, ZH SH. Gear Chamfering Profile Measurement System Based On Laserb And Machine Vision [C]. AMR, [EI] (2011)
4. LI Z Q, DONG X L, LI Q L, GUO C L. Research of small modulus gear vision measurement system [J].Coastal enterprise and science & technology (2010)
5. XU Z P, YU D M. Auto-calibration method of light and image plates [J]. Journal of Tianjin institute of light industry（1993）
6. C Zhou，Z Xu，C Cai，Y Wang, A high precision collaborative vision measurement of great chamfering profile[C],International Conference on Advances in Mechanical Engineering & Industrial Informatics (2015)

Stability and Bifurcation Analysis of Functionally Graded Materials Plate Under Different Loads and Boundary Conditions

Shuping Chen [1,a], Wei Zhang [2] and Yuxin Hao [3]

[1] *College of Mathematics, Xiamen University of Technology, Xiamen 361024, P. R. China*
[2] *College of Mechanical Engineering, Beijing University of Technology, Beijing 100124, P. R. China*
[3] *College of Mechanical Engineering, Beijing Information Science and Technology University, Beijing 100192, P. R. China*

Abstract. This paper presents the stability and bifurcation analysis of a simply-supported functionally graded materials (FGMs) rectangular plate subject to the transversal and in-plane excitations. A two-degree-of-freedom nonlinear system of the FGM plate is obtained via the Hamilton's principle and the Galerkin approach. The case of primary parametric resonance and 1:2 internal resonance is considered. The asymptotic perturbation method is utilized to obtain four-dimensional nonlinear averaged equation. With the aid of Matlab and normal form theory, the various types of dynamical behavior in the neighborhood of a kind of degenerated equilibrium point are investigated. It was found that static bifurcation and Hopf bifurcation exist for the FGM rectangular plate under certain conditions.

1 Introduction

The functionally graded materials (FGMs) are a kind of microscopically composite materials. Their properties can be spatially varied to achieve the desired structural, electrical functions, or thermal. It is emerging with great potential in mechanical science research because of their high performance and multi-function [1-3].

For FGM plate, many studies have concentrated on the nonlinear dynamic behavior analysis. Sladek et al. [4] applied the meshless local Petrov-Galerkin method to a circular plate with functionally graded piezoelectric material properties. Li et al. [5] studied the free vibration of functionally graded material sandwich rectangular plates with simply supported and clamped edges based on the three-dimensional linear theory of elasticity. Allahverdizadeh et al. [6] developed a semi-analytical approach for nonlinear free and forced axisymmetric vibration of a thin circular functionally graded plate. Gunes and Reddy [7] investigated the geometrically nonlinear analysis of functionally graded circular plates subjected to mechanical and thermal loads. Hao [8] investigated the nonlinear dynamics of a simply-supported FGM rectangular plate with the parametric and forcing excitations.

Many studies have considered the stability and bifurcation of nonlinear systems. Zhang [9] investigated the stability and bifurcation behaviors of a two-dimensional nonlinear viscoelastic panel in supersonic flow by using analytical and numerical methods. Anton et al. [10] applied the stochastic

[a] Corresponding author : shupingchen@126.com

normal form to an aeroelastic random dynamical system with uncertainties in the bifurcation parameter and the plunge and pitch nonlinear terms. Li et al. [11] analyzed the stabilities and bifurcations of a cantilevered plate with nonlinear motion constraints in an axial subsonic flow.

This paper focuses on research on the bifurcation and stability analysis of the FGM rectangular plate deflection with variation of material properties. Based on the theoretical model for the nonlinear oscillations of the FGM plate subjected to the transversal and in-plane excitations, the normal form theory and bifurcation theory are applied to discuss the local dynamical behaviors of the FGMs rectangular plate.

2 The functionally graded materials plate model

In this section, we use the normal form method to the rectangular plate made of FGMs with length a, width b and thickness h as presented in Fig.1. The plane is subjected to in-plane and transverse excitations. A Cartesian coordinate system is located in the middle surface of the FGMs rectangular plate. u, v and w are the displacements of an point within the plate in the x, y and z directions. $-\left(p_0 - p_1\cos\left(\Omega_2 t\right)\right)$ represents the in-plane excitation along the width. The transverse excitation is written as $F(x,y)\cos(\Omega_1 t)$. In this paper, the governing equations of transverse motion for the FGMs plate in dimensionless form can be expressed as follows: [8, 12, 13]

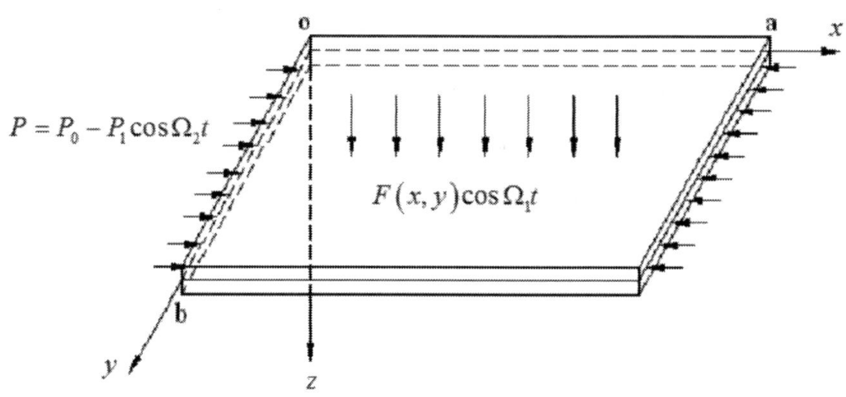

Figure 1. The model of a functionally graded materials rectangular plate

$$\ddot{w}_1 + \omega_1^2 w_1 + a_1\dot{w}_1 + a_2 w_1\cos\Omega_2 t + a_3 w_1^2 + a_4 w_2^2 + a_5 w_1 w_2^2 + a_6 w_1^3 + a_7 w_1 w_2 = f_1\cos\Omega_1 t, \tag{1a}$$

$$\ddot{w}_2 + \omega_2^2 w_2 + b_1\dot{w}_2 + b_2 w_2\cos\Omega_2 t + b_3 w_1 w_2 + b_4 w_1^2 + b_5 w_2^2 + b_6 w_2 w_1^2 + b_7 w_2^3 = f_2\cos\Omega_1 t, \tag{1b}$$

where a_i, b_i $(i = 1,\cdots,7)$ are non-dimensional coefficients and can be found in the Reference [13].

In this paper, we investigate the case of 1:2 internal resonance and primary parametric resonance for the FGMs rectangular plate,

$$\omega_1 = \frac{\Omega_1}{2} + \varepsilon^2\sigma_1,\ \omega_2 = \Omega_1 + \varepsilon^2\sigma_2,\ \Omega_1 = \Omega_2 = \Omega, \tag{2}$$

where σ_1 and σ_2 are two detuning parameters.

Using the Galerkin method and the asymptotic perturbation method, the four-dimensional averaged equation of a FGMs plate is defined by

$$\begin{aligned}\dot{x}_1 = &-\mu_1 x_1 + \left(\sigma_1 + \alpha_1 + \alpha_2 f_1\right)x_2 + 14\alpha_2 x_1^2 x_2 + 14\alpha_2 x_2^3 + \alpha_3 x_1 x_2^2 + \alpha_3 x_1^3 \\ &+\alpha_4 x_2 x_3^2 + \alpha_4 x_2 x_4^2 + \alpha_5 x_2 x_3 - \alpha_5 x_1 x_4,\end{aligned} \tag{3a}$$

$$\dot{x}_2 = (-\sigma_1 + \alpha_1 + \alpha_2 f_1)x_1 - \mu_1 x_2 - 14\alpha_2 x_1^3 - 14\alpha_2 x_2^2 x_1 - \alpha_4 x_1 x_3^2 - \alpha_4 x_1 x_4^2$$
$$+\alpha_3 x_1^2 x_2 + \alpha_3 x_2^3 + \alpha_5 x_1 x_3 + \alpha_5 x_2 x_4 \tag{3b}$$
$$\dot{x}_3 = \beta_1 f_1 x_2 - \mu_2 x_3 + \sigma_2 x_4 + \beta_2 x_1^2 x_4 + \beta_2 x_2^2 x_4 + \beta_3 x_4^3 + \beta_3 x_4 x_3^2 + \beta_4 x_3^2 x_2$$
$$+\beta_4 x_4^2 x_2 - \beta_5 x_2^3 - \beta_5 x_1^2 x_2, \tag{3c}$$
$$\dot{x}_4 = \beta_1 f_1 x_1 - \sigma_2 x_3 - \mu_2 x_4 - \beta_2 x_1^2 x_3 - \beta_2 x_2^2 x_3 - \beta_3 x_3^3 + \beta_3 x_3 x_4^2 + \beta_5 x_1^3$$
$$-\beta_5 x_2^2 x_1 - \beta_4 x_3^2 x_1 - \beta_4 x_4^2 x_1 + \frac{1}{4} f_2. \tag{3d}$$

3 Stability and bifurcation analysis of the FGMs plat

In this section, the normal form theory and bifurcation theory is utilized to analyze the dynamical behaviors of the FGMs rectangular plate. The expression for the characteristic polynomial of equation (3) can be given as

$$f(\lambda) = \lambda^4 + c_1\lambda^3 + c_2\lambda^2 + c_3\lambda + c_4, \tag{4}$$

where

$$c_1 = 2(\mu_1 + \mu_2),\ c_2 = \sigma_1^2 + \sigma_2^2 + \mu_1^2 + \mu_2^2 + 4\mu_1\mu_2 - \alpha^2,$$
$$c_3 = 2\mu_1\mu_2^2 + 2\mu_2\mu_1^2 + 2\mu_1\sigma_2^2 + 2\mu_2\sigma_1^2 - 2\mu_2\alpha^2,$$
$$c_4 = \mu_1^2\mu_2^2 + \mu_1^2\sigma_2^2 + \mu_2^2\sigma_1^2 + \sigma_1^2\sigma_2^2 - \mu_2^2\alpha^2 - \sigma_2^2\alpha^2, \tag{5}$$

when $\alpha = \alpha_1 + \alpha_2 f_1$, $\gamma_1 = \beta_1 f_1$.

Let

$$\mu_1 = 1,\ \mu_2 = \sigma_1 = 0,\ \sigma_2 = 1,\ \alpha = 1,\ \gamma_1 = 1,\ \alpha_2 = \alpha_3 = \alpha_4 = \alpha_5 = 1,\ \beta_2 = \beta_3 = \beta_4 = \beta_5 = 1, \tag{6}$$

Therefore, the eigenvalues can be given by

$$\lambda_1 = 0,\ \lambda_{2,3} = \pm i,\ \lambda_4 = -2. \tag{7}$$

Letting $\mu_1 = 1 + \varsigma_1$, $\mu_2 = \varsigma_2$, and using the normal form theory and the cylinder coordinate transformation

$$z_1 = y,\ z_2 = r\cos\theta,\ z_3 = r\sin\theta,\ z_4 = z_4, \tag{8}$$

The equations (3) can be rewritten as

$$\dot{y} = y\left(-\varsigma_1 - \frac{12}{5}r^2 - 2y^2\right), \tag{9a}$$

$$\dot{r} = r\left(-\varsigma_2 - \frac{119}{100}r^2 - \frac{3}{10}y^2\right). \tag{9b}$$

We can give some analyses to observe the dynamical behaviors of system (9).

(1) The original equilibrium solution

$$y = r = 0. \tag{10}$$

From equation (9), it is easy to see that the original equilibrium solution is stable when $\varsigma_1 > 0$ and $\varsigma_2 > 0$. Therefore, the invariant lines which represent the stability boundaries of the original equilibrium solution are L_1: $\varsigma_2 = 0$ and L_2: $\varsigma_1 = 0$.

(2) The static bifurcation solution

$$y^2 = -\frac{\varsigma_1}{2},\ r = 0. \tag{11}$$

It is obvious that the static bifurcation solution from the initial equilibrium solution takes place when $\varsigma_1 < 0$. It is to be noted that system (9) undergoes a static bifurcation on the curve:

$$T = \left\{ (\varsigma_1, \varsigma_2) : \varsigma_1 < 0,\ -\varsigma_2 + \frac{3}{20}\varsigma_1 < 0 \right\}.$$

(3) The incipient Hopf bifurcation solution

$$y = 0,\ r^2 = -\frac{100}{119}\varsigma_2. \tag{12}$$

A family of limit cycles are bifurcated when $\varsigma_2 < 0$. We note that the incipient Hopf bifurcation solution is stable whenever $\varsigma_2 < 0$ and $-\varsigma_1 + \frac{240}{119}\varsigma_2 < 0$. So the incipient Hopf bifurcation occurs on the line $\varsigma_2 = 0$ and we obtain a curve L_4: $-\varsigma_1 + \frac{240}{119}\varsigma_2 = 0$.

(4) The secondary Hopf bifurcation solution

$$y^2 = \frac{10}{83}\left(12\varsigma_2 - \frac{119}{20}\varsigma_1\right),\ r^2 = -\frac{100}{83}\left(\frac{3}{20}\varsigma_1 - \varsigma_2\right). \tag{13}$$

It is easy to see that the secondary Hopf bifurcation solution takes place when $-\varsigma_1 + \frac{240}{119}\varsigma_2 < 0$ and $-\varsigma_2 + \frac{3}{20}\varsigma_1 < 0$.

To reveal further the nonlinear dynamic behaviors of the functionally graded materials plate, the waveforms and the phase portraits are depicted. Fig. 2 shows that the trajectory converges to the origin when $(\varsigma_1, \varsigma_2) = (0.1, 0.2)$. It is also observed from Fig. 3 that trajectory converges to the static bifurcation solution when $(\varsigma_1, \varsigma_2) = (-0.2, 0.3)$.

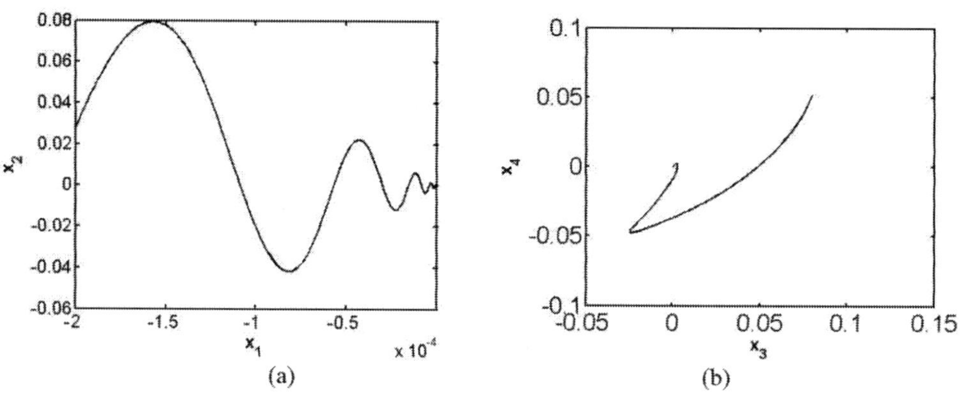

Figure 2. Trajectory projection starting form initial point $x_0 = (-0.0002, 0.028, 0.098, 0.032)$

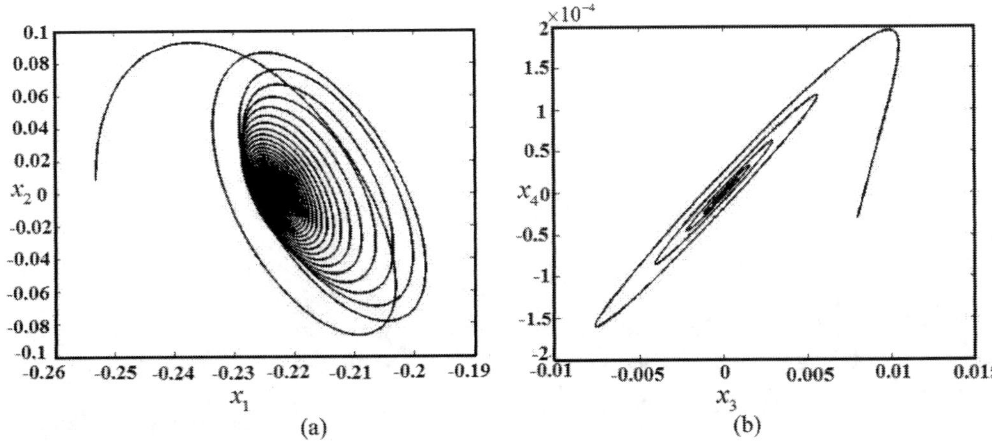

Figure 3. Trajectory projection starting form initial point $x_0 = (0.008, -0.05, 0.073, 0.08)$

4 Conclusion

In our paper, we investigate the bifurcation and stability behaviors of the FGMs rectangular plate subjected to transverse and the in-plane excitations. The nonlinear partial differential governing equations of motion are derived by using Hamilton's principle. The nonlinear vibration equations are deduced using Galerkin approach. Based on the normal form theory, the bifurcation theory is utilized to discuss the static bifurcation, incipient Hopf bifurcation and secondary Hopf bifurcation in the vicinity of degenerated equilibrium. The results provide some guidance for the design of the structure of FGMs plate. This technique, based on the method of normal form and bifurcation theory, can be used to deal with general engineering problems.

Acknowledgements

The authors gratefully acknowledge the support of the National Natural Science Foundations of China (NNSFC) through grant No.11302184, the Scientific Research Fund of Zhejiang Provincial Education Department through grant No. Y201121157 and the Scientific Research Foundation of Xiamen University of Technology through grant No. YKJ13007R.

References

1. S. Abotula, A. Kidane, V. B. Chalivendra, A. Shukla, Dynamic curving cracks in functionally graded materials under thermo-mechanical loading, Int. J. Solids Struct. **13** (2012) 1637-1655.
2. R. Gunes, J. N. Reddy, Nonlinear analysis of functionally graded circular plates under different loads and boundary conditions, Int. J. Struct. Stab. Dy. **8** (2008) 131-159.
3. D. V. Dung and, L. K. Hoa, Semi-analytical approach for analyzing the nonlinear dynamic torsional buckling of stiffened functionally graded material circular cylindrical shells surrounded by an elastic medium, Appl. Math. Model. **39** (2015) 6951-6967.
4. J. Sladek, V. Sladek, P. Stanak, C. Z. Zhang, M. Wünsche, Analysis of the bending of circular piezoelectric plates with functionally graded material properties by a MLPG method, Eng. Struct. **47** (2013) 81-89.
5. Q. Li, V. P. Lu, K. P. Kou, Three-dimensional vibration analysis of functionally graded material sandwich plates, J. Sound Vib. **311** (2008) 498-515.
6. A. Allahverdizadeh, M. H. Zaei and, M. N. Bahrami, Nonlinear free and forced vibration analysis of thin circular functionally graded plates, J. Sound Vib. **310** (2008) 966-984.

7. R. Gunes, J. N. Reddy, Nonlinear analysis of functionally graded circular plates under different loads and boundary conditions, Int. J. Struct. Stab. Dy. **8** (2008) 131-159.
8. Y. X. Hao, L. H. Chen, W. Zhang, J. G. Lei, Nonlinear oscillations, bifurcations and chaos of functionally graded materials plate, J. Sound Vib. **312** (2008)862-892.
9. X. H. Zhang, Local bifurcations of nonlinear viscoelastic penel in supersonic flow, Commun Nonlinear Sci Numer Simulat. **18** (2013)1931-1938.
10. C. Anton, J. Deng, Y. S. Wong, Hopf bifurcation analysis of an aeroelastic model using stochastic normal form, J. Sound Vib. **331** (2012)3866-3886.
11. P. Li, Y. R. Yang, H. J. Shi, Hopf and two-multiple semi-stable limit cycle bifurcations of a restrained plate subjected to subsonic flow, J. Sound Vib. **335** (2015) 286-303,.
12. W. Zhang, C. Z. Song, M. Ye, Further studies on nonlinear oscillations and chaos of a rectangular symmetric cross-by laminated plate under parametric excitation, Int. Bifurcat Chaos. **16** (2006) 325-347.
13. Y. X. Hao, W. Zhang, J. Yang, Analysis on nonlinear oscillations of a cantilever FGM rectangular plate based on third-order plate theory and asymptotic perturbation method, Compos Part B: Eng. **42** (2011)402-413.

Effect of Equivalence Ratio on Composition and performance of Biogas and Gasoline Exhaust from Spark Ignition Engine by Mathematical Modeling

Paramust Juntarakod[a]

Department of Mechatronic Engineering, Faculty of Engineering. Rajamangala University of Technology Isan, Khonkaen Campus 150 Srichan Road, NaiMuang, Muang, Khonkaen, 40000, Thailand

Abstract. This paper presents the numerical computationnal of pressure, temperature and exhaust characteristics of spark ignition engine with biogas as fuel. The solution of non-linear combustion equation systems have been computed, that based on a quasi-one-dimensional engine model, high order iteration method with the equilibrium constants method. Computer program was used to calculate the mole fractions of 10 combustion products when biogas and gasoline fuel are burnt along with variable equivalence ratios. In cylinder chamber model is based on the classical two-zone approach, wherein parameters like heat transfer from the cylinder, blow by energy loss and heat release rate are also considered and calculated. Biogas is defined as fuel produced from using anaerobic digestion of biodegradable or waste materials and the constituents are $C_5H_7O_2N$, CH_4, CO_2 N_2 H_2O of biogas and C_7H1_7 of gosoline. Which general fuel model is specified by way of its $C_aH_bO_cN_d$ values. The curve-fitted coefficients of energy were then employed to simulate air and fuels data along with frozen composition and practical chemical equilibrium routines from Gill data. The calculated data were used to plot the various pressure and temperature with the crank angle of each step of four stroke engine cycle and combustion products versus equivalence ratio. All results were compared with gasoline as reference fuel in the spark ignition engine according to the same numerical method.

1 Introduction

In policy of Thailand, under the responsibility of Energy Policy and Planning Office (EPPO) and Department of Alternative Energy Development and Efficiency (DEDE), Thailand, they have proven to be a great success. Therefore, on determing that within 2021, 600 megawatt (MW) of electricity, 1,000 ktoe of heat should be generated from biogas and Excluding Elephant grass 3,000 megawatt (MW), Tawarath [1]. In academic or research, produced biogas in anaerobic digestion plants is primarily composed of methane (CH_4), carbon dioxide (CO_2), nitrogen (N_2) with smaller amounts of hydrogen sulfide (H_2S), ammonia (NH_3) and saturated with (H_2O) water vapor. Most of the researches in biogas fuel on spark ignition engine were focused on experimental study for finding performance and emissions. For the value of biogas parameters for potential energy researches, the researches were conducted with five gas compositions; with CH_4 = 55-65 (Vol. %), CO_2=30-35 (Vol. %) and N_2 = 5 (Vol. %) respectively. In the present study, biogas was produced by anaerobic digestion of

[a] Corresponding author : paramust_kmitnb@hotmail.com

biodegradable materials, in this present used Buswell equation $C_5H_7O_2N$ such as 53% (Vol. %) of CH_4 and 47% (Vol. %) of CO_2, LHV is 19.1 MJ/kg, [2].

In experiments using simulated biogas by Crookes, Sáawomir [3, 4 and 5] to find combustion products, he used the assumption that the engine parameters have the compression ratios from 11-13:1, they found the suitable operation without knocking by using the constant equivalence ratio of 0.8 and the variation of combustion products. In their case of a biogas fuelled engine simulation, the compression ratios that can be employed considerably higher than normal engine and the CO_2 present help to suppress knocking. However, at high compression ratio, it was found to increase the HC and NO_x levels, but reducing N_2 and CO was found to be beneficial in lowering NO_x. In the present study, an effort is made by developing a new combustion mathematical model to simulate exhaust characteristics in biogas fuels by variation of equivalence ratio. The methods based on thermodynamic and equilibrium properties are constants, computer program using MATLAB which has been developed for the biogas-gasoline fuels to calculate the mole fractions of the emission gases by applying from Buttsworth [6]. Equilibrium constant method is based on thermodynamic measurements and empirical calculations [7, 8 and 9]. It is very accurate and precise in solving most of combustion and chemical kinetics problems. Some thermodynamic data for elements, combustion products and many pollutants are available in the new compilation published by Grill [7]. All results were compared by gasoline as fuel according to the same calculation method for all cases.

2 Description of the mathematical model

2.1 Spark Ignition Engine model

Mathematical model was applicable in any types of fuels provided the fuel induction technique was used or in other words the carburetion system with the used first law of thermodynamics, and this model was described as following equations Ferguson [8], Figure. 1. This model predicted the fuel inducted through the air inlet and serves as an alternative fuel (assumed biogas) to combust with fresh air as function of crank angle ($°\theta$) and thermo-fluids parameters. The mathematical models is to predict the temperature, pressure and engine performance by step by step of compression stroke, combustion stroke and expansion stroke, respectively, Mathematical model can be define as,

$$\frac{dP}{d\theta}=\frac{A+B+C}{D+E} \tag{1}$$

$$\frac{dT_b}{d\theta}=\frac{h\pi b^2/2+4v/bx_b^{1/2}\Delta T}{-vmcpb^X}+\frac{v_b}{c_{Pb}}\left(\frac{\partial \ln v_b}{\partial \ln T_b}\right)\left(\frac{dP}{d\theta}\right) \tag{2}$$

$$\frac{dT_u}{d\theta}=\frac{-h\left(\frac{\pi b^2}{2}+\frac{4v}{b}\right)1-x_b^{1/2}T_u-T_w}{vmcpb1-x_b}+\frac{v_u}{c_{Pb}}\left(\frac{\partial \ln v_u}{\partial \ln T_u}\right)\left(\frac{dP}{d\theta}\right) \tag{3}$$

Where, the variation team in eq (1) A, B, C, D and E reference base on Ferguson [8].

2.2 Assumption of the Carburetor model

The control volume selected to analyze the flow in the mixing chamber is shown in the left hand section of Figure. 1. In the analysis, the following assumptions were made,

1) The inner wall of the carburetor is adiabatic and Friction loss is negligible.
2) The primary stream and secondary stream have the same static pressure at the entrance of the mixing chamber, $P_{p(air)} = P_{s(fuel)} = P$.

Both primary stream (air) and secondary stream (biogas fuel) can be considered as perfect gases with constant specific heat ratio γ. So, $m = m_{p(Air)} + m_{s(fuel)}$. Using the definition of critical condition, it can be shown fig. (1) that,

$$m_p = \frac{P_{P0}A_t}{\sqrt{T_0}} = \sqrt{\frac{\gamma_P}{R_P}}\left(\frac{2}{\gamma_P+1}\right)^{\frac{\gamma_P+1}{\gamma_P-1}} \tag{4}$$

$$m_s = \frac{P_{s0}A_s}{\sqrt{T_0}} = \sqrt{\frac{\gamma_s}{R_s}}\left(\frac{2}{\gamma_s+1}\right)^{\frac{\gamma_s+1}{\gamma_s-1}} \tag{5}$$

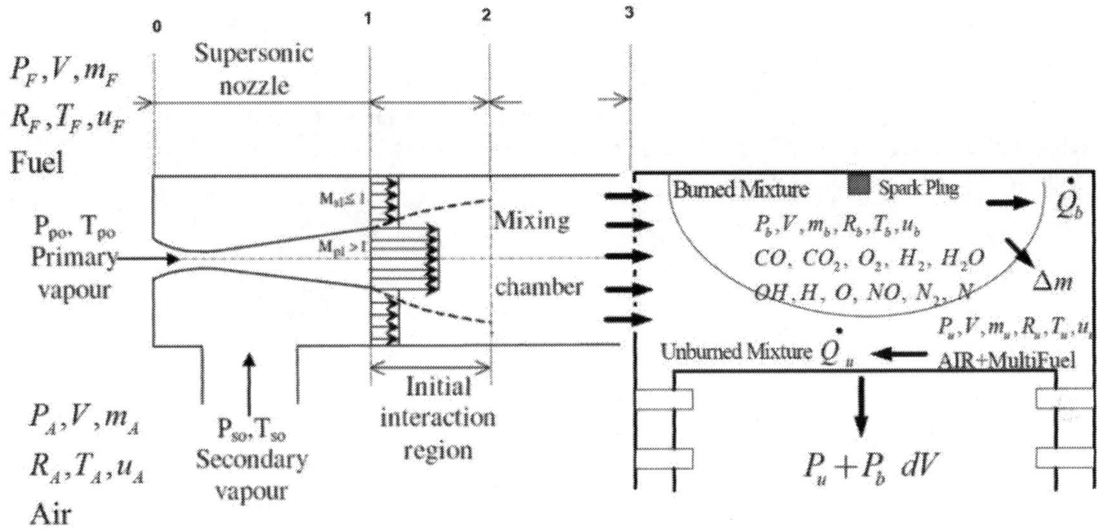

Figure 1. Definition of fresh air, fuel intake (Carburetor model), SIE by Mathematical Modeling

2.3 Practical of combustion and chemical equilibrium

Basis of the equilibrium combustion products with fuel-air mixture model is a solution to the atom balance equations from the chemical reaction equation of fuel and air forming. This mixture equation is given in eq (6) for the condition of equivalence ratio ($\phi < 3$). This process based on conservation of mass that was modified form Buttsworth [6] and Ferguson [8], where the following equation represents the chemical reaction with the relevant species involved which was added gasoline supplied as given below,

$$\begin{aligned} &\varepsilon\phi C_aH_\beta O_\gamma N_\delta \\ &+0.21O_2+0.79N_2 \end{aligned} \rightarrow \begin{aligned} &\rightarrow v_1 CO_2 + v_2H_2O + v_3N_2 + v_4O_2 + v_5CO \\ &+ v_6H_2O + v_7H + v_8O + v_9OH + v_{10}NO \end{aligned} \tag{6}$$

Where, $\varepsilon = 4,(4\alpha+\beta)$ and biogas, $C_5H_7O_2N$ [2], gasoline, C_7H_{17} [8].

The method for solving of non-linear equation system, in the High order iteration Method was selected to solve in eq (6) with equilibrium constants method of biogas and gasoline fuels, which were summarized in eqs. (7) - (11). The many non-linear equations are available in the free compilation published by Paramust [10].

$$f_i(y_3,y_4,y_5y_6)=0 \quad j=1,2,3,4 \tag{7}$$

$$f_1(y_3,y_4,y_5y_6)=c_1y_5y_4^{1/2}+c_2y_4y_6^{1/2}y_6+y_3+y_4+y_5+y_6+c_4y_6^{1/2}+y_5y_4^{1/2}+c_7y_4^{1/2}y_3^{1/2}+c_6y_3^{1/2}-1 \tag{8}$$

$$f_2(y_3,y_4,y_5y_6)=2\left(c_2y_4^{1/2}y_6\right)+2c_6+c_4y_6^{1/2}+c_3y_4^{1/2}y_6^{1/2}-d_1\left(c_1y_5y_4^{1/2}+y_5\right) \tag{9}$$

$$f_3(y_3,y_4,y_5y_6)=2\left(c_1y_5y_4^{1/2}\right)+c_2y_6^{1/2}y_6+2y_4+y_5+c_5y_4^{1/2}+c_3y_4^{1/2}y_6^{1/2}+c_7y_4^{1/2}y_3^{1/2}-d_2\left(c_1y_5y_4^{1/2}-y_5\right) \tag{10}$$

$$f_4(y_3,y_4,y_5y_6)=2\left(c_2y_4^{1/2}y_6\right)+c_7y_4^{1/2}y_3^{1/2}+c_6y_3^{1/2}-d_3\left(c_1y_5y_4^{1/2}+y_5\right) \tag{11}$$

3 Results and discussions

The simulation was developed under no-load operating conditions compared gasoline and biogas fuel . The pressure change from compression stroke to expansion was clearly shown in Fig. 2 (left). The unburned and burned gas temperature was shown in Fig. 2 (right). Both pressure and temperature history was very useful parameter to determine the performance (work output of engine) of the engine cycle. The unburned gas temperature was usually referred to the residual gas whereby the burned gas temperature was referring to the gaseous burned during the combustion. This fig can show the average work cycles of the cylinder engine during the compression to the expansion stroke. Fig. 3 is the effect of equivalence ratio on composition of Biogas and Gasoline, it shows the average emissions of the burned gas during the combustion where using the varied equivalence ratio between 0.5 to 2.0 by high iteration method. This is numerical results and not the experimental result. An assumption on engine specification to be used in the calculation is shown in Table 1. The two versions of the simulation were used to compute engine performance (area of pressure-volume diagram) for the above conditions.

Table 1. The engine specifications for computing and outputs[10]

Input parameters	**Values**
Bore(m)/Stroke (m)/ Stroke per 2 (rod Length)	0.068/0.0540.25
Compression ratio/ Press (kPa)/Temp ($^{\circ}K$)	8.5/100/350
Start and Finish of Combustion process	-35 $^{\circ}\theta$ BTDC and 40 $^{\circ}\theta$ ATDC
Engine speed (RPM)	1,000
Computing	**Output**
Biogas (B) $C_5H_7O_2N$,Gasoline (G) C_7H_{17}	Press and Temp vs. crank angle, (Figure.2)
Biogas (B) $C_5H_7O_2N$,Gasoline (G) C_7H_{17}	Mole fraction vs. equivalence ratio (Figure. 3)

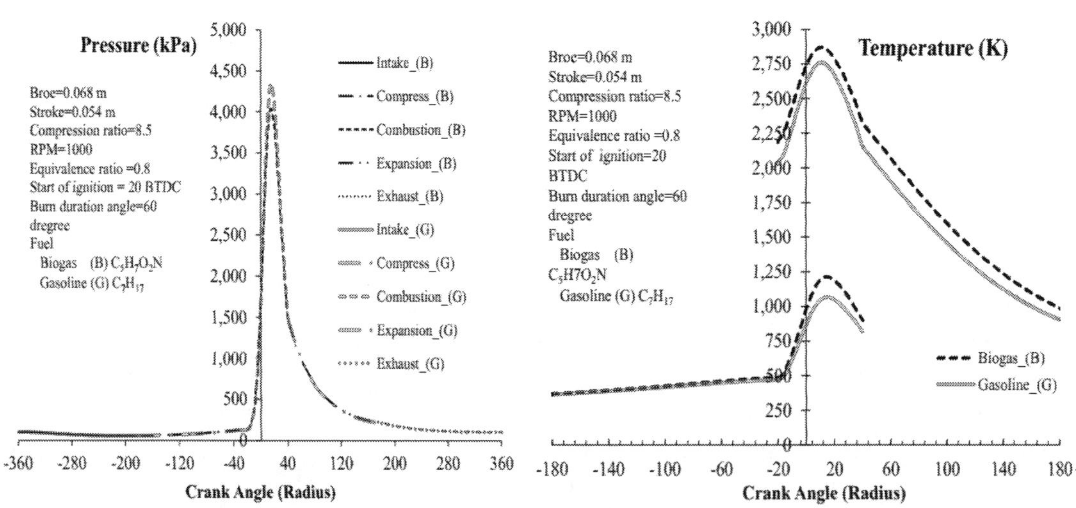

Figure 2. Pressure, Temperature VS Crank angle of all strokes, the engine specifications in Table 1.

Fig. 2, show that the general trend of the pressure results for all strokes agree with the pressure trend of the results with a low heating value of gasoline that have more internal energy more than biogas. So, the agreement of the pressure values depends on the value of energy each fuel. For the engine cycle, the peak pressure is clearly different, only during the combustion stroke. There is a different about 300 kPa (3 bars) at 15 °θ BTDC, between gasoline and biogas as fuel in the engine. As a result, the pressure is less than this, if the amount is less methane. In general, the case for gasoline as fuel is stable of performance engine because it is a constant value, but biogas is variable property, dependent on the amount of methane and type of reactor. This illustrates the sensitivity of the cylinder pressure to the value of the low heating value or internal energy. The overall gas temperature dependence on crank angle is similar for both the fuel (gasoline and biogas) of the simulation. The peak temperature is 2,700 °K for gasoline as fuel in combustion stroke of engine and the peak temperature is 2,850 °K for the case with biogas at 15 °θ BTDC as the same of pressure. This temperature of gasses in the engine cycle are involved the heat flux was actually determined by making use of the temperature and applying the Newton's law. Consider carefully, the characteristics of the temperature curve for biogas combustion engine is higher than the temperature of gasoline engine as fuel. This is due to the fact that the characteristic of gases properties and equilibrium constant properties for types of reaction in combustion process of biogas fuel are higher and rapidly more than gasoline. This is described in more detail of biogas are low energy and low types of hydrocarbons, but more reaction process than gasoline. This reason can be used to cooling of the engine, while changing fuel.

Results illustrating composition shifts with equivalence ratio are given in Fig. 3 (Biogas, Left) and Fig 3 (Gasoline, Right) in the combustion cylinder chamber of biogas and gasoline at the nearly adiabatic flame temperature 2,000 °K, pressure 4,000 kPa, respectively. The results to find, the mole fraction behavior relative to equivalence ratio is complex, the general mole fraction line trends are similar for this result as well. The product types of CO and H_2 generally increase with equivalence ratio are higher, while the O_2, NO, OH and O mole fraction decrease. But, CO_2 and H_2O of biogas have the clearly differential, while gasoline is similarly value. In additional case of The most significant emission is NO_x as according internal combustion engine theory, NO_x emission formation requires many ways such as O_2, NO, N, O, high enough temperature, low pressure, the maximum this value is generated when engine runs at stoichiometric (r=1) as following theory. In simulation of these cases used summation of mole fraction O_2, NO, N and O. To found, NO_x emissions increase with lower combustion of equivalence ratio is lean and decrease with higher combustion of equivalence ratio is rich. The reason be described in more affection of NO_x is the mole faction of O_2, higher specie of O_2 is lean side for more reaction (low equivalence ratio) and not enough reaction of O_2 at high equivalence ratio. So this model is consistent for the theory in all respects.

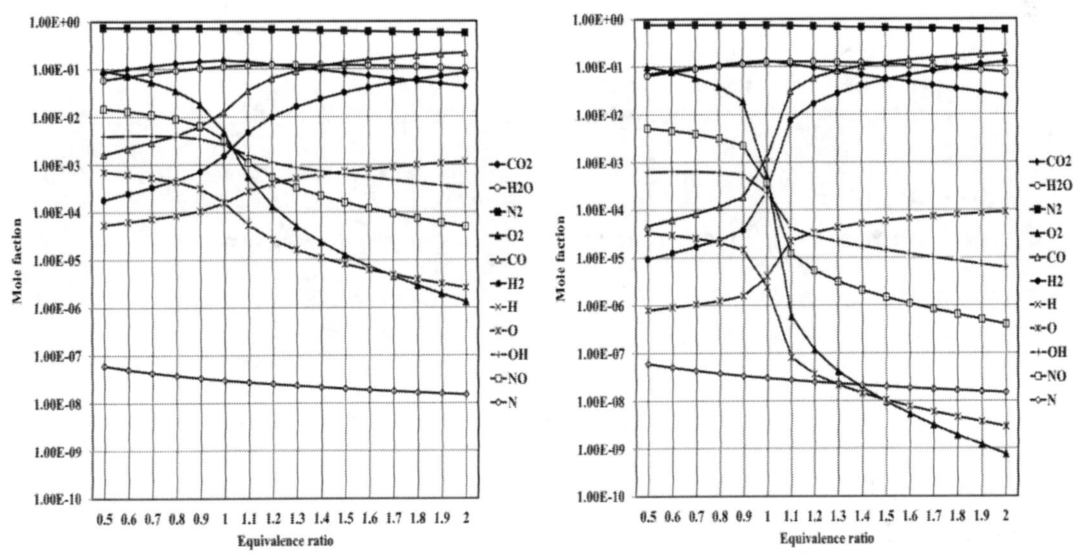

Figure 3. Mole fraction of Biogas (Left) and Gasoline (Right) Combustion with the Equivalence ratio 0.5-2.0

4 Conclusion

Mathematical Modeling program has been used successfully calculate the mole fractions of 11 combustion products when biogas or gasoline (fuel) are burnt along with the spark ignition engine at variable equivalence ratios. It can be observed that hydrogen (biogas, H=7 and gasoline H=17) substitution causes significant increase in H_2 and H emissions during rich combustion and OH and NO_2 emissions during lean combustion. As the equivalence ratio increases during rich combustion, mole fractions of CO and atom C decreases with increment of hydrogen in fuel substitution. N_2, atom N and CO_2 emissions decrease whereas no significant changes in O_2, NO and atom O emissions throughout all equivalence ratios as hydrogen is added in fuel to the combustion.

The simulation work achieves its goal of being a simple, fast, accurate engine numerical model for the changing conditions such as equivalence ratio, compression ratio, types of fuel and combustion products. The results obtained can be used as a first-degree approximation and is useful in numerous engineering applications including general design predictions. The numerical model of parameters can predict an array of thermodynamic properties, equilibrium constant as a function of temperature, and non-liner combustion equations and adapt to any mathematical method for solving. It be can used in to study at graduate program of engineering or mathematical sciences. The numerical model of engine parameters can predict an array of thermodynamic term outputs such as pressure, temperature, heat flux, work and easily adapt to any combustion chamber shape or any internal combustion engine. Due to its simplicity and computational efficiency, the model can also be used as a preliminary test for a wide range of alternate hydrocarbon fuels.

Acknowledgment

This paper was partially financed and given by Department of Mechatronic Engineering, Faculty of Engineering, Rajamangala University of Technology Isan, Khonkaen Campus.

References

1. T. Sutabutr, Alternative Energy Development Plan: AEDP 2012-2021, International Journal of Renewable Energy, **7**, 1 (2012)
2. E. Danso-Boateng, G. Shama a, A.D. Wheatley, S.J. Martin, R.G. Holdich, Hydrothermal carbonisation of sewage sludge: Effect of process conditions on product characteristics and methane production, Bioresource Technology 177, (2015)
3. J. Huang, R.J Crookes, Assessment of simulated biogas as a fuel for the spark ignition engine. Fuel, **77**, 15 (1998)
4. W. Sáawomir, Biogas as a fuel for diesel engines, Journal of KONES Powertrain and Transport, 19, 3 (2012)
5. Crookes R.J.: Comparative bio fuel performance in internal combustion engines. Int J Biomass Bioenergy, 30, (2006)
6. D. R. Buttsworth, Spark Ignition Internal Combustion Engine Modeling using Matlab, Faculty of Engineering & Surveying Technical Reports, University of Southern Queensland, (2002)
7. M. Grill, M. Chiodi, H. Berner, M. Bargende, Calculating the Thermodynamic Properties of Burnt Gas and Vapor Fuel for User-Defined Fuels, MTZ 05-2007 (2007)
8. Colin R. Ferguson, Internal Combustion Engines, Applied Thermosciences. New York: John Wiley and Sons, (1986)
9. Heywood. Joh. B. Internal Combustion Engine Fundamentals. New York McGraw Hill, (1988)
10. J. Paramust, T. Soontornchainacksaen. Calculation of Combustion Products by the New Iteration Method of Non-linear Equations, Journal of Mathematics and System Science, 3, 1 (2013)

Study on the Performance of a Proposed Fire Safe Elevator System used for Evacuation in Supertall Buildings

Na Cai[a] and Shifei Shen

Institute of Public Research, Department of Engineering Physics, Tsinghua University, Beijing 100084, PR China.

Abstract. Long evacuation time is a key fire safety concern when a supertall building is on fire. The elevator system can be an effective alternative. The performance of a design of fire safe elevator system combining the refuge place with fire safe elevator is studied. An example building based on this proposed design is considered. Smoke spread to the system is studied by the Computational Fluid Dynamics (CFD) code Fire Dynamics Simulator (FDS). Different arrangements of smoke extraction with pressurization systems are evaluated by analysing the smoke dispersion and pressure distributions in this fire safe elevator system. Numerical results were compared with that by theoretical equations. The results show that a smoke extraction system with a four-floor approach pressurization system can be an efficient method for smoke control in elevator system for supertall buildings.

1 Introduction

Buildings taller than 300 m (984 feet) are regarded as supertall buildings by the Council on Tall Buildings and Urban Habitat [1] in the USA. Long evacuation time is a key fire safety concern when a supertall building is on fire. Stairs alone are inadequate for full evacuation at any significant height, for example up to 2 hours as reported in the previous case [2]. Phased-evacuation or the "stay-in-place" approach by providing refuge storeys or refuge areas is a feasible way for fire safety management for supertall buildings. However, occupants are no longer willing to delay egress by staying in the building on fire since the World Trade Center incident. Elevators seem to be the most efficient alternative means and is arising interest in the use of elevator system for egress and access all over the world [3]. A case study conducted on Burj Khalifa, which is now the tallest building in the world, showed that the total evacuation time would be significantly reduced with elevator assistance [4]. The elevator system can be an effective alternative if it is reliable, accessible and safe during a major fire. Upgrading the fire safety provisions for the elevator system to stand a large fire is important.

A fire safe elevator design consisting of an elevator and refuge places was proposed for emergency evacuation of supertall buildings based on a concept design proposed by Chow [5]. This design combined the fire safe refuge places with a fire safe elevator for evacuation. In order to keep the elevators accessible during major fires, the fire safe elevator was provided through refuge places with at least 2-hour Fire Resistance Rating (FRR). The performance of this fire safe elevator system will be evaluated in this paper using an example supertall building by Computational Fluid Dynamics (CFD).

[a] Corresponding author: cainabj@163.com

2 An Example Building with Proposed Fire Safety Elevator System

An example building of height 306 m with 68 storeys was taken for illustrating the concept of fire safe elevator system. Each floor was of length 42 m, width 45 m and height 4.5 m. Figure. 1(a) shows the side view of the building model. The typical floor plan of the fire safety elevator with refuge area used in this study is shown in Figure. 1(b). Two fire safe elevators are added going through the refuge places. Floors with refuge places are located every eight storeys. Fire safe elevators of length 2.4 m and width 2.4 m with lobbies of length 2.4 m and width 2.4 m are located in the centres of the refuge places. The exterior sides of the refuge place are opened to provide adequate cross ventilation. Each refuge place also has an access to a staircase. It is further assumed that the FRR (at least 2 hours) of the enclosure materials of the elevator shafts and lobbies are the same as those of the refuge areas.

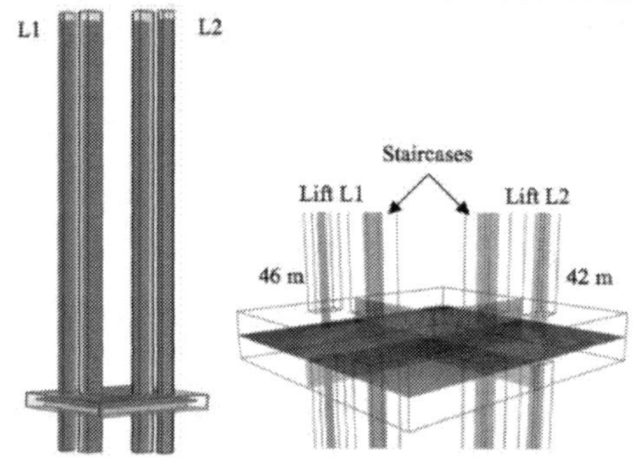

(a) Typical refuge floor plan of a financial center (b) Refuge walls with 2 hours FRP

Figure 1. The example case on supertall building

The elevator can stop at any storey in the traditional way. The smoke would easily spread into the elevator shaft and then spread throughout the building when the elevator stops at the fire floor. Therefore, in this proposed design, the elevators are controlled to stop only at those storeys with refuge places. Surrounded by refuge places, the elevator can be accessible during major fires. Refuge places can protect both the occupants and the elevator system. During the evacuation, occupants can go to a refuge place first and then evacuate through the elevator to the safe level.

However, when the fire breaks out at other storeys without refuge places, there are still chances that the smoke can infiltrate into the elevator shafts through door cracks. Under such circumstances, the stack effect and effect of wind action may be serious for the very tall elevator shafts and stairwells. Smoke control should be considered under such circumstances as follows:

- Traditional elevator system without refuge places and the elevator stops at a fire floor, so that the smoke would enter the elevator shaft and spread throughout the building.
- In the proposed elevator system with refuge places, the fire breaks out at other storeys without refuge places.

3 Numerical Experiments

The computing domain for the two floors involved is 54 m long, 57.6 m wide and 10 m high in a Cartesian co-ordinate, which is extended outside the enclosure to capture all movement of air. This area comprises a total of 691,200 grid cells 0.3 m by 0.3 m by 0.5 m high, as shown in Figure. 1(b). Elevator shafts with lobbies were modelled with 250,000 grid cells 0.3 m by 0.3 m by 0.5 m high. The staircases were modelled with 300,000 grid cells 0.3 m by 0.3 m by 0.5 m high. The total number of grid cells is 1,241,200. Each door has a one grid size leakage to simulate the door cracks and the openings due to the entry and exit of people. For the door located on the refuge floors, open boundaries were applied to the door cracks.

Two scenarios without fire were first carried out to study the pressure distribution in the elevator shaft under different ventilation conditions and boundary conditions, as shown in Figure. 2.

DOI: 10.1051/matecconf/20168201015

Scenario NNN:
No ventilation at the top of the shafts, without wind.
Scenario NVN:
Same as scenario NNN, but with ventilation at the top of the shafts.

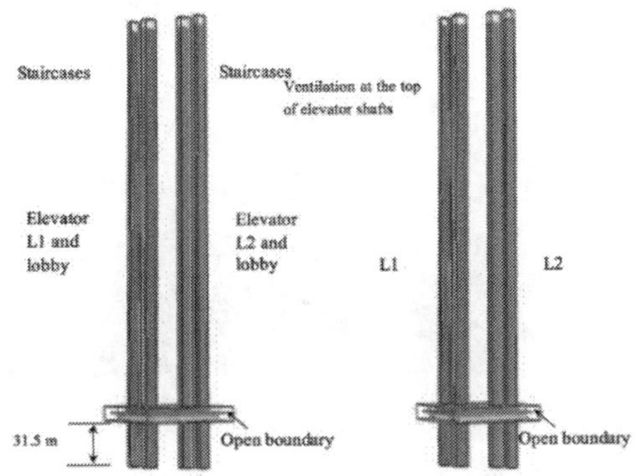

(a) Scenario NNN (b) Scenario NVN

Figure 2. Plans of scenarios without fire

Two scenarios with the fire located on the 31th floor (139.5 m above ground level) were carried out to investigate the smoke extraction and pressurization systems, as shown in Figure. 3. The fire source was placed on the floor below the storey with refuge places. It was located in front of the lobby door of elevator L1. The fire was assumed to be an ultra-fast t^2 fire with a heat release rate of 5 MW.

Scenario MNN:
Without smoke extraction and pressurization systems.
Scenario MSP2:
With smoke extraction system on the fire floor and pressurization system in the elevator shafts in the way of four-floor approach.

4 Scenarios without Fire

Theoretical and numerical studies were first carried out to investigate the pressure distribution in the elevator shaft in scenarios without fire. The effect of ventilation of the shaft, stack effect and wind effect were studied first.

Since the open refuge places are located every eight storeys, the elevator shaft can be considered as a shaft with a continuous opening. The location of the neutral plane can be given in terms of the height of the shaft Z (m), the absolute temperature of air in the shaft T_s (K) and the absolute temperature of outside air T_o (K) [6]:

$$Z_n = Z\left(\frac{1}{1+(T_s/T_o)^{1/3}}\right) \quad (1)$$

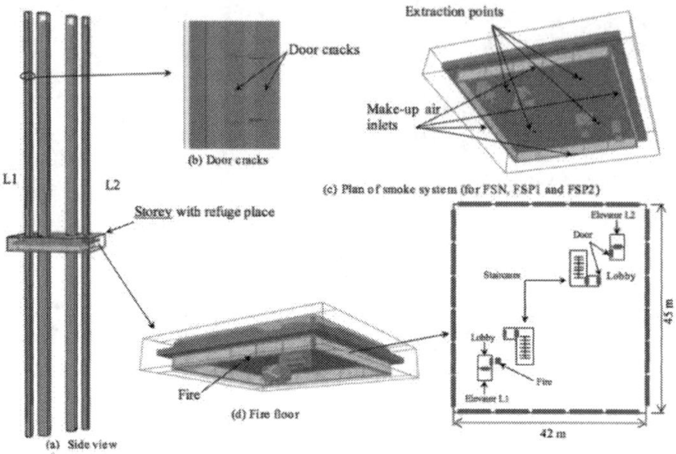

Figure 3. Fire scenarios with fire located on a low floor

If a shaft is ventilated, regardless of whether the vent is above or below the neutral plane, the neutral plane will be located between the height of the neutral plane of an unvented shaft and the vent elevation. The location of the neutral plane can then be expressed as [6]:

$$\frac{2}{3}A'(Z - Z_n)^{3/2} + A_v(Z_v - Z_n)^{1/2} = \frac{2}{3}A'Z_n^{3/2}(T_s - T_o)^{1/2} \tag{2}$$

where A' is the area of the opening per unit height (m^2).

Two scenarios were studied first. In scenario NNN, the elevator shaft was not ventilated, air could flow in and out only through the door cracks. In scenario NVN, the elevator shaft was ventilated by one grid size (0.15 m^2) opening on the top. For both scenarios, the outside air temperature was 0°C, and the air temperature of the shafts and inside floors was 20°C. As the floor plan is symmetric, the pressure distribution of the two elevator shafts should be similar. Therefore, only the numerical results of the shaft of elevator L1 were analysed in this section. The predicted pressure distributions at the central line of the shaft of elevator L1 are shown in Figure. 4. It can be observed that the location of the neutral plane of scenario NVN is higher than that of scenario NNN. This is consistent with the theoretical analysis by Klote and Milke[6].

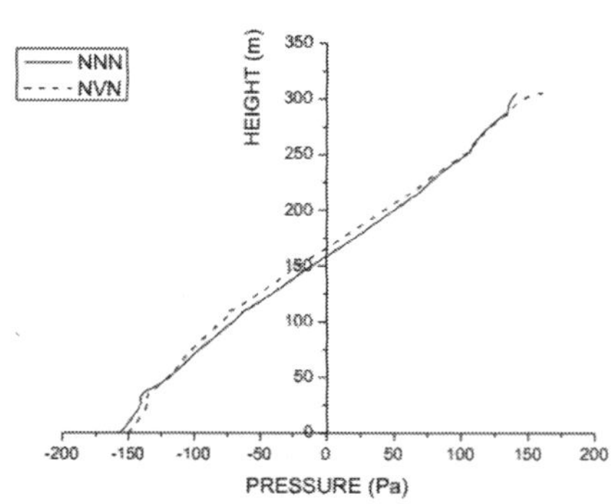

Figure 4. Pressure distributions at the central line of L1 for scenarios NNN and NVN

The theoretical location of the neutral plane for shaft with a continuous opening calculated by equation (1) is 151.5 m. FDS result for scenario NNN is 158.5 m, with an error of 4.6 % compared to theoretical result. The theoretical location of the neutral plane for ventilated shaft calculated by equation (2) is 169 m. FDS result for scenario NVN is 164 m, with an error of 3.0 %.

5. Scenarios with Fire

Smoke extraction system for the fire floor and pressurization system for the elevator shafts were studied in the fire scenarios. Smoke extraction system with a minimum extraction rate not less than eight air changes per hour of the total compartment volume is adopted. Pressurization systems are applied to stairwells, lobbies and elevator shafts. A certain range of pressure differentials should be achieved across all doors. A pressurization approach known as 'four-floor approach' was first proposed in zoned smoke control by [6], and was adopted by the Seattle jurisdiction in the USA[7]. The primary fire floor, the floor directly above, and two floors immediately below the fire floor would be considered. The pressure differences across the elevator doors on these four floors are assumed to be within a certain pressure range. This 'four-floor approach' pressurization system is investigated.

Figure 5 shows the smoke dispersion snaps at different time for scenarios MNN. Without ventilation system, the smoke entered the nearest vertical passage L1 shaft and quickly reached the top of the shaft. Using the pressurization in the way of four-floor approach, relatively large amount of pressurized air flow rates was required for the injection points located in the shafts on the four fire affected floors, including the fire floor, the floor above with refuge places and the two floors below the fire floor. The smoke can be confined in the area of the fire floor for scenario MSP2, as shown in Figure 6. All the vertical passages were free of smoke for as long as 1200 s (20 min) when the simulation was stopped manually. The vertical slices of pressure distribution are shown in Figure 7. Positive pressure differentials around 25 Pa and 45 Pa were kept in the elevator shafts and stairwells respectively.

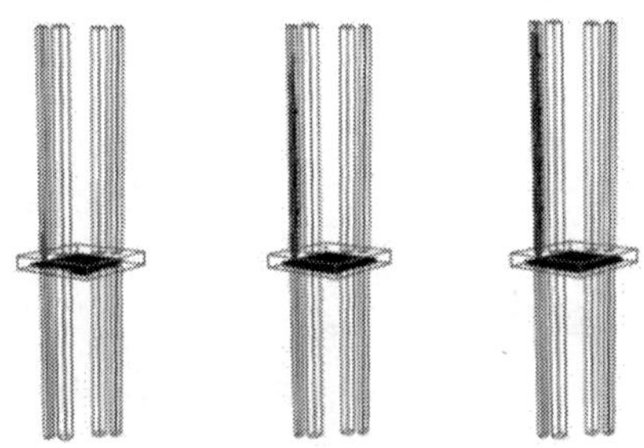

t = 100 s t = 200 s t = 1200 s

Figure 5. Smoke dispersion in scenarios MNN

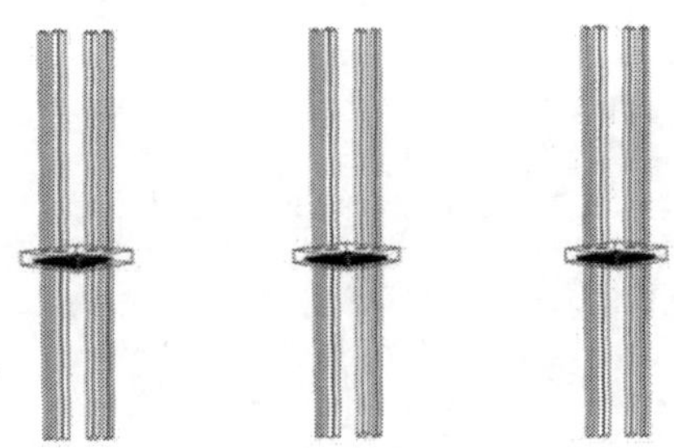

t = 100 s t = 200 s t = 1200 s

Figure 6. Smoke dispersion in scenarios MSP2

6 Conclusions

The performance of a fire safe elevator system combined fire safe elevators and refuge places for emergency evacuation of supertall buildings is evaluated for various scenarios in this paper. The fire safe elevators were designed to go through the refuge places, with all areas built with at least 2 hours FRR. For the example supertall building concerned, the pressure distributions in the elevator shaft in scenarios without a fire were investigated. Compared to the theoretical results, FDS can reasonably predict the location of the neutral plane and the vertical pressure distributions in the elevator shaft when no fire was presented. In the scenarios with fire, smoke spread from the fire breaking out at other storeys without refuge places was considered. The 'four-floor approach' of pressurization, which targets the primary fire floor, the floor directly above, and two floors immediately below the fire floor, can confine the smoke in the area of fire floor and keep all the vertical passages free of smoke for a sufficiently long time, at least for 20 minutes in this case. This 'four-floor approach' [6] pressurization system is an efficient way for smoke control for elevator shafts in supertall buildings.

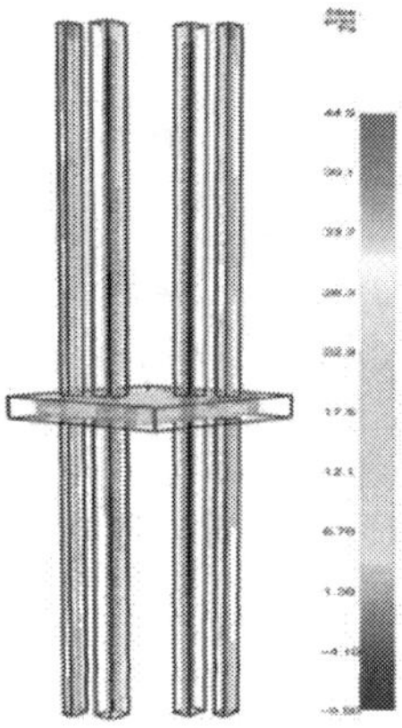

Figure 7. Pressure distribution in vertical passages at 600 s

Acknowledgement

The work described in this paper was supported by a grant from the China Postdoctoral Science Foundation for the project 'Study on the Influence of Elevator Evacuation on Smoke Movement and Control in the Fire Safe Elevator System used for Evacuation in Supertall Buildings' (Grant No. 2015M571045).

References

1. Council on Tall Building and Urban Habitat (CTBUH), Tall Buildings in Numbers. CTBUH Journal – International Journal on Tall Buildings and Urban Habitat 2, 46-47 (2011)
2. W. K. Chow, A retrospective survey on elevator evacuation of supertall buildings under fires. J. Appl. Fire. Sci 16, 315-327 (2006-2007)
3. J. Koshak, Elevator evacuation in emergency situations. In: Proceedings of Workshop on Use of Elevators in Fires and Other Emergencies, Atlanta, GA, March 2-4, (2004)
4. A. F. Vanney, Protection of Elevators for Building Evacuation: Code Requirements, Design Practices and Case Studies. Presented in CTBUH 2010 World Conference - India Remarking Sustainable Cities in the Vertical Age. Mumbai, India, 3-5 February (2010)
5. W. K. Chow, Proposed design on fire safe evacuation elevator for supertall buildings. International Journal on Architectural Science, 8, 54-59 (2011)
6. J. H. Klote, J. A. Milke, Principles of Smoke Management. American Society of Heating, Refrigerating and Air-Conditioning Engineers, Incorporated (2002)
7. R. S. Miller, Elevator shaft pressurization for smoke control in tall buildings: The Seattle approach. Build. Environ, 46, 2247-2254 (2011)

MATEC Web of Conferences **82**, 01016 (2016) DOI: 10.1051/matecconf/20168201016
D2ME 2016

The Application of Quantum Energy Saver on Engine

Xiong Fang[1,a], Wenwu Mao[2], Xisheng Shen[1] and Zhixin Chen[1]

[1]*Fishery Machinery and Instrument Research Institute, Chinese Academy of Fishery Sciences Research Laboratory of marine fishing vessels and equipment, Shanghai, China*
[2]*School of Engineering, Shanghai Ocean University, Shanghai, China*

Abstract. In order to reduce diesel fuel consumption, this paper conducts the research in view of a new type of quantum energy saving device, and then produce the sample and applied on automobile engine, Detect fuel use of an automobile by automobile fuel saving technology assessment methods from the department of transportation. Compare the changes of fuel use before and after installation of quantum energy saving device on the same car, and give the feedback of energy saving capability. The result shows, after installed quantum energy saver, both fuel consumption and the smoke of tail gas has decreased. The analysis and application of this paper carry out the conclusion that the quantum energy saver can play an important role in energy saving and emission reduction, and provide a reference for other related research.

The automobile has become one of the most important means of transportation in the world and it's the symbol of development of science and technology. All developed countries take the automobile industry as the mainstay industry of the national economy. Automobile industry related to national economy and has played a huge role in promoting the social economic development and scientific and technological development. At the same time the automobile is also a sign of the development level of social life. In 2015 Both China's auto production and sales reached 24.5 million. The market share has reached from 3.5% in 2000 to 28% and become an important part of the world's automotive industry [1].

Engine is the only source of vehicle power. An engine with reliable quality and low energy consumption will provide huge power to the auto industry of China and make it develop quickly, continuously and healthy.

For ordinary consumers, the cost of fuel accounted for 40%-50% of the whole cost, even more. With the decrease of oil production and the increase of fuel price, consumers' travel costs are higher and higher. In order to save fuel and reduce exhaust emissions, most of automobile and engine manufacturers have adopted many measures such as use small displacement engine with turbocharger technology, or adjust throttle to make the fuel burned more fully. But the methods described above have some defects, the turbocharger technology is unstable and maintenance cost is high, also the stability of throttle adjustment remains to be improved [2]. In this paper, by the research and experiment of a new type of quantum energy saver, achieve the purpose of energy saving and consumption reduction.

[a] Corresponding author :13472617125@163.com

1 Quantum Energy

In 1990, the German physicist Max Planck found in the research of electromagnetic radiation energy distribution: The emission and absorption of electromagnetic waves are not continuous, but one by one. Every one has energy; the energy could be a multiple of an elementary unit:

$$\varepsilon_0 = h\nu \tag{1}$$

where ε_0 is energy, h is Planck's constant, and v is the frequency of the radiation. He called the energy unit which is the minimum and cannot be divided as energy quantum. When explain the distribution of the energy of the electromagnetic radiation, Planck first creatively put forward the concept of "quantum" and introduce the Planck's constant; There are more than 100 years when "quantum" was firstly introduced. Further research shows, in the micro field, some physical quantities are also jumping in the smallest units, not continuous, and the smallest units are called the quantum of the physical quantity. Such as: energy quantum, light quantum, flux quantum, sound quantum and etc [3].

In the composition of the universe, there are different levels of the smallest units, such as: all kinds of molecules, atoms, neutrons, protons, electrons, etc. In the range of $10^{-10}m$ and smaller scale, the material movement has obviously shown "quantized" characteristics. And in which, the Planck constant always plays an important role.

At the end of the nineteenth Century, scientists attempted to understand the micro world classical mechanics and macro view, but they failed, which reflected the contradiction between people's subjective understanding and objective reality. In order to solve this contradiction, the only way is to change the view of microscopic particles and get knowledge from practice. From the classical physics, we know that all microscopic particles have the property of the particle and the property of the wave. In 1905, the German physicist Albert Einstein's, 1879 - 1955) introduced the concept of quantum and photons and also put forward the "wave-particle duality" of the light. The concept of particle are mutually exclusive, but in a unified entity and it is shown by De Broglie equation

$$\lambda = h / p \tag{2}$$

The De Broglie wavelength is the wavelength, λ, associated with a massive particle and is related to its momentum, p, through the Planck constant, h [4].

Entire quantum mechanics is based on the concept of wave-particle duality and uncertainty relationship is very important. [5] When it is out of the accuracy which is ruled by uncertainty relationship, we can follow the classical laws, otherwise, follow the quantum laws. Therefore, the uncertainty relation is the boundary of quantum mechanics and classical mechanics; it reflects the process from quantitative change to qualitative change.

2 Quantum technologies

Quantum theory has been perfected and developed from theoretical research to practical application. This kind of wave can be stored in different carrier by the laser sintering machine, which makes it produce energy value in different applications. Materials which have been conversed by quantum energy create billion/sec supermicro wave in the effect of light, force and heat. When the liquid flows through the quantum pipe, hydrocarbon chains are rearranged into a long molecular chain, which greatly improves the combustion and explosion, so it can reduce the fuel consumption.

3 Designs and application

3.1 Structure design of quantum energy saver

The new quantum energy saver is a cylinder which is composed of two symmetrical quantum tile pipes. There are two grooves on both sides of the cylinder and there is fastener in the middle. All grooves and fasten are used to fix the cylinder on the pipe, so fluid can flow through the pipe and affected by the cylinder. The tile tubes are used to enhance the physical and chemical properties of the fluid in the pipe and reach the goal of high efficiency and optimization. Structure is shown in Figure 1.

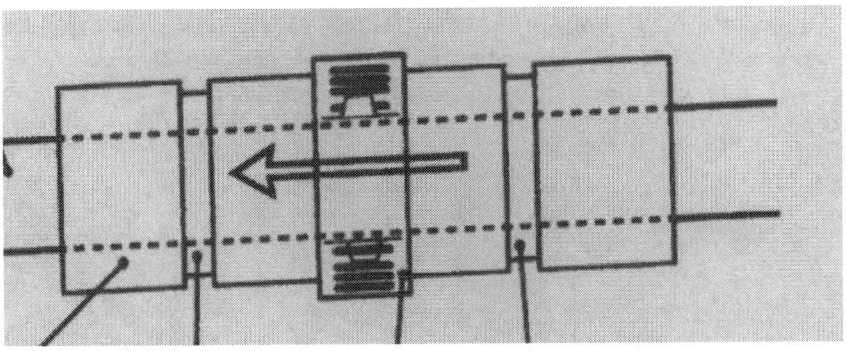

Figure 1. Structure of quantum energy saver

3.2 Application of engine

3.2.1 Application background and objects

In the new century, China's national standard of living has improved rapidly and automotive industry is developing quickly. As an important part of automobile industry, automobile components industry has been widely concerned. During 2000-2012, the engine industry in the proportion of GDP has increased a lot and especially in 2014. According to statistics, in 2014 the total number of engine production in China was 21.0816 million, increased 5.78%, total engine sales number was 21.1695 million, increased 9.64%. China has become a big country of automobile engine production and each year the number of automobile engine production accounts for about 1/4 of the world's total engine production [6]. So the energy saving and emission reduction of engine is a hot topic in current years. Major domestic and foreign automobile manufacturers, relevant research institutions have focused on this hope to be able to make a breakthrough. So we put the engine as the test object to verify fuel saving and emission reduction effect of quantum energy saver.

Figure 2. Dongfeng EQ5041XXY35D3AC automobile

3.2.2 Authentication

The object of this experiment is the Dongfeng EQ5041XXY35D3AC automobile, as shown in figure 2. The basis for the test was issued by the Ministry of transport GB/T14951-2007 《Automobile fuel saving technology assessment method》. The test site is in the Ministry of transport road traffic test site, engine laboratory. The main instruments of the test are non-contact speed analyzer, fuel flow meter, extinction type smoke meter, engine speed and so on. Test method was to put the quantum energy saver on the automobile engine oil inlet pipe, the location was near the end of the engine and then fasten belt and fixed the saver. Finally, drive the vehicle in constant speed and finish the fuel consumption test and free acceleration smoke (light reflection method) test, as shown in Figure 3.

Figure 3. Installation of quantum energy saver

3.3 Results

Before installing energy saver, tested fuel consumption in 30 km/h, 45 km/h, 60 km/h, 75 km/h and 90 km/h and then, compare with the data when saver was installed in all five speeds. The oil savings were 0.07L, 0.11L, 0.21L, 0.17L, 0.3L and the rates were 1.5%, 1.6%, 3%, 2%, 2.9%, integrated saving rate was 2.2%. The free acceleration smoke (light reflection method) purification rate was 4.3%. Results are shown in Table 1 and 2.

Table 1: Test data

Nominal speed（KM/h）	30	45	60	75	90
Actual Fuel consumption /speed data 0	4.91/29.8	6.06/45.0	7.34/59.9	8.74/75.0	11.0/89.8

Actual Fuel consumption /speed data 1	4.85/30.0	5.95/44.9	7.13/60.1	8.55/74.9	10.7/90.0
Calculation of fuel saving based on data 0 and data 1	0.07	0.11	0.21	0.17	0.3
Calculation of saving rate based on data 0 and data 1	1.5%	1.6%	3%	2%	2.9%
Note：Data 0 is the vehicle without quantum energy saver, data 1 is with the saver Units of fuel consumption and fuel saving：Kg/Km					

Table 2: Test data

	Free acceleration smoke（light reflection method）（m^{-1}）
Vehicle without quantum energy saver	1.61
Vehicle with quantum energy saver	1.54
Purification rate（%）	4.3

In the test of the automobile engine, quantum energy saver has obvious effect of fuel saving, There are no adverse effects and accidents during the test, the engine operated in good condition and energy saver was used in steady state.

4 Conclusions

According to the experiment above, use of new quantum energy saver can obviously save fuel and reduce tail gas emission. Data shows it saved most fuel when driven in 90KM/H and when driven in 60KM/H the fuel saving ratio is highest. The conclusion can be drawn that the quantum energy saver is particularly suitable for city roads and highways and the smoke emission rate of the exhaust gas is also improved.

Focus on the development direction of energy-saving and emission-reduction; design the new quantum energy saver by the research of quantum theory. There are many advantages such as simple structure, convenient to install and there is no adverse reaction.

If the new quantum energy saver can be promoted to the automotive industry and all automobiles install energy savers, it can save nearly 100 thousand tons of oil and billions of cost. Also provide support and protection for the country's sustainable development and green ecological environment.

References

1. L. J. Guan, Modern Industrial Economy and Informationization, **6**, 10-11 (2016)
2. J. Li, Z. L. Hei, Science & Technology Vision, No.12， 93-94 (2015)
3. R. X. Rui, H. T. Chen, International Electronic Elements, **21**, 15-16 (2013)
4. J. H. Gong, J. F. Gao, L. H. Duo, Journal of Teaching and Management, No.12, 99-100 (2006)
5. J. K. Fen, Discovery of Nature, **4**, No.1, 135-136 (1985)
6. Y. Wang, Dynamics of Industrial Economy in China, No.9, 41-43 (2014)

MATEC Web of Conferences **82**, 01017 (2016) DOI: 10.1051/matecconf/20168201017
D2ME 2016

Local Search Optimization for a Multi-state Series-parallel System

Ming-Ang YIN[1,a], Zhi-Li SUN[1], Jian WANG[1] and Yu GUO[1]

[1]*School of Mechanical Engineering and Automation, Northeastern University, Shenyang 110004, China*

Abstract: Generally, the redundancy allocation problem is characterized as a kind of NP-hard combinatorial optimization problem, which is difficult to solve with the classical method because it has exponentially increasing complexity with the augmented size. In this paper, we propose the efficient approach with two phases that is comprised of particle swarm optimization (PSO), which can efficiently represent the region search, and a local search, to solve the adjacent domains of best particle, which is great useful to improve optimization efficiency, and the parameter of local search is also studied.

Yun proposes a hybrid genetic algorithm with adaptive local search scheme[1]. Otsuki presents a local search algorithm which utilizes pruning and deepening techniques in the variable depth search framework[2]. Dengiz presents a genetic algorithm with specialized encoding, initialization and local search genetic operators to optimize communication network topologies[3]. Caniou chooses as the basis for these experiments the Adaptive Search method, an efficient sequential Local Search method for constraint satisfaction problems[4]. Gupta develops and compares different local search heuristics for the two-stage flow shop problem with makespan minimization as the primary criterion and the minimization of either the total flow time, total weighted flow time, or total weighted tardiness as the secondary criterion[5]. Cura presents a local search approach to the uncapacitated warehouse location problem that is quite simple and robust and is efficient in some cases[6].

Based on installation and maintenance cost are seen as the objective function, while availability of the system is seen as the constraint function, This paper mainly focuses on PSO combined with LS algorithm is applied to optimize such a problem, and a novel local search method is put forward.

1 Description

In a typical multi-state series-parallel system, N subsystems are put together in series, each of them is consisted of Mi ($i \in \{1,2,\ldots,N\}$) same devices in parallel. Configuration of one production line is shown in figure 1. With regard to the ith subsystem, there are Vi different versions available in market. To make things simplified, the ith subsystem in which devices with type of j (where j=1,2,…,Vi) is applied as an example below. CIN,ij represents installation charge and there are Kij states in total for the devices from brand new state to failure. State Transition Probability Matrix of this device is $\boldsymbol{Pij} \in RKij \times Kij$. $Xijl(t)$ represents the state of lth (where l=1,2,…,Mi) device at t moment. $Xijl(t)$=1

[a] Corresponding author : yinma@me.neu.edu.cn

represents that the *l*th device is brand new at *t* moment and $Xijl(t)=Kij$ represents that this device is in failure. Productivity of the devices will be $\gamma ij(w)$ if $Xijl(t)=w$, while production efficiency of invalid devices will be zero, that is $\gamma ij(Kij)=0$. Production efficiency of subsystem is the sum of that of all its components, while production efficiency of the whole system is the minimum of that of all its subsystems.

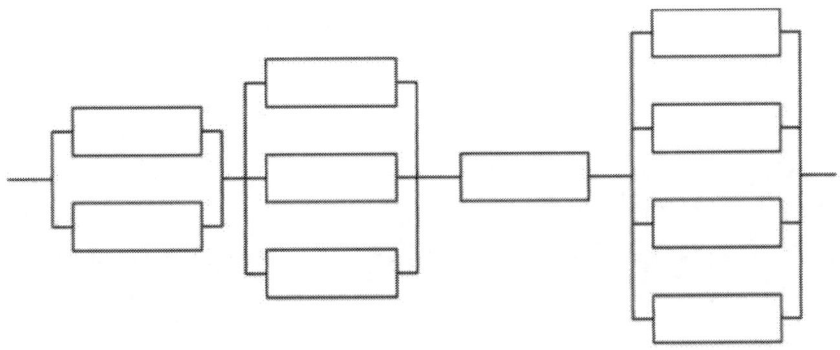

Figure 1. Schematic diagram of typical series-parallel system

2 Objective function

Formula (1) model of redundancy allocation problem (RAP) whose goal is to get the least acquisition and maintenance cost and meanwhile availability *As* of the system during the whole running time is higher than required value A_0.

$$\begin{cases} \text{Goal:} \quad \min C_S = C_{IN} + C_M \\ \text{Constraint:} \quad A_s \geq A_0 \end{cases}$$

Maintenance scheme will be confirmed according to detection results after detection, and furthermore, according to the second hypothesis that maintenance is instantaneous, so maintenance of system occurs at detection point, mean detection and maintenance cost of subsystem in unit time can be calculated through renewal theory[7].

$$C_{M,ij}\left(\theta_{ij}\right)=\sum_{i=1}^{N}\left(C_{INS,ij}+E\left(C_{ST,ij}\left(\theta_{ij}\right)\right)+E\left(C_{CM,ij}\left(\theta_{ij}\right)\right)+E\left(C_{PM,ij}\left(\theta_{ij}\right)\right)\right)\Big/E\left(T_{SR,ij}\left(\theta_{ij}\right)\right) \quad (2)$$

Acquisition expenses *CIN* of system can be expressed as

$$C_{IN}=\sum_{i=1}^{N} M_{ij}\times C_{IN,ij}$$

The system efficiency threshold value is W_R, so the system availability can be expressed as

$$A=\sum_{\{v|W_v\geq W_R\}} A_v(\theta)$$

One important problem in swarm optimization analysis is how to search in feasible region, penalty function can be used to solve this problem effectively. So, objective function is defined as the sum of objective function and penalty function. Calculation method of penalty function is put forward below.

$$C_P=\begin{cases} 0 & \text{,If } A_S \geq A_0 \\ C_\gamma \tan^{1.5}\left[-0.5\pi\left(1+A_S/A_0\right)\right], & \text{else} \end{cases}$$

Wherein $C\gamma$ is the penalty coefficient, objective function can be changed to

$$C_S=C_{IN}+C_M+C_P$$

3 Coding of RAP problem

According to the description of above model, the whole series-parallel system can be coded [***X***, ***Y***], ***X*** is the position vector and ***Y*** is the velocity vector. With regard to position vector X=[X_1,X_2,…,X_N], wherein , X_i represents the feature of the subsystem, M_i the number of devices of the *i*th subsystem, j_i the type selected in the subsystem and the serial number that the detection and maintenance strategy selected. Too much devices will add the acquisition expenses of the system and it's impossible to achieve the goal of low cost, so there is need to set an upper limit ***XB*** to restrain the number of devices in each subsystem.

4 Local search strategy

To compensate for the effectiveness that discretization works on particle's position and velocity vector, local search method is applied on individual optimal solution ***XPB*** and global optimal solution ***XGB***. To reduce the calculation cost, local search is used on particle from ***XPB*** that take up very little proportion η in every n iteration, while for ***XGB***, if there's a better one, local search will be applied on it.

Main idea of local search is that a solution with lower cost and higher availability. Research object of local search is characteristic of the subsystem which includes number and model of devices in the subsystem and maintenance strategy. Number and version of devices in the subsystem should be enumerated firstly, maintenance strategy of the subsystem will be worked out by restricting maintenance cost, of which the upper limit is the minimum cost of the subsystem before this local search, value of Ai will be ranked from high to low and the first Ns particles will be selected. Characteristics of other subsystems stay unchangeable when the subsystem undergoes local search. The particle with lowest cost is the optimal one to be output after all subsystems have been searched.

5 The detail of LS algorithm steps

Table 1 shows a part of one step LS, where N_L=3, other condition is not changed. The initial particle is [2 5 34 4 5 49 2 4 19 4 4 19]. Every sub step is local-searches one subsystem. Under the 1 version 1 device, finding the maintenance strategy which cost is less than the original one. Because of cost or the availability, it can be hardly to find the 4-verion-six-devices situation. The enumeration of version, number and maintenance strategy is calculated before optimization, so the ls needs just call objective function for a few times. At one ls, objective function is called $N \times M_i \times V_i \times N_L$=4×6×4×3=288 times at most.

Table 1. Details of Local Search

information	particle
Original particle	2 5 34 4 5 49 2 4 19 4 4 19
LS maintenance strategy for subsystem 1 , with 1 version 1 device	1 1 1 4 5 49 2 4 19 4 4 19 1 1 2 4 5 49 2 4 19 4 4 19 1 1 3 4 5 49 2 4 19 4 4 19

LS maintenance strategy for subsystem 1, with 1 version 2 devices	1 2 2 4 5 49 2 4 19 4 4 19 1 2 3 4 5 49 2 4 19 4 4 19 1 2 6 4 5 49 2 4 19 4 4 19
LS maintenance strategy for subsystem 1, with 4 version 3 devices	4 3 40 4 5 49 2 4 19 4 4 19 4 3 46 4 5 49 2 4 19 4 4 19 4 3 17 4 5 49 2 4 19 4 4 19

6 Example

An auto parts factory plans to put a vehicle air condition valve plug production line into production which includes four main processes. Units in every process are uniform. There are four versions of units in every process. Working performance of units from each factory varies from each other. Performance degradation process of all units all follow discrete time Markov process. Discrete interval is 1 week. State transition probability matrix of units produced from these four factories in the first process can be expressed as

$$\boldsymbol{P}_{11}=\begin{bmatrix}0.6 & 0.2 & 0.15 & 0.04 & 0.01\\ 0 & 0.6 & 0.3 & 0.08 & 0.02\\ 0 & 0 & 0.75 & 0.2 & 0.05\\ 0 & 0 & 0 & 0.75 & 0.25\\ 0 & 0 & 0 & 0 & 1\end{bmatrix}\quad \boldsymbol{P}_{12}=\begin{bmatrix}0.6 & 0.2 & 0.15 & 0.03 & 0.015 & 0.005\\ 0 & 0.6 & 0.3 & 0.06 & 0.025 & 0.015\\ 0 & 0 & 0.75 & 0.2 & 0.03 & 0.02\\ 0 & 0 & 0 & 0.75 & 0.15 & 0.1\\ 0 & 0 & 0 & 0 & 0.7 & 0.3\\ 0 & 0 & 0 & 0 & 0 & 1\end{bmatrix}$$

$$\boldsymbol{P}_{13}=\begin{bmatrix}0.7 & 0.15 & 0.1 & 0.04 & 0.01\\ 0 & 0.65 & 0.25 & 0.08 & 0.02\\ 0 & 0 & 0.6 & 0.25 & 0.15\\ 0 & 0 & 0 & 0.5 & 0.5\\ 0 & 0 & 0 & 0 & 1\end{bmatrix}$$

$$\boldsymbol{P}_{14}=\begin{bmatrix}0.7 & 0.15 & 0.1 & 0.03 & 0.015 & 0.005\\ 0 & 0.7 & 0.2 & 0.06 & 0.025 & 0.015\\ 0 & 0 & 0.75 & 0.2 & 0.03 & 0.02\\ 0 & 0 & 0 & 0.75 & 0.15 & 0.1\\ 0 & 0 & 0 & 0 & 0.8 & 0.2\\ 0 & 0 & 0 & 0 & 0 & 1\end{bmatrix}$$

To reduce redundant description, state transition probability matrix of units in other processes are the same with that of the first process such as $\boldsymbol{P}_{21}=\boldsymbol{P}_{11}$, productivity and installation cost of each unit is listed in Table 2. Number of inspection threshold of units $N_{INS,ij}$ in each process is 2, then maintenance strategy includes $L_{PM,ij}$, $L_{OM,ij}$ and two inspection thresholds $L_{INS,d,ij}$, where d=1,2

Table 2. Productivity, installation and maintenance cost of different versions in each subsystem

i	j	$\gamma_{ij}(1)$	$\gamma_{ij}(2)$	$\gamma_{ij}(3)$	$\gamma_{ij}(4)$	$\gamma_{ij}(5)$	$\gamma_{ij}(6)$	$C_{IN,ij}$	$C_{PR,ij}$	$C_{CR,ij}$	$C_{H,ij}$	$C_{INS,ij}$
1	1	0.2	0.15	0.1	0.05	0	-	1.56	0.686	0.312	0.156	0.052
	2	0.4	0.32	0.24	0.16	0.08	0	1.86	1.085	0.62	0.155	0.031
	3	0.24	0.2	0.16	0.08	0	-	2.16	0.756	0.36	0.216	0.036
	4	0.8	0.7	0.6	0.4	0.2	0	3.9	1.869	1.068	0.134	0.13
2	1	0.08	0.06	0.04	0.02	0	-	1.548	0.539	0.31	0.155	0.052
	2	0.25	0.2	0.15	0.1	0.05	0	2.748	1.603	0.916	0.229	0.046
	3	0.15	0.125	0.1	0.05	0	-	2.901	1.015	0.484	0.29	0.048
	4	0.6	0.525	0.45	0.3	0.15	0	4.101	4.55	1.64	0.205	0.21
3	1	0.24	0.18	0.12	0.06	0	-	0.642	0.364	0.128	0.064	0.021
	2	0.45	0.36	0.27	0.18	0.09	0	1.152	0.672	0.384	0.096	0.019
	3	0.54	0.45	0.36	0.18	0	-	1.602	0.56	0.267	0.16	0.027
	4	1.6	1.4	1.2	0.8	0.4	0	5.4	1.288	0.737	0.092	0.092
4	1	0.1	0.075	0.05	0.025	0	-	2.049	0.714	0.41	0.205	0.068
	2	0.125	0.1	0.075	0.005	0.025	0	1.935	1.127	0.645	0.161	0.032
	3	0.09	0.075	0.06	0.03	0	-	2.091	0.735	0.349	0.209	0.035
	4	0.56	0.49	0.42	0.28	0.14	0	3.57	3.92	1.428	0.179	0.18

Proportion and productivity requirements of busy season, normal season and slack season in one year are listed in Table 3.

Table 3. Productivity requirements of the system

W_{0p}	0.40	0.30	0.20
$T_p(\%)$	20	30	50

Some other parameters neeed in optimal algorithm are listed in Table 4.

Table 4. Parameters used in optimization algorithm

X_B	t_{max}	N_P	C_γ	η	n	N_S
6	50	500	2000	2%	10	5

Figure 2 shows that 20 optimization results, for W_R=15, A0=0.999 and other parameters without changed. For comparison, the times of optimization result and the dispersion are listed. The best particle is [4 2 3 4 3 10 4 2 27 4 3 3], the system cost is 32.358, and system availability is 0.9993.

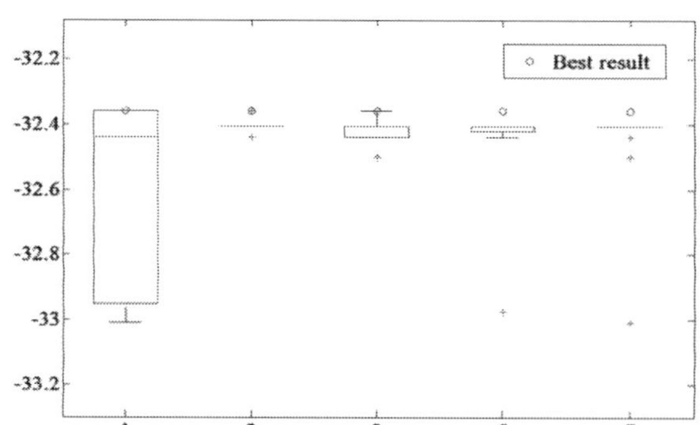

Figure 2. 20 times Optimization results with different N_L

In figure 2, the blue cycle is the best result. The upper quartile is the best result, as N_L=1. Meanwhile for other condition, the best result only can be the upper boundary or singularity, even cannot be optimized. The LS with N_L=1 is generally superior to the other N_L since there are more trials out of 20 runs achieving the lowest cost, but the variation of the 20 solution is lager. It is an efficiency method to get the best solution to run more times optimization.

There is a LS strategy proposed in Literature [8], and the details are listed in figure 3. Five LS strategies are listed in figure 3. For improving the efficiency of algorithm, combination of five LS strategies is used.

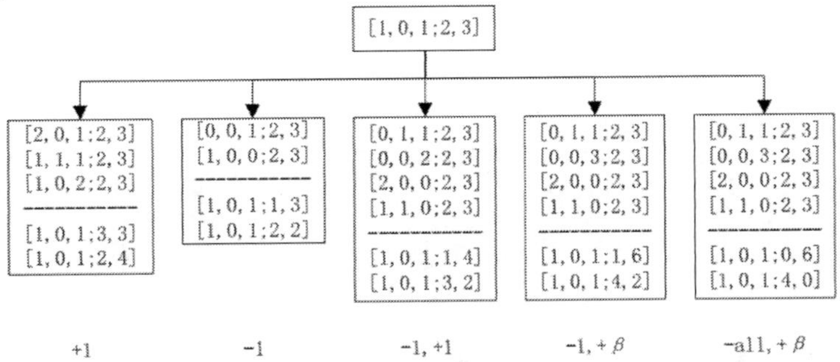

Figure 3. LS from reference [8]

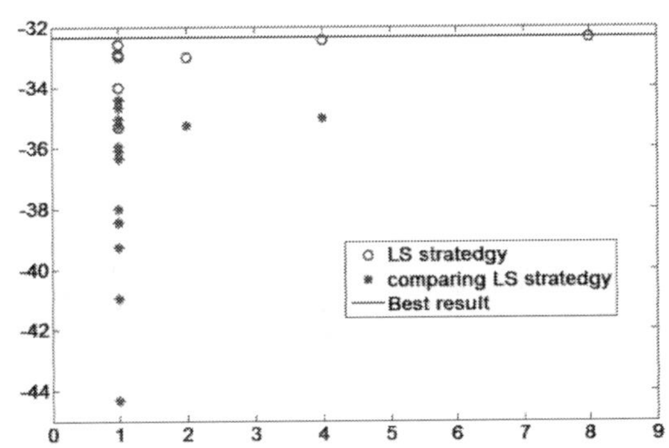

Figure 4. Comparison of two LS methods

Figure 4 shows comparison of two different ls strategies. The ls strategy with little cost more availability shows apparent solution improvement. It also lists times of achieving the best solution. In 20 runs, there 8 runs optimize the best result -32.358, and 4 runs optimize the second-best result, so it is show that the ls is efficiency and stability.

Acknowledgement

This study was supported by the Fundamental Research Funds for the Central Universities of China (Grant No. N140303010).

References

1. Y. S. Yun. Comput Ind Eng, **51**, 128-141 (2006)
2. T. Otsuki, K. Aihara. J Heuristics, 1-19 (2014)
3. B. Dengiz, F. Altiparmak. Ieee T Evolut Comput, **1**, 650-657 (1998)
4. Y. Caniou, P. Codognet, F. Richoux, et al. Constraints, **20**, 30-56 (2014)

5. J. N. D. Gupta, K. Hennig, F. Werner. Comput Oper Res, **29**, 123-149 (2002)
6. T. Cura. Comput Ind Eng, **59**, 1000-1009(2010)
7. Y. Zhou, Z. Zhang, T. R. Lin, et al. Reliab Eng Syst Safe, **111**, 248-259(2013)
8. Y. Wang, L. Li. Ieee T Syst Man Cy A, **42**, 464-474(2012)

MATEC Web of Conferences **82**, 01018 (2016) DOI: 10.1051/matecconf/20168201018
D2ME 2016

The Improvement of Carburater Efficiency Using Biogas-based Venturi

Ni Ketut Lasmi[a], Alamta Singarimbun and Last Wahyu Srigutomo

[1]*Physics of Earth and Complex System, Fac. of Mathematics and Natural Sciences, Institut Teknologi Bandung, Jl. Ganesa 10 Bandung, Indonesia 40132*

Abstract. The elimination of the fossil fuel subsidy by the Indonesian government has caused an increase in fuel prices, and a solution to find a relatively cheap and environmentally friendly alternative energy is needed. Biogas is one of the sources of renewable energy that has a potential to be developed, especially in farming area where the abundant animal excrement is not yet optimally used and causes environmental problems. Addressing this issue, we have developed an innovation by making a biogas and air mixer instrument through venturi pipe, using the basic theory of fluid mechanism in order to increase the use of biogas as an electricity source. Usually, biogas-based electric generators use dual fuel system such as fossil fuel and biogas to perform combustion due to the low octane contained in the biogas. By replacing the readily available manufactured venturi with the modified venturi, optimal combustion can be reached with using only single fuel of biogas. The results of the experiments show that the biogas debit on carburetor increases from 13 to 439 watts consuming biogas fuel from 0.22 to 4.96 liter/minute, respectively. The amount of combusted biogas depends on the value of the load power. Within the scope of our results, the maximum voltage reached is about 211.13 – 211.76 volts which is feasible to use for 220 volts electrical appliances

1 Introducton

In the last couple of years, the use of renewable energy has been increasing in Indonesia. If it is well organized and properly used, it can be a supplement for the limited fossil energy. One of the renewable energy sources that can be used is biogas [1] which can be transformed into bioelectric. Bioelectric is an electricity that is obtained from biogas resulted by cow excrement, liquid and solid waste through anaerobic digestion [2]. The composition of biogas varies depending on the source of the biogas, for example cow excrement [3]. Biogas is one of the energy technology solutions to overcome society's problem of electricity. This technology can be immediately applied to society in order to fulfil the national energy need.

The use of purified biogas as machine fuel will decrease greenhouse gas emission and provide other natural benefits, because combustion of biogas releases smaller amount of nitrogen oxide, hydrocarbon and carbon monoxide compared to the combustion of benzene and diesel oil. Compared to natural gases, biogas combustion releases smaller amount of hydrocarbon, carbon monoxide and carbon dioxide, but bigger amount of nitrogen oxide [4]. Another benefit is that a machine that uses

[a] Corresponding author : k.lasmi@yahoo.com

biogas fuel sounds softer than machine with diesel oil. Biogas conversion is done by changing the fuel in generator that uses mixture of benzene and biogas by single fuel biogas generator. Because the octane of biogas if very low, the available commercial venturi is modified by a new venturi having different sizes. The purpose of making this instrument is to increase the efficiency of biogas as an environmentally friendly electricity source [5, 6], and affordable by all citizen in Indonesia.

2. Materials and Methods

2.1 Research location and design of biogas mixer instrument

This research is conducted at Pondok Pesantren Baituraahman, Bojong Dusun village, Cikoneng Ciparai sub-district, Bandung, which is the location of biogas digester. This research is carried out through several steps: mixing the biogas and air in the modified venturi, measurement of CH4 emission by the self-made sensor after the biogas passing through 2 kW generator as shown in Fig.1 [8]. To produce the biogas, we utilize six biogas reactors which are airless installation enabling the cow excrement to be decomposed optimally thus decreasing the methane emission.

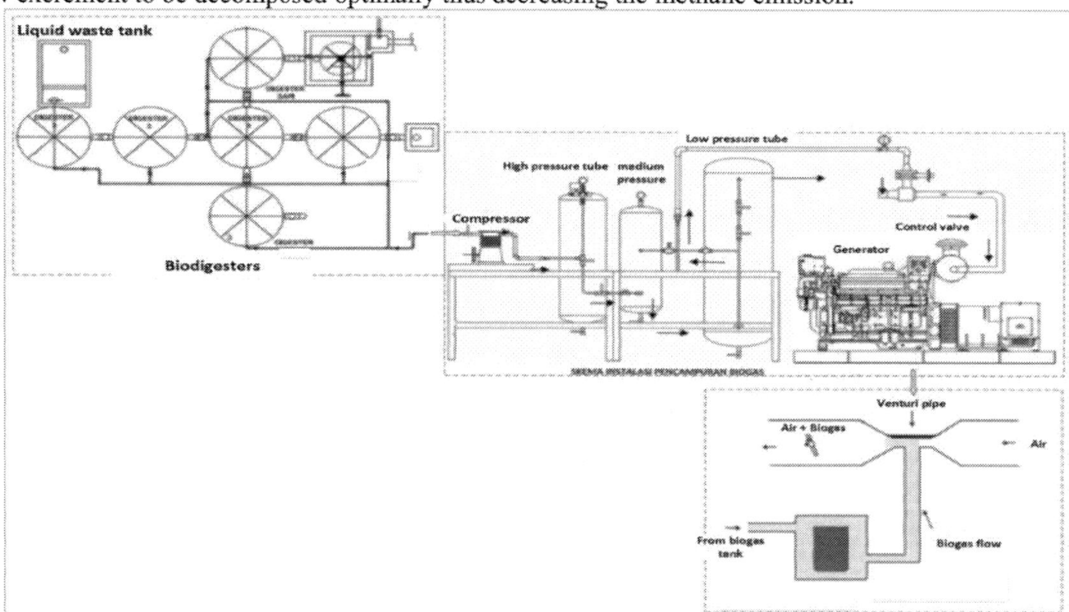

Figure 1. Scheme of biogas mixer instrument

2.2 Generator modification

Modification on the generator was done to increase the biogas generator performance. There are two modified components: carburetor and the venturi. The function of carburetor is to mix air and biogas with a certain ratio in the combustion room. The modification of carburater is done by replacing the original factory manufactured venturi made from lathed copper rod having 6 mm inlet and 3 mm outlet diameters by the modified brass venturi having 5.0 mm inlet and 3.5 mm outlet diameters and 150.0 mm long (Fig. 2 and Fig. 3). The venturi is then attached to the carburetor and serves as the air-biogas mixer.

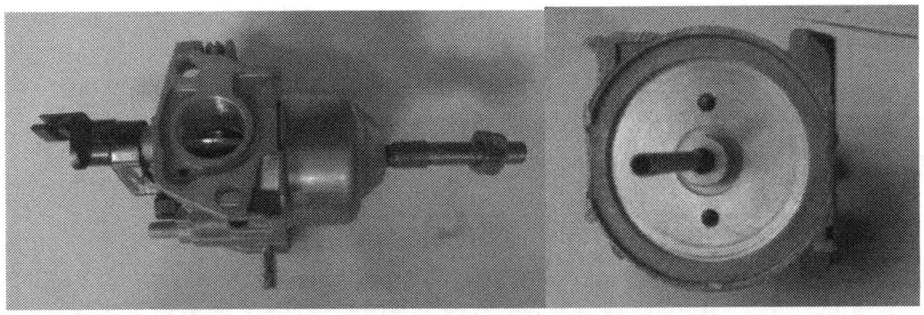

Figure 2. Modified carburater.

Figure 3. (Up) original factory manufactured venturi, (Down) modified venturi

2.3 Formulation of output power using venture

Fluid velocity in a tube can be measured by Bernoulli's and continuity equations under assumptions that the fluid follows ideal fluid basic equations [9]. To calculate the pressure difference of biogas- air mixture at the inlet and outlet of the venturi, the continuity equation is used [10, 11, 12] yielding

$$Q_{ideal} = A_1 v_1 = A_2 v_2 \quad (1)$$

In its general form Bernoulli's equation can be expressed as [13, 14],

$$\frac{p}{\rho} + \frac{v^2}{2} + gz = \text{constant} \quad (2)$$

whereas for two different diamaters the equation can be written as [15]

$$\frac{p_1}{\rho} + \frac{v_1^2}{2} + gz_1 = \frac{p_2}{\rho} + \frac{v_2^2}{2} + gz_2 \quad (3)$$

If the flow is horizontal ($z_1 = z_2$), then

$$\frac{p_1}{\rho} + \frac{v_1^2}{2} = \frac{p_2}{\rho} + \frac{v_2^2}{2} \quad (4)$$

with p1: pressure at point 2 (Pa), p_2: pressure at point 2 (Pa), v_1: velocity at point 1 (m/s), v_2: velocity at point 2 (m/s), and: ρ:density (kg/m^3).

The non-horizontal effect in the flow can be easily incorporated by inserting height difference (z_1–z_2) into the equation. The combination of continuity equation (1) and Bernoulli's equation (4) results the theoretic flow velocity equation if $A_2 < A_1$:

$$Q = A_2 \sqrt{\frac{2(p_1 - p_2)}{\rho\left[1 - \left(\frac{A_2}{A_1}\right)^2\right]}} \quad (5)$$

where Q: flow velocity (m3/s), A_1: area of part 1 (m^2), A_2: area of part 2 (m^2), and p_1–p_2 = p: pressure difference (Pa). The flow velocity can be written as

$$Q = c_v Q_{ideal} = c_v A_T \sqrt{\frac{2(p_1 - p_2)}{\rho(1-\beta^4)}} \qquad (6)$$

where $A_T = (\pi d^2)/4$ is the area of venturi neck (m^2), c_v: discharge venturi coefficient, and β = d/D: neck diameter ratio.

The result of theoretic flow velocity will be higher than the flow velocity in real measurement because of the difference of "real world" with assumptions that are used in Bernoulli equation. The difference can approximately be from 1 – 40% [14]. If the diameters of both pipes are known, the fluid flow velocity in venturi pipe can be measured by flow meter as well as the pressure.

Experimentally, the measurement of generator performance is expressed as

$$Q_{measured} = \frac{Q_1 + Q_2 + ... + Q_n}{n} \quad \text{(liter/minute)} \qquad (7)$$

The actual heat of biogas depends on CH4. Heat is very vital to determine the machine capability. To measure heat, the standard thermodynamics formula can be used

$$H_{u,actual} = \frac{V_{CH_4}}{V_{tot}} \rho_{CH_4,actual} H_{u,normal} \quad \text{(kJ/m}^3\text{)} \qquad (8)$$

with

$$\rho_{CH_4,actual} = \rho_{CH_4,standard} \frac{P_{actual}}{P_{standard}} \frac{T_{actual}}{T_{standard}}, \quad \text{(kg/m}^3\text{)} \qquad (9)$$

CH4,standard = standard methane density (=0.72 kg/m^3)
Pactual,standard = total pressure and standard pressure of biogas, (mbar)
Tactual,standard = standard temperature (273 K) and actual temperature from the biogas fuel
The flow velocity of biogas can be measured with the equation:

$$Q = vA = \frac{m}{Vol} \quad \text{(kg/m}^3\text{)}, \qquad (10)$$

where ρ: density (kg/m3), m: mass (kg), and Vol: volume (m^3), v: gas flow velocity (m/s), dan A: cross section (m^2). The flow velocity [13] is written as:

$$Q = \frac{m}{t} \quad \text{(kg/h)}, \qquad (11)$$

The biogas consumption is calculated by the equation:

$$E = QH_{u,normal} \quad \text{(kJ/kg)}, \qquad (12)$$

The output power can be measured as:

$$P = VI \quad \text{(watt)}, \qquad (13)$$

where V: output voltage (volt) and I: current (ampere).

3. Results and Discussion

The result is based on the experimental data obtained by direct observation in location. The data are taken by using different loads. The biogas combustion quality spreading out from the venturi exhibited straight splash pattern which is different from the original manufactured venturi, resulting a stable output current.

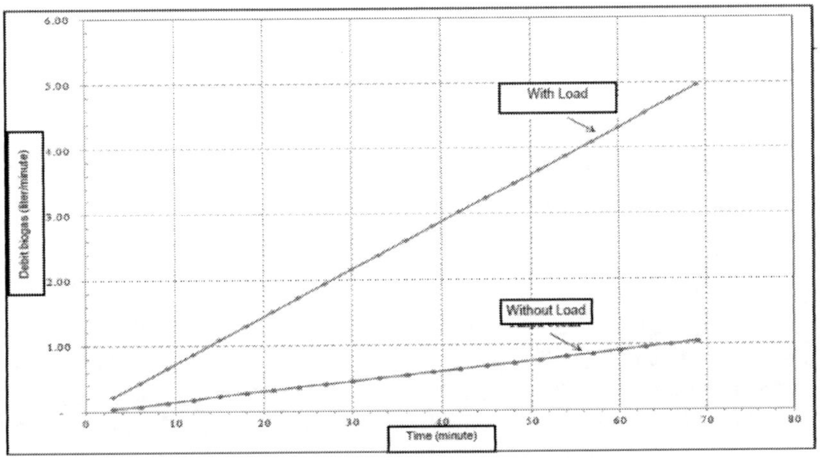

Figure 4. Graphics of biogas debit vs. time (with and without load)

The plot between biogas debit vs. time (with and without load) is depicted in Figure 4. This figure shows that there is a significant relation between biogas debit and time that is changing linearly. The longer the time, the higher the biogas debit is and vice versa. This indicates that the modified venturi provides efficient injection of biogas resulting perfect combustion.

Plot of the relation between the output power and the biogas debit is shown by Fig. 5 indicating the liniear relation between the load power and biogas debit. The plot between the actual power with biogas debit is also shown in the same figure indicating the improvement in combustion efficiency as the result of venturi modification. The power generated by the generator increases with time inline with the icrease in biogas consumption. The actual power outputted varies from minimum 21.12 watts up to 658.73 watts with biogas fuel consumption from 0.22 liter/minute to 4.96 liter/minute. The differences between the load power and the measured power are approximately 33.36% - 38.43%. The application of modified venturi affects the generated power, energy, fuel consumption, heat, fuel mass and generator efficiency. The more biogas fuel is combusted, the more output power that is obtained. The electric voltage obtained by the generator reached value of 220 volts when applied without load and reached approximately 211.13 V – 211.76 volts when load is applied which is still a stable voltage range.

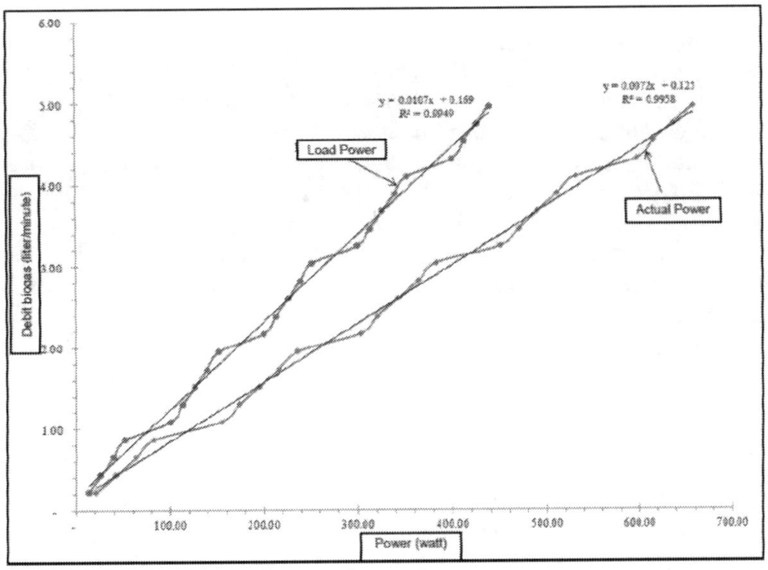

Figure 5. Graphic of relation between fuel debit with output power

4. Conclusion

The result shows that the increase in biogas debit is linearly related to the increase in output power. The actual power outputted varies from minimum 21.12 watts up to 658.73 watts with biogas fuel consumption from 0.22 liter/minute to 4.96 liter/minute. The electric voltage obtained by the generator reached value of 220 volts when applied without load and reached approximately 211.13 V – 211.76 volts when load is applied which is still a stable voltage range.

Acknowledgements

A sincere thank to Pondok Pesantren Baituraahman and the farmers in Cikoneng Kecamatan Ciparai Kabupaten Bandung and Lembaga Ilmu Pengetahun Indonesia (LIPI) Bandung for their help and support during the data collecting.

References

1. Herberg,T.C., Enskat,M.and Abmann, D., Conference Issue Paper. Renewables 2004 – International Conference for Renewable Energies, Bonn, Germany, 2014
2. Gunnerson, C.G. and Stuckey,D.C.,Anaerobic Digestion: Principles and Practises for Biogas System. The World Bank Washington, D.C. USA, 1986
3. Deublein,D. and Steinhauser,A., Biogas from Waste and Renewable Resources, Wiley-VCH Verlag Gmbh and Co. KgaA, Germany, 2008
4. Vijay, V.K., Chandra, R., Subbarao, P.M.V. and Kapdi, S.S., Biogas Purification and Bottling into CNG Cylinders: Producing Bio-CNG from Biomass for Rural Automotive Applications, The 2nd Joint International Conference on “Sustainable Energy and Environment 2006”, Bangkok, 2006
5. Porpatham, E., Ramesh,A. and Nagalingam, B.,Investigation on the effect of concentration of methane in biogas when used as a fuel for a spark ignition engine. Internal Combustion Engines Laboratory, Department of Mechanical Engineering, Indian Institute of Technology Madras, Chennai – 600 036, India, 2007
6. Prastya, R., Susilo,B. and Lutfi, M., Pengaruh Penggunaan Bahan Bakar Biogas terhadap Emisi Gas Buang Mesin Generator Set. Jurnal Keteknikan Pertanian Tropis dan Biosistem. Vol. 1 No. 2, Juni 2013, pp. 77-84
7. Lasmi, N.K., Singarimbun, A., Souisa, M., Rajagukguk, J. and Amunidin, A.,Desain Instrumen Pencampur Biogas dan Udara Untuk Menghasilkan Daya Listrik, Proceeding Seminar Nasional Energi Indonesia (Senendo), Universitas Atma Jaya, Yogjakarta, 2013, pp.111–114
8. Lasmi, N.K., Singarimbun, A. and Srigutomo, W., Pengukuran Perbedaan Tekanan dan Laju Aliran Biogas Pada Pipa Venturi Menggunakan Generator Berbahan Bakar Biogas Untuk Menghasilkan Energi Listrik, Prosiding Simposium Nasional Inovasi dan Pembelajaran Sains 2015 (SNIPS 2015), 8 dan 9 Juni 2015, Bandung Indonesia, 2015,pp. 29–32
9. Serway and Faugh, CBS Collega Physics, College Publishing, 1985, pp. 45-46
10. Munson, B.R., Young, D.F., and Okiishi, Th.H., Fundamentals of Fluid Mechanics, fourth edition. John Wiley and Sons, Inc. USA, 2002
11. White, F.M.,Fluid Mechanics, Second Edition. McGraw-Hill, Ltd, 1986
12. Abbas, H.A., Hasan, M. and Lucas, G.P., Experimental and theoretical study of the gas-water two phase flow through a conductance multiphase Venturi meter in vertical annular (wet gas) flow. University of Huddersfield Repository, Nuclear Engineering and Design. 241 (2011) 1998-2005
13. Zucker, R.D. and Biblarz, O., Fundamentals of Gas Dynamics, second edition, Published by John Wiley and Sons, Inc., Hobeken, New Jersey, 2002,26,41, pp. 64-66
14. Munson, Bruce R., Young, Donald, F., Okiishi, andTheodore, H., Mekanika Fluida, Jilid I, Edisi Keempat, Jakarta: Erlangga, 2004

MATEC Web of Conferences **82**, 01019 (2016)   DOI: 10.1051/matecconf/20168201019
D2ME 2016

Surface Roughness Prediction in Grinding: a Probabilistic Approach

Krishna Kumar Saxena[1], Sanjay Agarwal[2] and Raj Das[1,a]

[1]*Department of Mechanical Engineering, Centre for Advanced Composite Materials, University of Auckland, Auckland 1010, New Zealand, ksax995@aucklanduni.ac.nz r.das@auckland.ac.nz*
[2]*Department of Mechanical Engineering, B.I.E.T., Jhansi, Uttar Pradesh, India, [a] sanjay72ag@rediffmail.com*

Abstract. Surface quality of machined components is one of the most important criteria for the assessment of grinding processes. The importance of surface finish of a product depends upon its functional requirements. Since surface finish is governed by many factors, its experimental determination is laborious and time consuming. So the establishment of a model for the reliable prediction of surface roughness is still a key problem for grinding. In this study, a new analytical surface roughness model is developed on the basis of the stochastic nature of grinding processes. The model is governed mainly by the random geometry and the random distribution of cutting edges on the wheel surface having random grain protrusion heights. A simple relationship between the surface roughness and the chip thickness was obtained, which was validated by the experimental results using AISI 4340 steel in surface grinding.

1 Introduction

Advanced engineering materials have been extensively used in industrial applications during the last two decades [1, 2]. However, the actual utilization of advanced materials with high strength and hardness has been quite limited mainly because of the machining difficulties and associated high cost of machining these materials by grinding whilst ensuring the workpiece quality. A technological basis to achieve more efficient utilization of the grinding process requires an understanding of the interaction between the abrasive and the workpiece, which has direct bearing on the surface roughness, produced using electron discharge machining. Although extensive research [3-11] has been carried out to predict the surface roughness of the ground workpieces, a complete understanding of surface roughness in grinding is yet to be achieved.

Theoretical methods of surface roughness evaluation, as reported in the literature, can be classified into empirical and analytical methods. In the empirical methods, the surface roughness model is developed as a function of kinematic conditions [3-4]. Although the development of empirical models requires less efforts compared to analytical one, they have limited applicability; however, still these models are used in the field of grinding technology. Hence, the scope of these models is limited. The analytical models are always preferred over empirical models as these models are deductively derived from fundamental principles. So, the main advantage of the analytical model is that the results can easily be transferred to different grinding conditions and processes. Hence, these results can be made

[a] Corresponding author : r.das@auckland.ac.nz

applicable to a wide range of process conditions. The analytical models for surface roughness have always been characterized by the description of the microstructure of the grinding wheel, in one-dimensional form, taking the grain distance, the width of cutting edge and the grain diameter into account [5-7]. However, these models did not consider the different heights of cutting edges and assumed that the distance between the cutting edges was uniform. Tonshoff et al. [8] described the state of art in the modelling and simulation of grinding processes comparing different approaches to modelling. This work identified a simple basic model where all the parameters such as wheel topography, workpiece material properties and grinding parameters have been lumped into an empirical constant.

Several analytical models, based on the stochastic nature of grinding processes, were proposed [9-11] to simulate the surface profile generated during grinding. Assuming a particular probability distribution of these random cutting points, output surface profiles were generated for a known input surface profile and input grinding conditions. Experiments conducted by Lal and Shaw [12] with single abrasive grain under fine grinding conditions indicated that the grain tip could be better approximated by a circular arc. Therefore, it is evident that the groove produced by an individual grain can be better approximated by an arc of a circle. Based on this concept, Agarwal and Rao [13-15] developed an analytical model for the prediction of surface roughness in ceramic grinding. A probabilistic approach, for geometric analysis of the grooves generated, was used to describe the surface roughness, in which, the random variable had been defined by undeformed chip thickness probability density function. A simple relationship between the surface roughness and the undeformed chip thickness was found.

In this paper, an analytical model has been envisaged to evaluate surface roughness from the chip thickness probability density function. A simple relationship between the surface roughness and the chip thickness has been established with the chip thickness as a random variable. The model is then verified using experimental data from the surface grinding of AISI 4340 steel with aluminium oxide abrasive.

2 Model development

A schematic diagram showing the transformation of the grain tip to the work piece is given in Fig. 1. At any transverse section, the shape of the groove generated by any grain is shown in Fig. 2. Since an individual grain has many tiny cutting points on its surface and the speed ratio is high, the groove produced by an individual grain can be assumed to be an arc of a circle. It can be further assumed that the material is either ploughed with little side pile-up or removed in the form of chips whenever grain-work piece interference occurs.

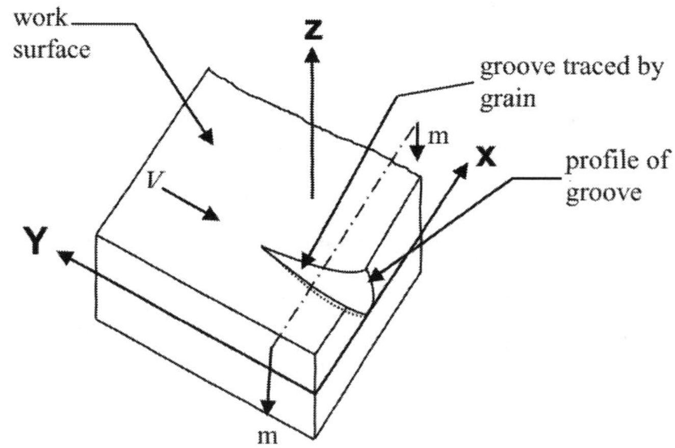

Figure 1. Schematic view of the workpiece in Cartesian coordinate system

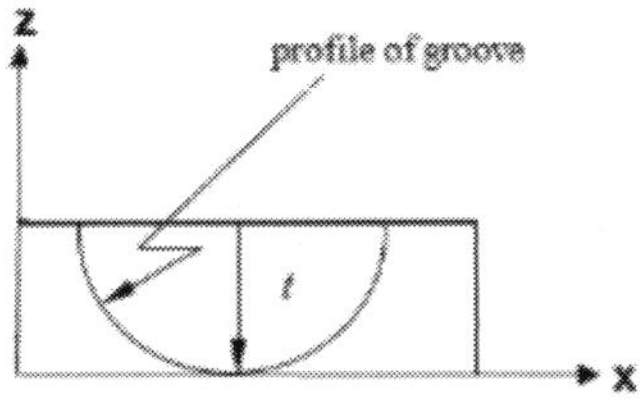

Figure 2. Sectional view showing the shape of groove generated.

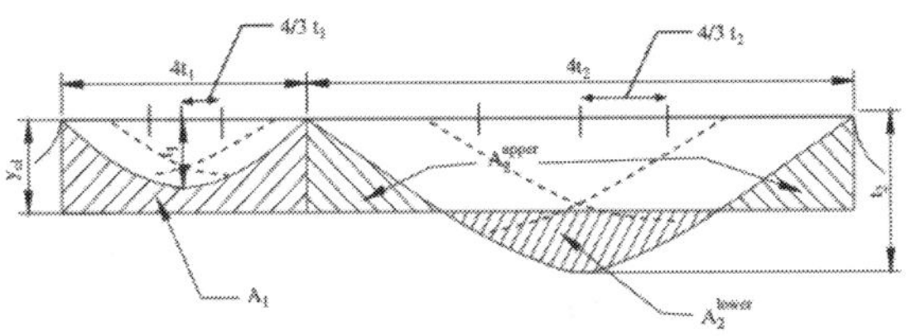

Figure 3. Profile of grooves generated.

Since the position of a grain tip will be random due to random occurrence of grains on the wheel surface, a probability density function is required to describe the surface roughness for all the grains engaged. Thus, the undeformed chip thickness '*t*' can be better described by Rayleigh's probability density function '$f(t)$' [11] as;

$$f(t) = \frac{t}{\beta^2} \exp\left\{-\frac{1}{2}\left(\frac{t}{\beta}\right)^2\right\} \quad \text{for } t \geq 0 \tag{1}$$

$$= 0 \qquad \text{for } t < 0$$

where β is a parameter that completely defines the probability density function and it depends upon the cutting conditions, the direction of grinding wheel and the properties of work piece materials. The expected value and the variance of the above function will be given as where β is a parameter that completely defines the probability density function and it depends upon the cutting conditions, the direction of grinding wheel and the properties of work piece materials. The expected value and the variance of the above function will be given as E (t) = $\beta\sqrt{\pi/2}$ and sd(t) = $\beta\sqrt{(2-\frac{\pi}{2})}$. The surface roughness R_a which is defined as the arithmetic average of the absolute values of the deviations of the surface profile height from the mean line within the sampling length l. *Ra* can be expressed as

$$R_a = \frac{1}{l}\int_0^l |y - y_{cl}|\, dl$$

where y_{cl} denotes the centre line drawn in such a way that the areas above and below it are equal. It can also be expressed statistically as

$$R_a = \frac{1}{l}\int_{y_{min}}^{y_{max}} |\, y - y_{cl} \,| \cdot p(y) \cdot dy \tag{2}$$

where y_{max} and y_{min} are the lowest and highest peak height of the surface profile and p(y) is the probability that height y of grain has a particular value y and may be expressed as

$$p(y) = \lim_{\Delta y \to 0} \frac{probability\{y - (1/2)\Delta y \leq y \leq y + (1/2)\Delta y\}}{\Delta y}$$

The surface roughness R_a can be calculated using probability density function defined in Eq. (1). The complete description of surface generated is very difficult due to the complex behavior of different grains producing grooves because of the random grain-work interaction. Thus, certain assumptions have to be made while predicting the surface roughness. The assumptions are as follows:

(1) The profile of the grooves generated is same and completely defined by the depth of engagement or undeformed chip thickness '*t*'

(2) There is no groove overlapping
(3) An individual grain has many tiny cutting points in its surface, therefore, for simplicity, the grains are approximated as uniform spheres of diameter d_g (=2t), randomly distributed throughout the wheel volume.

On the average, the expected value of an interference area through a sphere is about one fourth of the area of a circle.

As per definition of surface roughness, the area above and below the centre line must be equal i.e. the total expected area can be written as

$$E\{A(t)\} = 0 \tag{3}$$

The above expression can be represented in terms of the probability density function f (t) as

$$\int_0^\infty A(t).f(t).dt = 0 \tag{4}$$

There are two types of grooves generated depending upon their depth of engagement is either less or greater than centre line y_{cl} . For the case when the depth of engagement is less than y_{cl} can be expressed as

$$E\{A(t_1)\} = \int_0^{y_{cl}} A(t_1).f(t).dt \, , \tag{4a}$$

Similarly, for the groove with depth of engagement is greater than y_{cl} can be expressed as

$$E\{A(t_2)\} = \int_{y_{cl}}^\infty A(t_2).f(t).dt \tag{4b}$$

Substituting the values from eq. (4a) and (4b), in eq. (3), equation becomes

$$\int_0^{y_{cl}} A_1 \; f(t)\,dt + \int_{y_{cl}}^\infty \left(A_2^{upper} - A_2^{lower}\right) f(t)\,dt = 0 \tag{5}$$

or,
$$p_1 \; E(A_1) + p_2 \; \{E(A_2^{upper}) - E(A_2^{lower})\} = 0 \tag{6}$$

where p_1 and p_2 are the probabilities defined in terms of the chip thickness probability density function $f(t)$ as

$$p_1 = \int_0^{y_{cl}} f(t) \; dt \quad \text{for } t \langle y_{cl} \tag{7}$$

$$p_2 = \int_{y_{cl}}^\infty f(t)\,dt \text{ for } t > y_{cl} \tag{8}$$

The expected area value, for the groove with depth less than centre line contributing to surface roughness R_a can be calculated as

$$E\{A(t_1)\}=\left(\frac{2}{2-\sqrt{3}}\right)E(t_1)y_{cl}-\left(\frac{\pi}{6}-\frac{\sqrt{3}}{4}\right)\left(\frac{2}{2-\sqrt{3}}\right)^2 E\left(t_1^2\right) \tag{9}$$

and the expected value of area, for the groove with depth greater than the centre line contributing to surface roughness R_a will be given as

$$E\left(A_2^{upper}\right)=\left(\frac{2}{2-\sqrt{3}}\right)^2 E\left(t_2^2\right)Sin^{-1}\left(\frac{\left(2-\sqrt{3}\right)r}{2t_2}\right)-\left(\frac{\sqrt{3}}{2}\left(\frac{2}{2-\sqrt{3}}\right)t_2+y_{cl}\right)r \tag{10}$$

$$E\left(A_2^{lower}\right)=\left(\frac{2}{2-\sqrt{3}}\right)E(t_2)y_{cl}-\left(\frac{\pi}{6}-\frac{\sqrt{3}}{4}\right)\left(\frac{2}{2-\sqrt{3}}\right)^2 E\left(t_2^2\right)+$$
$$\left(\left(\frac{2}{2-\sqrt{3}}\right)^2 E\left(t_2^2\right)Sin^{-1}\left(\left(\frac{\left(2-\sqrt{3}\right)r}{2t_2}\right)\right)-E\left(\left(\frac{\sqrt{3}}{2}\left(\frac{2}{2-\sqrt{3}}\right)t_2+y_{cl}\right)r\right)\right) \tag{11}$$

Where $r=\sqrt{\left\{\frac{t_2^2}{\left(2-\sqrt{3}\right)^2}-\frac{2\sqrt{3}}{\left(2-\sqrt{3}\right)}t_2\,y_{cl}-y_{cl}^2\right\}}$ and $E\left(A_2^{lower}\right)$ and $E\left(A_2^{upper}\right)$ are the areas below and above the centre line as shown in Fig. 3. Rewriting the Eq. (6) after substituting the expected values from Eq. (9), (10), (11) as

$$2\,y_{cl}\{p_1\,E(t_1)+p_2\,E(t_2)\}=\frac{\pi}{2}\{p_1\,E\left(t_1^1\right)+p_2\,E\left(t_2^2\right)\} \tag{12}$$

To calculate the expected values in the above equation requires another probability density function for the cases where the chip thickness is smaller and greater than the centre line y_{cl}. Thus, the conditional probability density function $f_1(t)$ and $f_2(t)$ defined for each region as

$$f_1(t)=f_1\left(t\,|0\le t<y_{cl}\right)=f(t)\Big/\int_0^{y_{cl}}f(t)\,dt \tag{13}$$

for chip thickness smaller than y_{cl} and for rest of the chip thickness

$$f_2(t)=f_2\left(t|y_{cl}\le t<\infty\right)=f(t)\Big/\int_{y_{cl}}^{\infty}f(t)\,dt \tag{14}$$

Substituting the Eq. (7), (8), (13) and (14) in Eq. (6) to find y_{cl}. After simplification the expression for centre line can be expressed as

$$y_{cl}=\left(\frac{\pi}{6}-\frac{\sqrt{3}}{4}\right)\left(\frac{2}{2-\sqrt{3}}\right)\left\{\frac{E\left(t^2\right)}{E(t)}\right\} \tag{15}$$

substituting the expected values in the above equation, and after mathematical simplification, the value of the centre line will be

$$y_{cl}=\sqrt{\frac{\pi}{2}}.\beta \tag{16}$$

For the calculation of the surface roughness, both types of grooves are required. Since the contribution of two types of grooves is different, thus, the total expected value of surface roughness can be calculated as;

$$E(R_a) = p_1\ E(R_{a1}) + p_2\ E(R_{a2}) \tag{17}$$

Where $E(R_{a1})$ and $E(R_{a2})$ are the expected value of the surface roughness for depth of engagement smaller or greater than y_{cl}. Thus, the values of the expected values can be calculated by the definition of the surface roughness according to which the surface roughness can be calculated by adding the area between the profile and the centre line and divide it by the total profile length. Hence from Fig. 3, the values can be written as

$$E(R_{a1}) = E\left(\frac{A_1}{2t_1}\right) \tag{18}$$

$$E(R_{a2}) = E\left(\frac{A_2^{upper} + A_2^{lower}}{2t_2}\right) \tag{19}$$

Rewriting the Eq. (18) & (19) after substituting the expressions of A_1, A_2^{upper} & A_2^{lower} from Eq. (9), (10) and (11) as

$$E(R_{a1}) = y_{cl} - \left(\frac{\pi}{6} - \frac{\sqrt{3}}{4}\right)\left(\frac{2}{2 - \sqrt{3}}\right)^2 E(t_1) \tag{20}$$

$$E(R_{a2}) = y_{cl} - \left(\frac{\pi}{6} - \frac{\sqrt{3}}{4}\right)\left(\frac{2}{2-\sqrt{3}}\right)E(t_2) + 2\left(\left(\frac{2}{2-\sqrt{3}}\right)E(t_2)Sin^{-1}\left(\frac{(2-\sqrt{3})r}{2t_2}\right)\right)$$
$$- E\left(\left(\frac{\sqrt{3}}{2}\left(\frac{2}{2-\sqrt{3}}\right)t_2 + y_{cl}\right)\left(\frac{2-\sqrt{3}}{2}\right)t_2 r\right) \tag{21}$$

Substituting the expected values of $E(t_1)$, $E(t_2)$ from Eq. (8), (10), (13) and (16) in Eq. (20), (21) and then from eq. (17), the expected value of surface roughness can be expressed as

$$E(R_a) = 1.16\,\beta \tag{22}$$

It can also expressed also in terms of the chip thickness expected value $E(t)$ by replacing β in terms of $E(t)$, as

$$E(R_a) = 0.92\,E(t) \tag{23}$$

Eq.(23) shows a proportional relationship between the surface roughness and the chip thickness expected values under the assumption that the profile of grove generated by an individual grain to be a semi-circular in shape and the phenomena like back transferring of material, grain overlapping are absent.

3 Prediction of surface roughness

The model has to be evaluated to determine its validity and versatility. The validity of the model is assessed through a comparison between the predicted value and measured value of the surface roughness within the predefined range of parameters. The versatility is based upon the usefulness of the model in predicting the kinematic conditions which would be required in order to produce specified surface finish. In order to compare the theoretical results obtained by making use of the model developed, experiments has been conducted in the horizontal surface grinding configuration. The work material taken was AISI 4340 steel with hardness 65HRC. The amount of material removed was the same for each experiment to maintain a consistent state of the wheel surface. The other conditions taken for experiment purpose are as follows; wheel speed 36.6 m/sec, wheel diameter 250 mm, wheel width 19 mm. The kinematic parameters for each experiment are depth of cut (a_e) and the speed ratio (v_s/v_w) where v_w (5, 10 and 15 m/min) is the feed rate and v_s is the wheel speed, as shown in Table 1.

Table 1 Experimental Conditions

Exp. No.	1	2	3	4	5	6	7	8	9
a_e (μm)	5	5	5	10	10	10	15	15	15
v_s/v_w	440	220	146	440	220	146	440	220	146
R_a (μm)	0.184	0.271	0.329	0.206	0.297	0.367	0.220	0.331	0.394

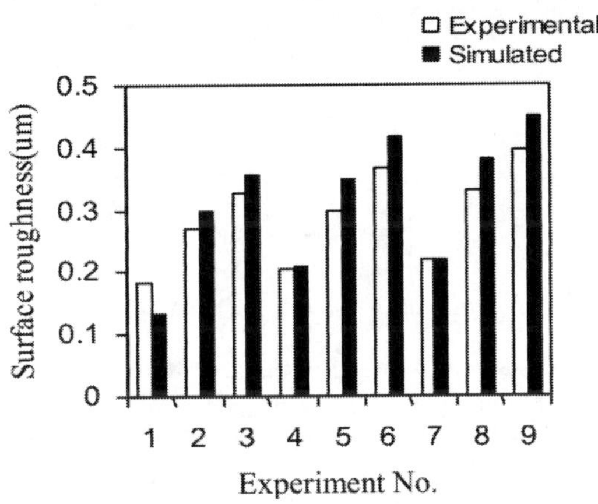

Figure 4. Experimental and simulated values of surface roughness for surface grinding.

Surface roughness measurements were made using Talysurf-VI (cut-off length was 0.8mm) at five different places on the 20 × 5 mm^2 cross-section of the workpiece after grinding and the arithmetic mean of the values of the measurements has been reported in the experimental results as shown in Table 1. The experiments are replicated five times (as shown in Table 1) to mask the variability of the process. By making use of the model of chip thickness developed in [16], the chip thickness expected value was calculated for each experiment.

To validate the model, the centre line average of surface roughness of all ground work pieces of the each set of experiment after a single pass was obtained using Talysurf-6 surface roughness measuring instrument. The results are plotted along with the experiment values of the surface roughness as shown in Fig. 4. It can be easily observed that the predicted value of the surface roughness shows a good agreement with the experimental data obtained from the different kinematic conditions in horizontal surface grinding.

4 Results and discussion

The number of variables that influence the surface roughness of ground surfaces are numerous. These include the wheel, work and machine characteristics and the operating characteristics. Since the surface roughness model developed above is based on the chip thickness model which includes many parameters such as kinematic conditions, material properties etc. Therefore, this model can be used to calculate the surface roughness under different conditions of these parameters. The most common kinematic parameters are depth of cut and the speed ratio. Further the deviation of the surface roughness of the new models from the actual values, for various feeds and depth of cut is shown in Fig. 5. Typical results showing the effect of the grinding conditions on the surface roughness. It is seen that with increase in the values of feeds and depth of cut, surface-finish deteoriates. This is probably due to the forces generated at higher feeds and depth of cut that results in higher deflections.

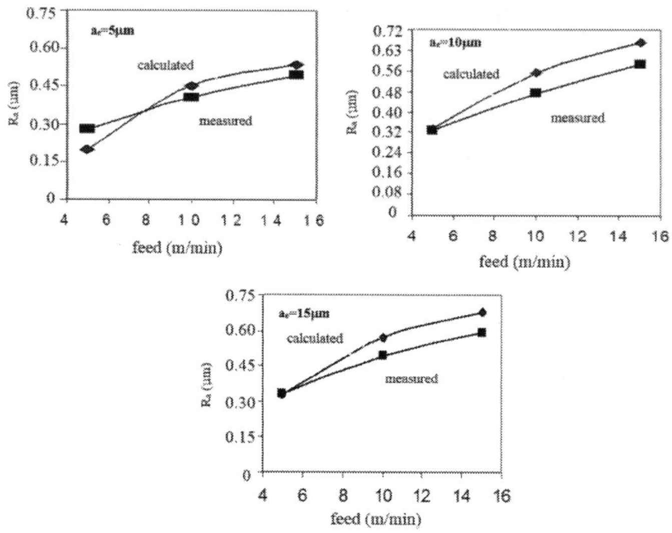

Figure 5. Surface roughness presented against feed at various depths of cut.

Also, the value of surface roughness is slightly higher than the measured value for various values of feeds and depth of cut. The probable reason for this is the assumption that the profile of grooves generated is perfectly semi-circle which is not always the case. It can be observed from the Fig.5 that the maximum deviation in the prediction of surface roughness with chip thickness model is about 1.15 times from the actual value. Another important parameter that will have significant effect on the surface roughness is the undeformed chip thickness. With increasing chip thickness, surface roughness will increase. Thus it appears that the chip thickness has pronounced effect causing a deteoriation of surface finish. This aspect is indicated by both theoretical and experimental results as shown in Fig. 4. Further, there appears to be reasonably close agreement between the experimental and simulated results.

5 Conclusions

In this paper, a simplified model is proposed for predicting the surface roughness of a work piece during the grinding process, taking into consideration the random distribution of the grain protrusion heights and assuming the profile of a groove generated by an individual grain to be an arc of a circle. The model is based on the relationship between the variables of the surface roughness and the undeformed chip thickness. The variables such as wheel microstructure, the kinematic and dynamic grinding conditions, and the material properties were included in the model through undeformed chip thickness model.The predicted surface roughness appears to yield results which agree reasonably well with the experimental values for different kinematic conditions in the horizontal surface grinding of AISI 4340 steel. Thus, the analytical model developed above is established to be a reliable and efficient technique for the prediction of surface roughness.

References

1. S. Agarwal, P.V. Rao, Int. J.Mach. Tools Mf. **50-12**, 1077 (2010)
2. S. Agarwal, P.V. Rao, Int. J.Mach. Tools Mf. **48-6**, 698 (2008)

3. S. Malkin, *Grinding Technology: Theory and Applications of Machining with Abrasives* (Ellis Horwood, Chichester, West Sussex, 1989)
4. T. Suto, T. Sata, Bull. Jap. Soc. Prec. Engng. **15-1**, 27 (1981)
5. K. Sato, Reports of Tokohu University, **20-1,** 59 (1955)
6. T. Orioka, Bull. Japan Soc. Grinding Engrs, 27 (1961)
7. G.K. Lal, M.C. Shaw, Trans. ASME, J. Engng for Industry. 1119 (1975)
8. H. Tonshoff, J. Peters, I. Inasaki, T. Paul, Ann. CIRP. **41-2,** 677 (1992)
9. X. Zhou, F.Xi, Int. J. Mach. Tools Mf. **42**, 969 (2002)
10. S.S. Law, S.M. Wu, Trans. ASME, J. Engng for Industry. **95,** 972 (1973)
11. R.L. Hecker, S. Liang, Int. J. Mach. Tools Mf. **43,** 755 (2003)
12. G.K. Lal, M.C. Shaw, Proc. Int. Grindng Conf., Carnegie-Mellon University, Pittsburgh, USA. 107 (1972)
13. S. Agarwal, P.V. Rao, Int. J. Mach. Tools Mf. **45-6,** 609 (2005)
14. S. Agarwal, P.V. Rao, Proc. of IMechE, Part B, J. Eng. Mf. **219,** 811 (2005)
15. S. Agarwal, P.V. Rao, Int. J. Mach. Tools Mf. **50-2,** 1035 (2010)
16. S. Agarwal, P.V. Rao, Int. J. Mach. Tools Mf. 5**6,** 59 (2012)

Combined Ultrasonic Elliptical Vibration and Chemical Mechanical Polishing of Monocrystalline Silicon

Defu LIU[a], Tao CHEN and Riming YAN

College of Mechanical and Electrical Engineering, Central South University, Changsha, Hunan province, 410083, China

Abstract. An ultrasonic elliptical vibration assisted chemical mechanical polishing(UEV-CMP) is employed to achieve high material removal rate and high surface quality in the finishing of hard and brittle materials such as monocrystalline silicon, which combines the functions of conventional CMP and ultrasonic machining. In theultrasonic elliptical vibration aided chemical mechanical polishingexperimental setup developed by ourselves, the workpiece attached at the end of horn can vibrate simultaneously in both horizontal and vertical directions. Polishing experiments are carried out involving monocrystalline silicon to confirm the performance of the proposed UEV-CMP. The experimental results reveal that the ultrasonic elliptical vibration can increase significantly the material removal rate and reduce dramatically the surface roughness of monocrystalline silicon. It is found that the removal rate of monocrystalline silicon polished by UEV-CMP is increased by approximately 110% relative to that of conventional CMP because a passive layer on the monocrystalline silicon surface, formed by the chemical action of the polishing slurry, will be removed not only by the mechanical action of CMP but also by ultrasonic vibration action. It indicates that the high efficiency and high quality CMP of monocrystalline silicon can be performed with the proposed UEV-CMP technique.

1 Introduction

Chemical mechanical polishing (CMP) is a popular finishing process by using a combination of chemical reactions and mechanical forces [1-2]. CMP is widely used to finish materials with great hardness, high wear resistance and chemical inertness including monocrystalline silicon, sapphire, copper, gallium nitride (GaN) or silicon carbides (SiC), which are employed frequently in integrated optoelectronic and semiconductor devices. With the rapid development of integrated optoelectronic

[a] Corresponding author: liudefu@csu.edu.cn

and semiconductor industries, higher polishing quality and efficiency of CMP are requested.

However, the present research results show that the conventional CMP technology is hard to obtain high material removal rate (MRR) and perfect polished surface simultaneously due to the material's properties [3-4]. To acquire both high MRR and great polished surface quality simultaneously, ultrasonic vibration is introduced into conventional chemical mechanical polishing process, which has been proved to bevery useful to improve the polishing quality and efficiency for the hard and brittle materials [5-7], such as wafer and sapphire.

Wenhu Xu,*etl.* [2] developed an ultrasonic flexural vibration assisted chemical mechanical polishing (UFV-CMP) system to improve both MRR and surface quality of sapphire substrates. The MRR of UFV-CMP was 2.5 times larger than that of conventional CMP. The roughness and flatness of the sapphire surfaces polished by UFV-CMP were 0.83Å and 0.12μm respectively, which were much better than that of conventional CMP. Ming-Yi Tsai, *etl.* [6] introduced an ultrasonic vibration assisted chemical mechanical polishing (UV-CMP) method to finish copper substrates. It was found that material removal rate of the copper substrates increased by approximately 50-90% relative to that of conventional CMP, and the roughness of the substrate surfaces was also improved dramatically. Li Liang, *etl.* [7] applied ultrasonic vibration with the frequency of 27.99kHzto the CMP of wafer. According to the large amounts of experiments on the chemical mechanical polishing assisted by ultrasonic vibration, it was found that morphologies of silicon wafers polished by UV-CMP were superior to that of conventional CMP. A novel model of UV-CMP for silicon wafer was proposed, which could be used to explain the experimental results.

In this paper, an innovative chemical mechanical polishing method named "ultrasonic elliptical vibration assisted chemical mechanical polishing (UEV-CMP)" aimed at achieving high material removal rate and high surface quality simultaneously is presented and applied in the finishing of monocrystalline silicon. The UEV-CMP experimental setup, which can make work piece vibrate simultaneously in both horizontal direction and vertical direction with the same ultrasonic frequency, is developed. To investigate the material removal characteristics of UEV-CMP, the experiments about the effects of ultrasonic vibration on chemical corrosion of reagent and mechanical impact of abrasive particles are conducted respectively. The combined ultrasonic elliptical vibration and CMP process is expected to improve the polishing efficiency and quality for monocrystalline silicon.

2 Experimental setup

The ultrasonic elliptical vibration can be synthesized by two ultrasonic vibrations in orthogonal directions with the same frequency [8]. A schematic of the ultrasonic elliptical vibration transducer combined by a longitudinal vibrator and a bending vibrator is illustrated in Fig. 1. The transducer consists of one end cap, 2 ring-shape piezo-electric actuators, one middle cap, 4 half ring-shape actuators and one front cap, which are bonded together by a bolt. The exciting alternating current with high frequency generated by an ultrasonic generator is transported to the ring-shape PZT actuators and the half ring-shape PZT actuators, which generate longitudinal vibration and bending vibration respectively under the converse piezoelectric effect. The direction of longitudinal vibration and the direction of bending vibration are orthogonal. The longitudinal ultrasonic vibration wave signals generated by the ring shape actuators and the bending ultrasonic vibration wave signals generated by the half-ring shape actuators can be transferred to the workpiece by the ultrasonic horn. At the same time, the vibration amplitude of the workpiece fixed at the end of the horn is amplified by the horn.

Ultrasonic elliptical vibration assisted chemical mechanical polishing experiments are performed on a UEV-CMP machine designed by ourselves. The schematic of the UEV-CMP experimental setup designed for the polishing experiments is illustrated in Fig. 2. The experimental setup mainly consists of an ultrasonic elliptical vibration apparatus illustrated in Fig. 1, and a polishing machine. The polishing pressure can be adjusted by changing the weight on the node point of horn. The polishing slurry can be provided to the interface between polishing pad and workpiece. This experimental setup is convenient to work as ultrasonic vibration assisted CMP or conventional CMP governed by the operation of the two switches to control the ultrasonic generator. As shown in Fig. 2, when both the switch 1 and the switch 2 are turned off, the workpiece does not vibrates, this polishing process is called conventional CMP; when the switch 1 is turned on and the switch 2 is turned off, the workpiece just vibrates ultrasonically along horizontal direction, this polishing process is called ultrasonic horizontal vibration assisted CMP (UHV-CMP); when the switch 1 is turned off and the switch 2 is turned on, the workpiece just vibrates ultrasonically along vertical direction, this polishing process is called ultrasonic vertical vibration assisted CMP (UVV-CMP); when both the switch 1 and the switch 2 are turned on, the workpiece vibrates ultrasonically along an elliptical trajectory, this polishing process is called ultrasonic elliptical vibration assisted CMP (UEV-CMP). The ultrasonic generator converts 50Hz electrical supply to high frequency (20 kHz ~ 50 kHz) AC output. Under the circumstances, the workpiece fixed at the end of horn also vibrates at the same frequency. When the output power of the ultrasonic generator is 180W, the amplitudes of the workpiece are 3μm and 2.5μm in horizontal direction and vertical direction respectively measured by laser Doppler vibration meter LK-G5001V.

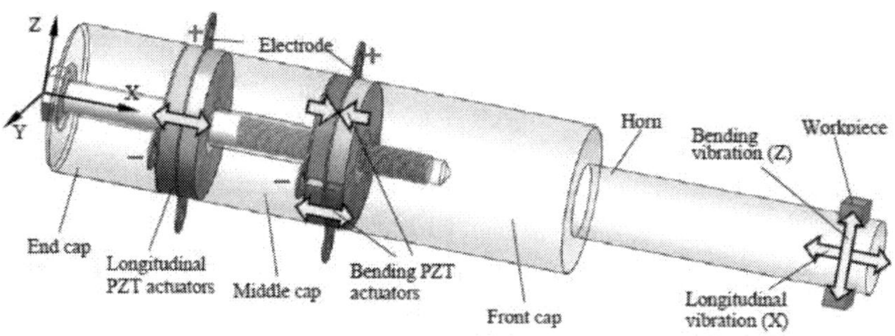

Figure 1. A schematic of ultrasonic elliptical vibration device

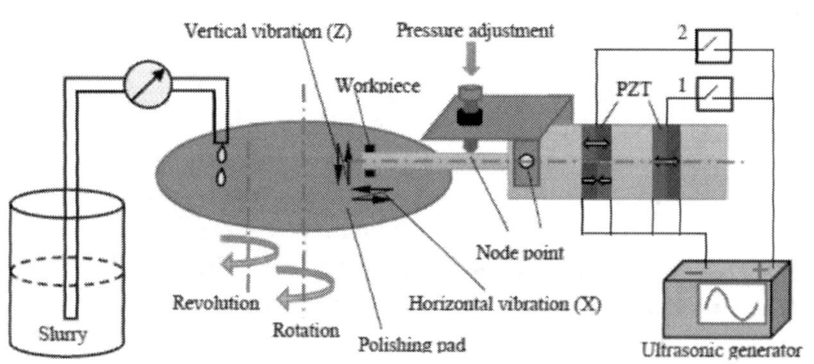

Figure 2. Schematic of ultrasonic elliptical vibration assisted CMP experimental setup

3 Experimental details

Ultrasonic elliptical vibration assisted chemical mechanical polishing (UEV-CMP) experiments are conducted to evaluate the polishing quality. The polishing process parametersare listed in Table 1. The workpiece size is 10mm×10mm. The workpiece material is monocrystalline silicon. The polishing padis made of polyurethane. The material of the abrasive particles is silica (SiO_2), and the abrasive size of particles is 50nm. The polishing experiments are divided 4 different groups. 20 workpieces are polished in each experiment of conventional CMP, UHV-CMP, UVV-CMP, and UEV-CMP. The MRR is determined on the basis of weight loss of workpieces before and after polishing. The surface roughness values and morphologies of the polished silicon workpieces are measured and observed by Veecoprofilometer equipment.

Table 1. Polishing parameters of CMP

Polishing parameters	Value	Polishing parameters	Value
Polishing time /min	30	Rotation speed of pad /($r{\cdot}min^{-1}$)	3
Abrasive size of particles /nm	50	Revolution speed of pad /($r{\cdot}min^{-1}$)	30
Slurry density /($g{\cdot}cm^{-3}$)	0.96	Polishing pressure /kPa	51.43
Concentration of polishing particles	2%	Ultrasonic vibration frequency /kHz	35.3
pH value of slurry	9	Horizontal ultrasonic amplitude /μm	3
Slurry feed rate /($ml{\cdot}min^{-1}$)	15	Vertical ultrasonic amplitude /μm	2.5

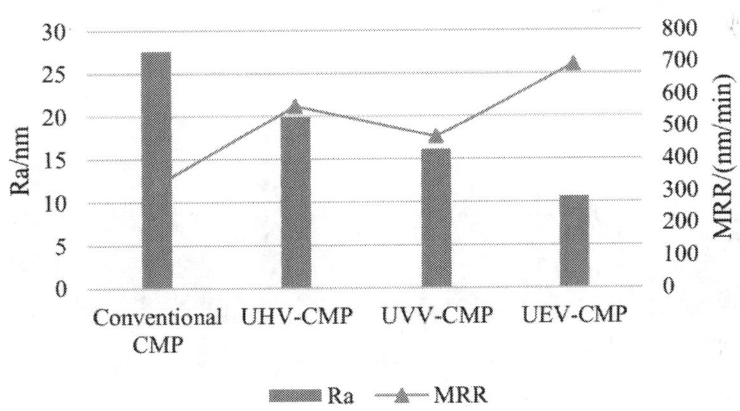

Figure 3. The relations between MRR, roughness and polishing process

The MRRs of the workpieces polished by conventional CMP, UHV-CMP, UVV-CMP and UEV-CMP are 320.1, 565.2, 469.7 and 693.6 nm/min respectively, as shown in Fig. 3. It can be known that the MRR of UEV-CMP increases significantly compared to that of conventional CMP. The relation between surface roughness and polishing process is also plotted in Figure 3. The average surface roughness Ra values of 20 workpieces polished by conventional CMP, UHV-CMP, UVV-CMP, UEV-CMP respectively are 27.61, 19.94, 16.14 and 10.61nm respectively. It can be seen from Fig. 4

that the surface roughness of the workpieces polished by conventional CMP, UHV-CMP, UVV-CMP, UEV-CMP respectively decrease significantly. Hence, ultrasonic vibration has significant effect on the MRR and surface roughness of workpiece polished by CMP. Especially, both the MRR and surface quality are improved greatly when the workpiece is polished by UEV-CMP.

4 Material removal mechanism

To clearly evaluate the material removal mechanism of ultrasonic elliptical vibration assisted chemical mechanical polishing, the effects of ultrasonic vibration on chemical corrosion of polishing solutionand mechanical impact of abrasive particles are studied respectively by design of experiments.

4.1 The effects of ultrasonic vibration onchemical corrosion

Experiments of ultrasonic vibration assisted chemical corrosion on silicon are conducted to evaluate the effects of ultrasonic vibration on chemical corrosion of polishing solution. The size and material of the specimens used in the chemical corrosion experiments are same with that of the workpiece used in the above polishing experiments. The original specimens are cleaned by ultrasonic vibration in absolute ethyl alcohol and deionized water respectively. The chemical reagent used in experimentsis NaOH with concentration of 2mol/L. The specimen is fixed at the end of the horn of the ultrasonic elliptical vibration device shown in Fig. 1and they are immersed in the NaOH solution. The experiments are performed for 30minutes with or without assistant of ultrasonic elliptical vibration respectively for comparing the effects. The frequency of the ultrasonic elliptical vibration is 35.3 KHz. Fig. 4 shows the process of chemical corrosion experiments.

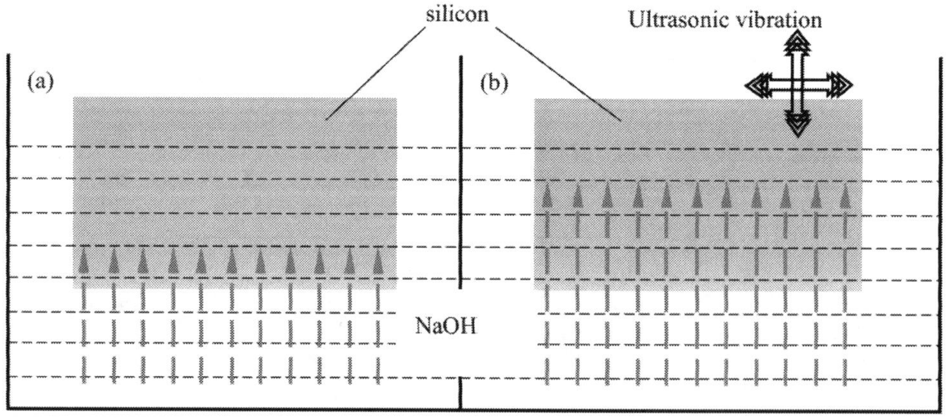

Figure 4. Chemical corrosion experiments (a)without assistant of ultrasonic vibration and (b) with assistant of ultrasonic vibration

The surface morphologies of the original specimen, the specimen corroded by NaOH solution without assistant of ultrasonic vibration, and the specimen corroded by NaOH solution with assistant of ultrasonic vibration are shown in Fig. 5. The roughness value of the original surface is only 8.62nm. However, the roughness value of the specimen surface corroded by NaOH solution without assistant of ultrasonic vibration is20.28nm, while the roughness value of the specimen surface

DOI: 10.1051/matecconf/20168202001

corroded by NaOH solution with assistant of ultrasonic elliptical vibration increases to 191.66nm. The results indicate that the ultrasonic elliptical vibration assisted chemical corrosion can increase the speed of chemical reaction significantly.

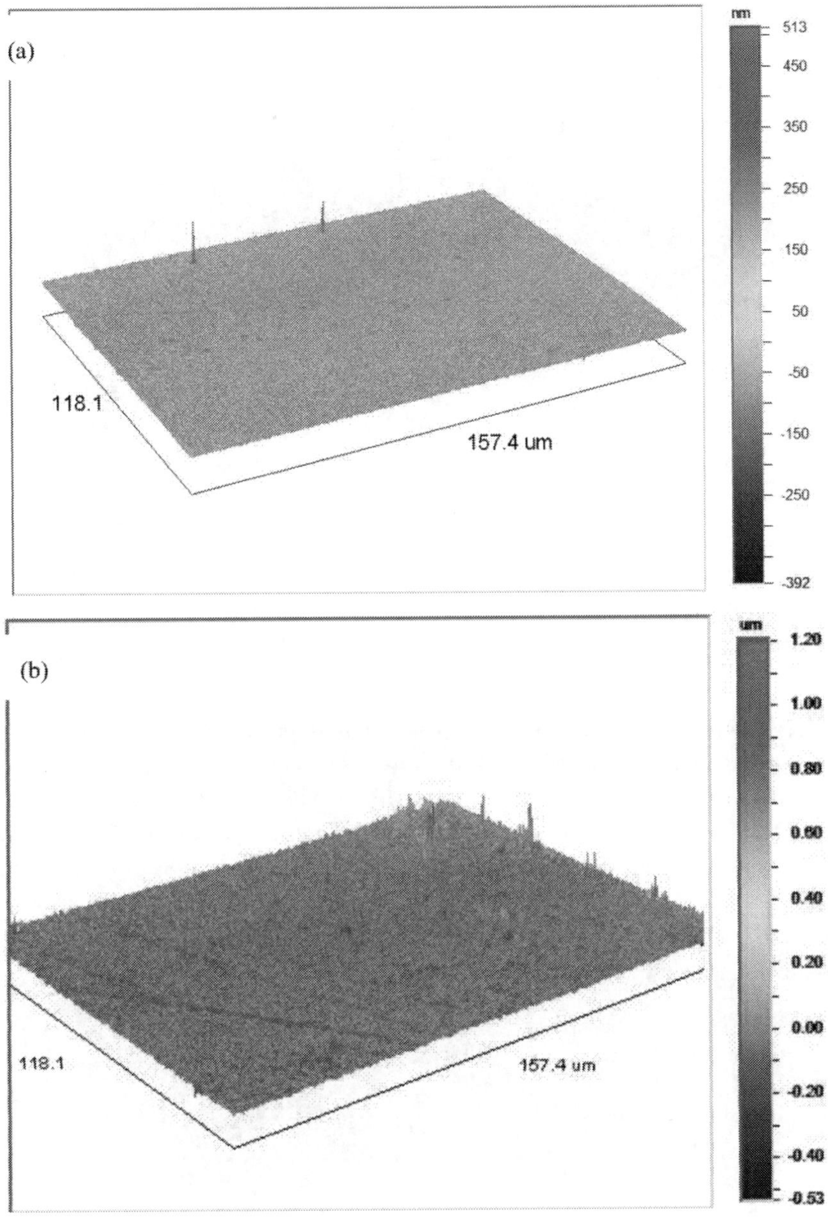

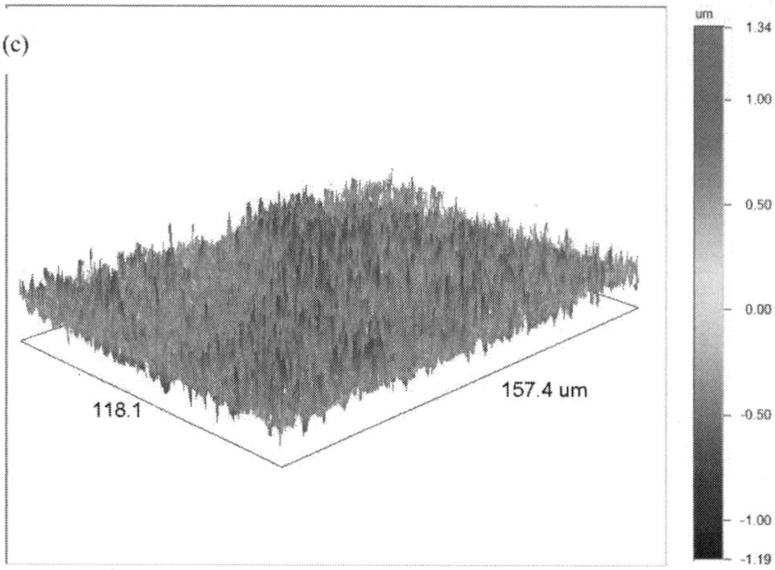

Figure 5. (a) The original surface of specimen, (b) The surface corroded by NaOH solution without assistant of ultrasonic vibration, and (c) the surface corroded by NaOH solution with assistant of ultrasonic elliptical vibration

4.2 The effects of ultrasonic vibration on mechanical impact of abrasive particles

The experiments of impact of abrasive particles on specimen are designed to evaluate the effects of ultrasonic vibration on the mechanical impact of particles on specimen. The material of specimens is silicon, and the size of the specimen is also 10mm×10mm. The specimens are put into the polishing slurry withα-Al_2O_3, SiO_2 and CeO_2nanoparticlesrespectively, as shown in Fig. 6. Ultrasonic vibration is imposed on the specimen through the ultrasonic vibration device shown in Fig. 1. The frequency of the ultrasonic vibration is also 35.3 KHz. The physical parameters of the specimen and the abrasive particlesα-Al_2O_3, SiO_2 and $CeO_2$are listed in Table 2. It can be observed from Table 2that α-Al_2O_3, $SiO_2$are harder than the specimen, while CeO_2 is softer than the specimen.

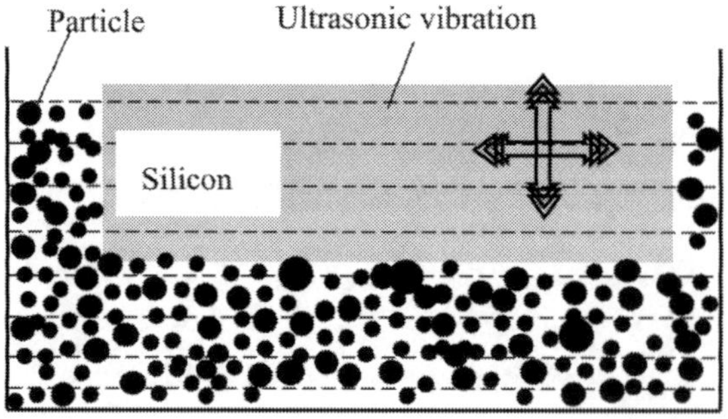

Figure 6. The effects of ultrasonic vibration on mechanical impact of particles

Table 2. The parameters of materials involved in experiments of mechanical impact of particles

	Specimen	**α-Al_2O_3**	**SiO_2**	**CeO_2**
Hardness /GPa	9.8	21	11	8.9
Density /g·cm^{-3}	2.329	3.9	2.2	7.65
Modulus of Elasticity /GPa	112.4	370	73	165
Poisson's Ratio	0.28	0.22	0.17	0.5

The surface morphologies of specimens impacted byα-Al_2O_3, SiO_2 and CeO_2 particles with assistant of ultrasonic vibration are shown in Fig. 7, Fig. 8andFig. 9respectively. The experimental results indicate that the hard particles (α-Al_2O_3, SiO_2) cause obviouspitson the specimen surface, while the soft particles (CeO_2) do not cause this phenomenon. Moreover, the pits caused byα-Al_2O_3particles are more, larger and deeper than those caused bySiO_2 particle, which means the particle is harder, the removed effects is more obvious. The results meet well with the impact wear theory. Therefore, it can be supposed that the hard particles can cause plastic scratch effects on soft materials under assistant of ultrasonic vibration, while the soft particles cannot.

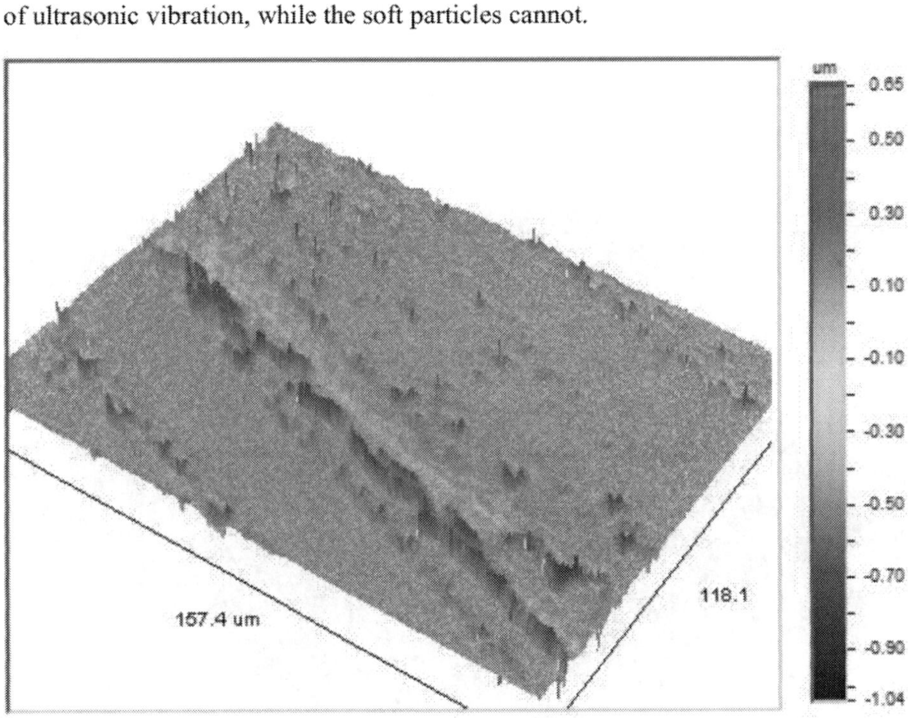

Figure 7. Surface morphology of specimen impacted by α-Al_2O_3 particles with assistant of ultrasonic vibration, Ra=47.83nm, ×40

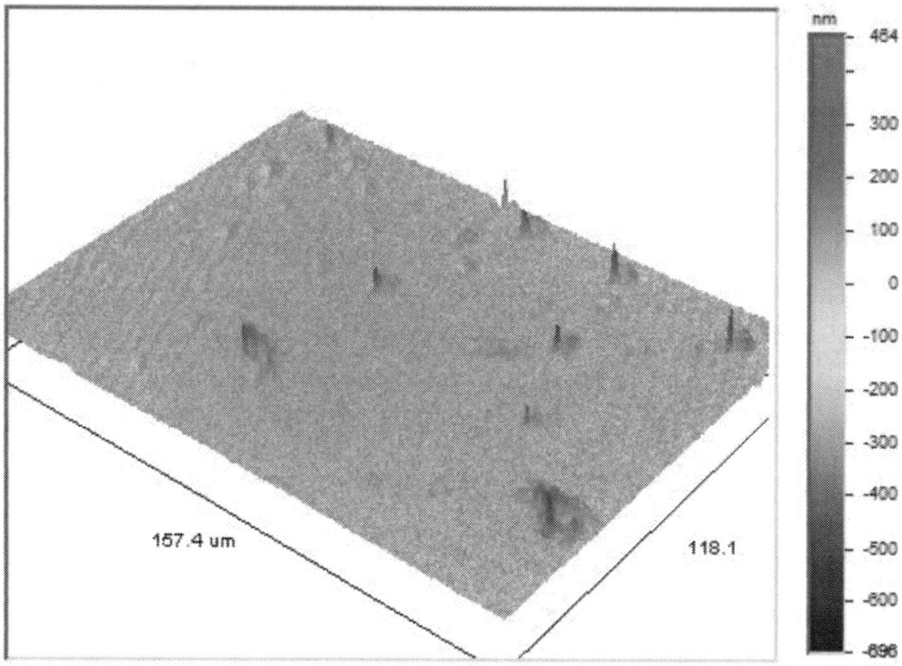

Figure 8. Surface morphology of specimen impacted by SiO_2 particles with assistant of ultrasonic vibration, Ra=15.14nm, ×40

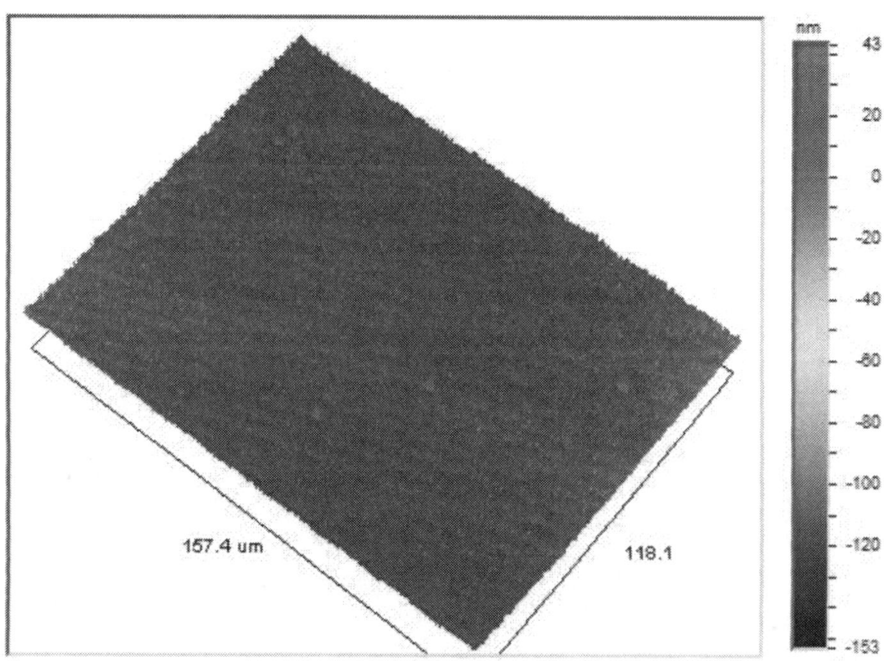

(a)

Figure 9. Surface morphology of specimen impacted by CeO_2 particles with assistant of ultrasonic vibration, Ra=12.68nm, ×40

5 Conclusions

This paper presents an experimental investigation of the characteristics of a monocrystalline silicon polished by ultrasonic elliptical vibration assisted chemical mechanical polishing process. UEV-CMP is able to improve the MRR and polished surface quality of monocrystalline silicon greatly compared with conventional CMP. The MRR of silicon samples polished by UEV-CMP is 693.6 nm/min, which is twice larger than that obtained by conventional CMP. The polished surface roughness value of UEV-CMP is 10.61nm, which is much better than 27.61nm obtained by conventional CMP. For UEV-CMP, a soft layer on workpiece surface formed by the chemical action of the polishing slurry will be removed not only by the mechanical action of conventional CMP but also by the ultrasonic vibration action of UEV-CMP. The combined actions of ultrasonic elliptical vibration and conventional CMP can facilitate better polished surface and higher MRR.

Acknowledgements

The authors would like to thank the National Natural Science Foundation of China (Grant 51275534) and Natural Science Foundation of Hunan province of China (Grant 2015JJ2153) for their support to this work.

References

1. H. S. Lee, H. D. Jeong. CIRP Annals-Manu. Tech. **58**, 485-490 (2009)
2. H. Gong, G. Pan, Y. Zhou, X. Shi, C. Zou, S. Zhang. Appl. Surf. Sci. **338**, 85-91 (2015)
3. H. Zhu, L.A. Tessaroto, R. Sabia, V.A. Greenhut, M. Smith, D.E. Niesz. Appl. Surf. Sci. **236**, 120-130 (2004)
4. X. Hu, Z. Song, Z. Pan, W. Liu, L. Wu.Appl. Surf. Sci. **255**, 8230-8234 (2009)
5. W. Xu, X. Lu, G. Pan, Y. Lei, J. Luo. Appl. Surf. Sci. **256**, 3936-3940 (2010)
6. M. Tsai, W. Yang. Int. J. of Mach. Tools and Manu. **53**, 69-76 (2012)
7. L. Li, Q. He, M. Zheng, Z. Liu.Ultrasonics **56**, 530-538 (2012)
8. H. Suzuki, T. Moriwaki, T. Okino, Y. Ando. CIRP Annals-Manu. Tech. **55**, 385-388 (2006)

MATEC Web of Conferences **82**, 02002 (2016) DOI: 10.1051/matecconf/20168202002
D2ME 2016

Degradation of Phenol in Wastewater with Ozone Produced by Self-design Ozone Generator

Shibo Wang[1], Dongwei Zhong[1], Guangfei Qu[1,a], Ping Ning[1], Jiankang Quan[1] and Xiaotian Chen[1]

[1]*Kunming University of Science & Technology, Environmental Science and Engineering Department, 650500 Yunnan, China*

Abstract. The optimized tube's structure of the self-design ozone generator was made with the double dielectric inner electrode and small metal cones were embedded in the outside electrode. This ozone generator was used for the degradation of phenol in wastewater. The research was studied from ozone gas flow rate, reaction time, the initial pH and concentration of the phenol in wastewater. In addition, the article also discusses the reaction mechanism of ozone degraded the phenol in wastewater. The results illustrate that the ozone concentration of self-design ozone generator under the pure oxygen flux (1.0 L/min) were 7.06 mg/L, compared with the unoptimized ozone, the optimized ozone generator's efficiency was improved 56.89%; the phenol removal rate was increased as the initial pH increased in wastewater, when pH was 12, the phenol removal rate was 80.17%; GC-MS analysis was adopted to analyze and determine the phenol intermediates. The results illustrate that the major oxidation degradation products of phenol may be p-benzoquinone benzenequinone, 4-cyclopentene-1, 3-dione and diacetone alcohol.

1 Introduction

Phenolic compounds were biodegradable and widely existed in industrial waste water which has a high toxicity. Phenol is a typical kinds of phenolic compounds with high biotoxicity. [1] The discharge of phenol in wastewater under strict control in water pollution management. So developing an effective method to deal with phenol in wastewater has important meaning to environmental protection and peoples' health.

At present, several methods was reported to treat the phenol in wastewater. Wastewater with high concentration of phenol is usually treated by an extraction process to reuse phenol. [2,3] In addition, some advanced technologies had been adopted to deal with wastewater with a low concentration of phenol, such as chemical processes, [4,5] ozonation, [5-8] ultrasonic processes, [9,10] fenton reaction, [11] compound oxidation process. [12,13] It is effective for ozone to sterilize, decrease chromaticity and smell, increase the biodegradability of refractory organics and improve the flocculent effect with its strong oxidation characteristics which is due to the high oxidation potential (2.08 V). [14-16] However, in the process of ozone generating, the high consumption of energy and low concentration of ozone limited the application in wastewater treatment in industry, so to improve the efficiency of ozone generation is the key to gain a broad prospect of applications in wastewater treatment of

[a] Corresponding author: qgflab@sina.com

ozonation technology. At present, the research in ozone generator mainly focuses on the following aspects: (1) The power system turns high voltage power frequency to medium voltage high frequency or uses the high-voltage bipolar pulsed corona etc.; (2) With thin and high dielectric constant of the dielectric materials; (3) Using the narrow discharge gap; (4) An improved cooling system with dual cooling system.

On the basis of development and insufficiency of ozone generator and treatment of phenol in wastewater. In this study, the phenol was selected as a model of toxic organic contaminant in solutions. The self-design ozone generator was optimized for its tube structure and electrode, which was applied as ozone source to removal phenol in wastewater. Finally, the ozone concentration of self-design ozone generator and the influence of operational conditions such as the initial concentration and pH of the water, the concentration of ozone on the phenol degradation and the intermediate products were evaluated.

2 Experiments and methods

2.1 Reagent and instrument

Reagent: Phenol was purchased from Tianjin Bodi Chemical Reagent Co., Ltd. 4-aminoantipyrine was purchased from Chengdu Kelong Chemical Reagent Co., Ltd. Sulfamic acid (AR grade, purity≥ 99.5%). Potassium ferricyanide was purchased from Tianjin Fengfan Chemical Reagent Co., Ltd. Potassium iodide was purchased from Tianjin Guangfu Fine Chemical Research Institute. Co., Ltd.

Main equipments: electronic scales (FA/JA); hot type constant temperature magnetic stirrer (DF-101s) ; ozone analyzer (PTM400-O3) ; self-design ozone generator; Gas Chromatography-Mass Spectrometer (PerkinElmer Clarus SQ 8,Clarus 680).

2.1.1 Self-design ozone generator

Ozone generator was assembled according to the standard of CJ/T322-2010 《Water Treatment By Ozone Generator》 , the structure of ozone generator's tube had been ameliorated on the premise of other parts didn't change, the materials of tube's dielectric and electrode is quartz glass and stainless steel, respectively. The optimized tube's structure of the self-design ozone generator was made with the double dielectric inner electrode and small metal cones were embedded in the outside electrode, and the bottom radius and height of the small metal cone is 0.1 cm and 0.1 cm, respectively, as the pattern in Figure 1. The double-dielectric barrier discharge was occurrence in the inner electrode of ozone generator.

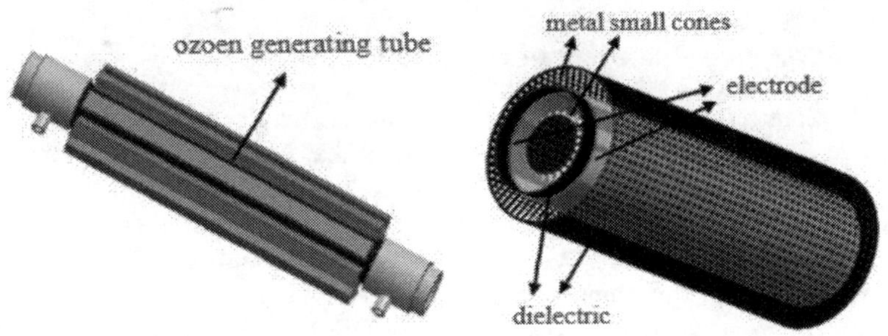

Figure 1. Ozone generating tube and its inner structure.

2.1.2 Preparation of the phenol wastewater

Dissolve phenol (analytically pure) in distilled water, and dilute with distilled water to 2000 ml in the brown glass flask.

2.1.3 Preparation of the 4-aminoantipyrine solution

Dissolve 2 g 4-aminoantipyrine in distilled water, and dilute with distilled water to 100 ml. The prepared solution was kept in the refrigerator to avoid the loss of potency.

2.1.4 Preparation of the potassium ferricyanide solution

Dissolve 8 g potassium ferricyanide in distilled water, and dilute with distilled water to 100 ml. The prepared solution was kept in the refrigerator to avoid the loss of potency.

2.2 Experimental method

The graph of ozone generator's laboratory equipment was shown in Figure 2. Ozone generator used pure oxygen to produce ozone under DBD ionization, and ozonization degraded wastewater with high concentration of phenol in 20℃. The samples of phenol wastewater were examined and analyzed for a period of time. Ozone concentration was detected by ozone analyzer, and the concentration of phenol was detected by direct spectrophotometric method. A variety of degraded products were identified by GC-MS analysis.

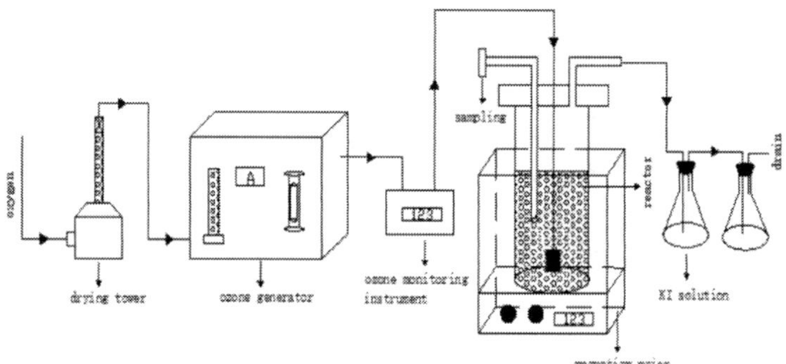

Figure 2. Stimulated phenol wastewater treatment device.

3 Results and discussion

3.1 Effect of pure oxygen flux on ozone concentration

The variation of ozone concentration with pure oxygen flux is presented in Figure 3 and other operating parameters remain the same. It could be found that with the pure oxygen flux increased the concentration of ozone first increased then decreased. As the concentration of pure oxygen was 1.0 L/min, the highest concentration of ozone was 7.06 mg/L. Compared with the traditional ozone generator, the concentration of ozone was only 4.5 mg/L in the same condition. The efficiency of the optimized ozone generator was improved 56.89%. The phenomenon might due to the corona discharge capacity of ozone generator was limited by DBD. The double dielectric barrier discharge could expand the discharge space of the ozone generator's tube thus raised discharge efficiency. Small metal cones were embedded in outside electrode which is advanced to increase the discharge efficiency and produced more oxygen atoms then improved the ozone concentration. The results indicated that the optimized self-design ozone generator was effective.

 DOI: 10.1051/matecconf/20168202002

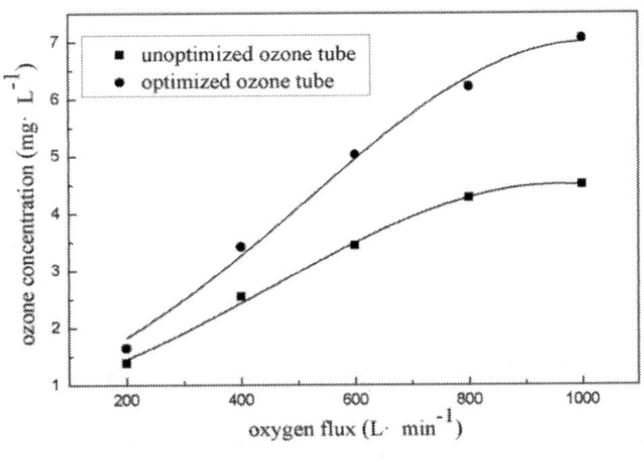

Figure 3. Effect of the oxygen flux on ozone concentration.

3.2 Effect of mixed ozone gas flux on phenol wastewater treatment

Prepared 2 L wastewater with 6 g phenol as phenol wastewater, the effective of mixed ozone gas flux (0.5 L/min, 1.0 L/min and 1.5 L/min) on ozone treated phenol wastewater was evaluated as the other operation conditions remained the same and the reaction continued for 240 min. The removal ratio of phenol in wastewater by different mixed ozone gas flux was showed in Figure 4. It could be fined that the best condition of the treatment was the mixed ozone gas flux was 1.5 L/min. After the reaction for 240 min, the removal ratio of phenol reached 75.8% under the reactive condition with 1.5 L/min of mixed ozone gas flux. This phenomenon was consistent with the highest concentration of ozone in the optimized ozone generator with the pure oxygen flux in 1.5 L/min. It was founded that the solubility of ozone in wastewater was related to the on concentration of ozone, the solubility of ozone increases with the concentration of ozone increased. As the time goes by, the removal ratio of phenol in wastewater increased, which was corresponded to the first-order kinetics equation in the reaction for 240 min.

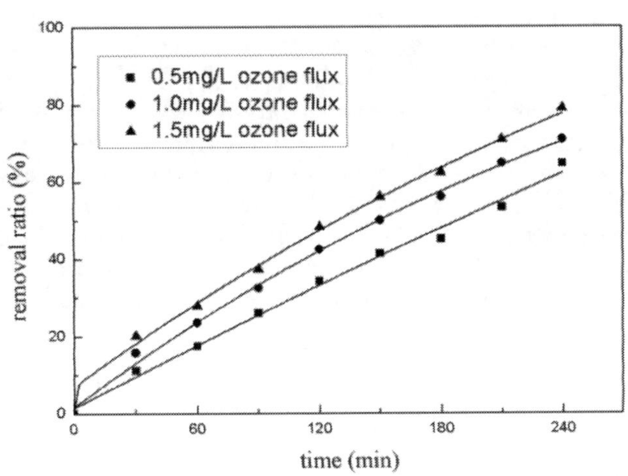

Figure 4. Phenol removal rate in wastewater by different ozone flux (reaction for 240 min, the concentration of phenol wastewater was 3000 mg/L).

3.3 Effect of the initial concentration of phenol wastewater on phenol wastewater treatment

The initial concentration of wastewater with phenol removal rate was presented in Figure 5, in the condition of mixed ozone gas flux was 1.5 L/min. When the concentration of wastewater was 100 mg/L, the phenol removed 99.9% in the oxide time for 30 min. While the concentration of wastewater was 3000 mg/L, the phenol removed 98.6% in oxide time for 480 min. In Figure 5, the results showed that the phenol removal rate increased as the reaction time increased, and the removal rate of phenol was near 100%. It illustrated that the removal efficiency of ozone generator produced by self-design was favorable for phenol wastewater treatment in a short time, and the removal rate of phenol was near 100%, even the wastewater with high concentration of phenol.

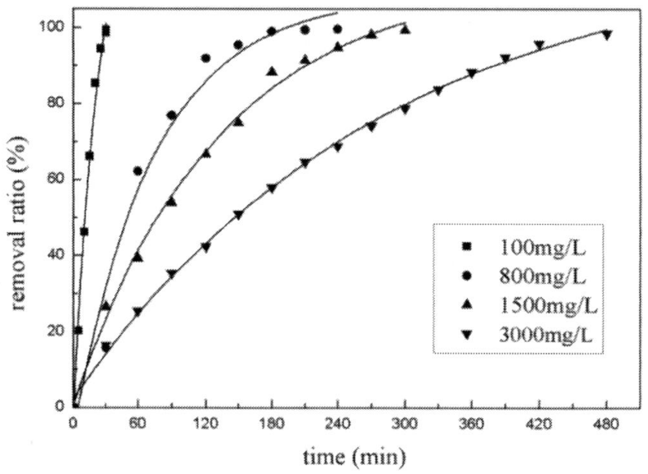

Figure 5. Reaction time on phenol removal rate at different concentration of phenol wastewater.

3.4 Effect of pH on phenol wastewater treatment

The relation shape between phenol removal rate and pH of initial wastewater was shown in Figure 6, and other operating parameters remain the same. The removal rate of phenol increased with the increasing pH value. The results illustrated that the removal of phenol was better in alkaline condition, and the phenol removal ratio was 80.2% with a pH of 12. However, when the pH was 2 the removal ratio of phenol was 64.6%. It might because of the neutral reaction of phenol with alkalis, according to the weak acid property of phenol, moreover, improved the initial pH of wastewater was favorable for the production of hydroxyl radical (•OH). [17] In this condition, compared with the reaction rate constant of hydroxyl radical and the ozonation, the former was seven times bigger than the latter one, this change causeed the phenol removal ratio increased. Fortunately, the removal rate of phenol was 64.6% when the pH was 2. It demonstrated that the self-design ozone generator could effectively treat the phenol wastewater whatever the pH condition was.

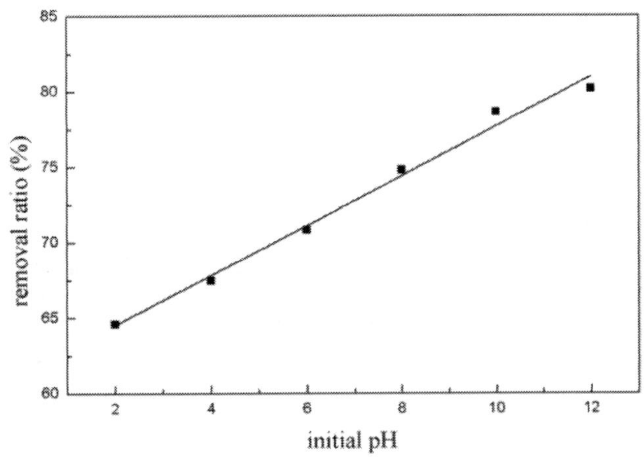

Figure 6. Initial pH on phenol removal rate (the concentration of phenol was 300 mg/L, reaction for 240 min).

3.5 Degradation analysis

The degradation products of oxidized phenol in wastewater reactor were analyzed by gas chromatography-mass spectrometer (GC-MS). The concentration of the phenol wastewater was 100 mg/L.As the phenol was totally degraded by ozone for 30 min, so during wastewater treatment process there are 4 periods time(5 min, 10 min, 20 min and 30 min) were measured by GC-MS. The peak of phenol appeared at about 10.4 min in the Figure 7 (a, b, c). Compared the Figure 7 (a) with (b), the peak of p-benzoquinone just appeared at about 9.1 min in the Figure 7 (b) and disappeared in the Figure 7(c). The peak of 4-cyclopentene-1, 3-dione appeared at about 13.13 min in the Figure 7 (a, b, c, d), and the intensity of the peak increased as the reaction time increased. The peak of diacetone alcohol appeared at about 7.1 min in the Figure 7 (a, b, c and d), and the intensity of the peak also increased as the reaction time increased. The rest of peaks was impure peaks. In the Figure 7 (c), the peak of p-benzoquinone and phenol did not appear and only left the peak of diacetone alcohol and the peak of 4-cyclopentene-1, 3-dione in the Figure 7 (d). Besides, the peak of 4-cyclopentene-1, 3-dione gradually decreased and the peak of diacetone alcohol gradually increased, these results showed that the phenol and p-benzoquinone were oxidized by ozone and new intermediate products emerged. This conclusion was in consistent with the phenomenon that the golden solution turned to be colorless transparent solution after reaction for about 20 min, which was shown in Figure 8. Presumably, the phenyl structure of the p-benzoquinone were destroyed, p-benzoquinone was degraded into diacetone alcohol by ozone or hydroxyl radical, and then, the further oxidation products of diacetone alcohol might be the diacetone alcohol. To sum up, p-benzoquinone, 4-cyclopentene-1, 3-dionethe and diacetone alcohol were the intermediate products, which were degenerated from phenol by ozone in wastewater solution. And this degradation mechanism could be expressed by an equation was shown in Figure 8.

DOI: 10.1051/mateconf/20168202002

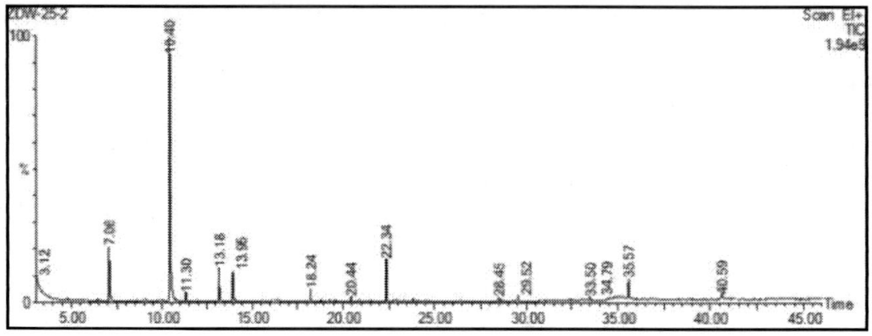

(a)

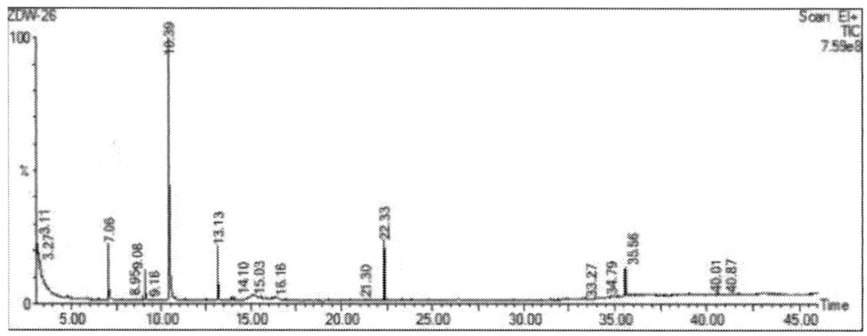

(b)

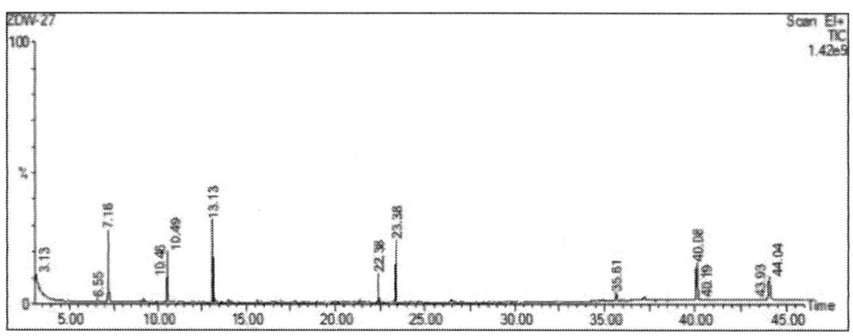

(c)

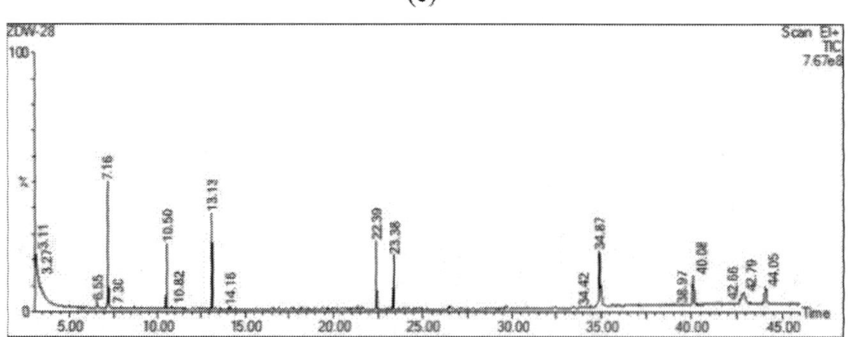

(d)

Figure 7. GC-MS spectrum of the degradation of phenol at 4 reaction periods, 5 min (a), 10 min (b), 20 min (c) and 30 min (d).

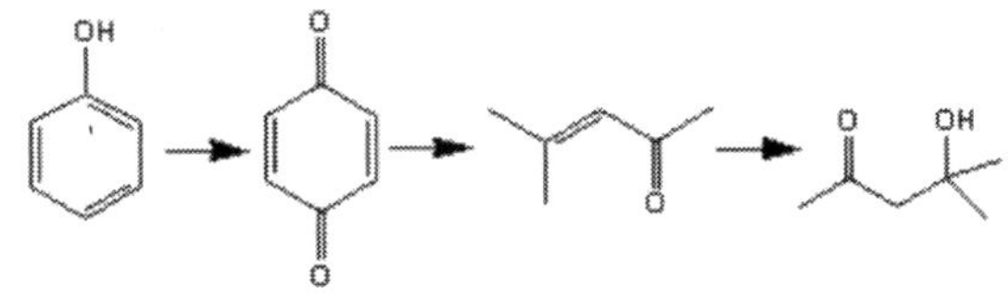

Figure 8. The degradation mechanism of phenol.

4 Conclusions

1. As the concentration of pure oxygen was 1.0 L/min, the highest concentration of ozone might be 7.06 mg/L. The concentration of ozone was increased by optimizing ozone generator in a range of pure oxygen gas flux. The concentration of ozone depended on discharge space and efficiency, as all the other conditions remained the same. It is effective for the optimization that the inner electrode of double dielectric and small metal cones were embedded in the outside electrode.

2. The concentration of ozone and the initial concentration of phenol have significant effect on the degradation of phenol. The increase in the initial concentration of phenol as the removal time increased, and as the time went on, the phenol removal ratio in wastewater was increased nearly 100%, whichwas in keeping with the first-order kinetics equation.

3. The self-design ozone generator can effectively treat phenol whatever the pH conditions was.

4. The intermediate products of the oxidation degradation of phenol were detected by GC-MS. According to a preliminary inspection, the intermediate products were p-benzoquinone, 4-cyclopentene-1, 3-dione and diacetone alcohol.

Acknowledgement

The authors acknowledge the research grant provided by the national science and technology major project for water pollution control and management (Project No. 2014ZX07105-001). And the major science and technology program of Yunnan Province, China（Project No. 2012ZB005）.

References

1. Z.Q. Zeng, H.Q. Zou, X. Li, M. Arowo, B. C. Sun, J.F. Chen, G.W. Chu, L. Shao, Chem. Eng. **229**, 404(2013)
2. C. Zidi, R. Tayeb, M. Dhahbi, Hazard. Mater **194**, 62(2011)
3. H Jiang, Y Fang, Y Fu, Q Guo, Hazard. Mater **101**, 179(2003)
4. S.T. Christoskova, M. Stoyanova, Wat. Res. **35**, 2073(2001)
5. H. Wright, A.N. James, Bioresour. Technol. **70**, 69(1999)
6. S.H. Lin, C.H. Wang, Ind. Eng. Chem. Res. **42**, 1648(2003)
7. K. Turhan, S. Uzman, Desalination **229**, 257(2008)
8. E.B. Azevedo, F.R.A. Neto, M. Dezotti, Hazard. Mater. **128**, 182(2006)
9. K.C. Namkung, A.E. Burgess, D.H. Bremner, H. Staines, Ultrason. Sonochem. **15**, 171(2008)
10. L.K. Weavers, N. Malmstadt, M.R. Hoffmann, Environ. Sci. Technol. **34**, 1280(2000)
11. J.A. Zazo, J.A. Casas, A.F. Mohedano, J.J. Rodriguez, Wat. Res. **43**, 4063(2009)
12. S. Esplugas, J. GimA Nez, S. Contreras, E. Pascual, M. RodrA I Guez, Wat. Res. **36**, 1034(2002)
13. J. M. Monteagudo, A. DurA N, M. Aguirre, I. San Mart, Hazard. Mater. **185**, 131(2011)
14. J. Hoigne, *Springer-Veriag Beriin Heideiberg*, **84**(1998)
15. I.V. Came, A. Bermond, Wat. Res. **32**, 3208(1998)
16. J.N. Wu, R. Klaas, Advances in Environmental Research **4**, 339(2000)
17. D.M. Feng, N.Y. Gao, X.C. Wang, Y. Ma, C. Zhou, J. Cent. Sout. Univ. **44**, 2130(2013)

Bamboo Fibre-reinforced Semi-Metallic Brake Friction Materials for Automotive Applications

R. J. Talib[1,2,a], K. Ramlan[3], N. I. Ismail[1,2], M. F. Ismail[1,2] and A. O. Eliasidi[4]

[1]*Faculty of Mechanical Engineering, Universiti Teknologi MARA Pulau Pinang, 13500 Permatang Pauh, Malaysia*
[2]*ARTeC, Universiti Teknologi MARA Pulau Pinang, 13500 Permatang Pauh, Penang, Malayisa*
[3]*Faculty of Mechanical Engineering, University Technology MARA, 40450 Shah Alam, Malaysia*
[4]*Advanced Materials Centre, SIRIM Berhad, Jalan High-Tech 2/3, Kulim Hi-Tech Park, 09000 Kulim, Malaysia*

Abstract. Three friction material formulations composed of bamboo fiber along with binder, friction modifiers and filler have been prepared through powder metallurgy process. Sample F1 and F2 are composed of 10 wt. % of copper and barium, respectively, while the other ingredients in both formulations have the same wt. %. The wt. % of bamboo fiber in sample F3 is, however, increased by 100%, while the compositions of the other ingredients are proportionally decreased. The samples were examined for their porosity, hardness, and friction and wear properties using hot bath, Rockwell hardness tester, and CHASE friction dynamometer, respectively. The test results are compared with those of a commercial sample as the benchmark. Normal and hot frictions of all the three samples developed comply with the requirements specified by Automotive Manufacturer Equipment Companies Agency (AMECA). However, sample F3 which is composed of 20 wt. % of bamboo fiber does not comply with the minimum requirement of friction coefficient. Whereas, sample F2, which is composed of 10 wt. % of bamboo fiber and 10 wt. % of barium, has lower friction coefficient than the commercial sample, and has a sudden drop in friction coefficient at a temperature of 500°F. Out of three developed samples, sample F1, which is composed of 10 wt. % of bamboo fiber and 10 wt. % of copper, complies with all the requirements and has higher friction coefficient than the commercial sample, and has higher fade resistance. Thus, it could be postulated that bamboo fiber could be used as a reinforcing fiber with composition of 10 wt. %.

1 Introduction

Friction materials are a heterogeneous material and are composed four components, namely; (i) the reinforcing fibres, (ii) binders, (iii) fillers, and (iv) frictional additives. Reinforcing fiber is use to provide the necessary mechanical strength, rigidity, integrity, and thermal stability at high temperatures. The selection of ingredients and weight percentages used in the friction formulation will significantly affect the friction and wear characteristics, braking performance, and friction-induced noise [1, 2]. In the braking process, kinetic energy of a moving vehicle is converted into thermal energy. The heat generated is dissipated to the surroundings through the mechanism of convection, conduction, and radiation. The ability of brake disc to dissipate this heat significantly affects the performance and wear life of the friction materials.

The phenolic resin had the strongest influence on tribological behaviours on the semi-metallic brake friction materials [3]. There is no simple correlation between friction values and wear rates with mechanical, chemical and thermal properties of the friction materials [4, 5]. In the process of developing a new formulation, the researchers should compromise some of the properties in order to

[a] Corresponding author : drtalib@ppinang.uitm.edu.my

get the best formulation. The main criterion for acceptance of a new formulation is based on the results of performance test rather than the mechanical test [6]. The mechanical properties are test is used to control the quality of the formulations that has been developed. The effect of bamboo fiber weight percentage in the semi-metallic brake material formulation on the mechanical and tribological characteristics will be the main focus in this study.

2 Experimental

Three friction material formulations composed of bamboo fiber along with binder, friction modifiers and filler have been prepared through powder metallurgy process. The powders were mixed for 10 minutes in tabular mixer and then warm compacted under a pressure of 150 kg/cm^2 at a temperature of 190°C with holding time of 3 minutes. The samples were compacted in a assmebly die with a sample dimension of 25 mm x 25 mm x 7 mm. Post curing was carried out in an oven at a temperature of 180°C for 3 hours. Sample F1 and F2 are composed of 10 wt. % of copper and 10 wt. % of barium, respectively, while the other ingredients in both formulations have the same wt. %. The wt. % of bamboo fiber in sample F3 is, however, increased by 100%, while the compositions of the other ingredients are proportionally decreased and composed both copper and barium having the same wt. %. Table 1 shows the composition for each formulation. Sample COM is a commercially available semi-metallic brake pad is used as a benchmark. Each sample was subjected to porosity, hardness, and friction tests. Table 2 lists the test equipment and test procedures employed in this work.

Table 1. Ingredients in the formulation

Ingredients	FC (%)	FB(%)	FBC (%)
Bamboo	10	10	20
Resin	15	15	13.3
Graphite	15	15	13.3
Steelwool	25	25	22.2
Iron Oxide	25	25	22.2
Copper	10	0	4.4
Barium	0	10	4.4
Total	100	100	100

Table 2. List of equipment and standards

Test procedures	Equipment	Standard
Porosity	Hotbath	JIS D 4418 1996 [7]
Rockwell hardness	Rockwell hardness tester	MS 474. Part 2. 2003 [8]
jFriction and wear test	CHASE dynamometer	SAE J661 Feb 1997 [9]

Friction coefficient and wear were results were obtained using Link Model 600 CHASE machine. The test procedure using CHASE machine is described in Society of Automotive Engineer standard SAE J661 Feb 1997 [9]. In this test, the sample was pressed against a rotating brake drum with a constant rotating speed of 417 rpm under the load of 647 N and subjected to test program. Briefly, each sample was subjected to seven test runs with the following sequences; (i) baseline, (ii) first fade, (iii) first recovery, (iv) wear, (v) second fade, (vi) second recovery, and (viii) baseline rerun. The samples thickness were measured and weighed before and after testing.

Society of Automotive Engineer defined the normal friction coefficient as average of the four readings taken at 200, 250, 300 and 400°F on the second fade curve [10]. The hot friction coefficient is defined as the average of the ten readings taken at 400 and 300°F on the first recovery; 450, 500, 550, 600 and 650°F of the second fade; and 500, 400 and 300°F of the second recovery run.

Automotive Manufacturer Equipment Companies Agency, USA has set the friction coefficient requirements as follow [11]; (a) shall have normal friction coefficient of 0.25 and higher or a hot of 0.15 and above, (b) shall have friction coefficient above 0.15 between 200 and 550 ^{O}F inclusive in second fade, or between 300 and 200 ^{O}F during the secondary fade.

3 Results and Discussions

3.1 Mechanical Properties

Table 3 shows the mechanical and tribological test results while Figure 1 shows the effect of material compositions on the porosity and hardness properties of the sample. It was observed that there is no simple correlation between the material compositions with the mechanical properties of semi-metallic brake friction materials as shown in Table 3 and Figure 1 and Figure 2. Sample COM has the lowest porosity and the highest hardness. Sample FB which composed of barium has the lowest porosity among the three developed sample and supposed to have higher hardness. Sample FBC which composed of higher bamboo fibre wt. percentage has the highest hardness even though having higher porosity than sample FB. The friction materials are heterogeneous materials, thus the mechanical and tribological properties cannot be predicted [1]. Thus, each formulation develop need to be tested in the lab as well on the road before it could be allowed to be used on the road.

Table 3. Test results

Sample	Porosity (%)	Rockwell hardness (HR_S)	Normal friction	Hot Friction	Thickness loss (%)
COM	14.88	82.4	0.38	0.33	0.56
FC	24.78	56.8	0.41	0.38	2.8
FB	20.98	45.6	0.32	0.27	0.91
FBC	23.79	78.7	0.36	0.29	2.7

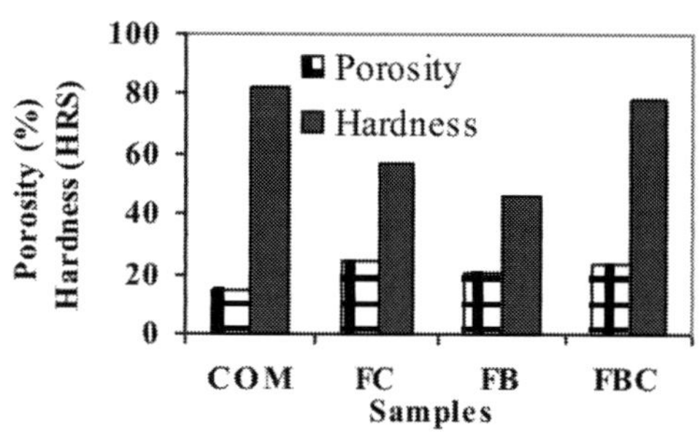

Figure 1. Porosity and Hardness Vs material composition

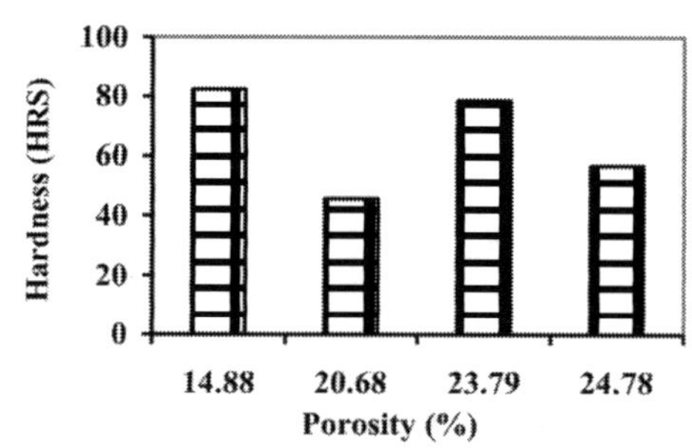

Figure 2. Hardness Vs porosity

3.2 Friction Coefficient

Normal and hot frictions of all the three samples developed comply with the requirements specified by AMECA (Table 3). However, sample FBC does not comply with the minimum requirement of friction coefficient which less than 0.15 in second fade test condition as shown in Figure 3. This phenomenon could be due to higher wt. percentage of bamboo fibre in the composition. Second fade run generated higher temperature during braking which will elucidate the brake fade of the sample being tested. Based on test results, it could be postulated that 10% wt. percentage of bamboo fibre is the optimum wt. percentage in the composition. Whereas, sample FB has lower friction coeffient than the commercial sample, and has a sudden drop of friction coeffient at a temperature of 500°F. Sample FC complies with all the requirements and has higher friction coeffient than the commercial sample, more stable friction coefficient, has higher fade resistance and therefore could be seen as the optimized material composition under this work. The addition copper in the composition has improved the frcition coefficient due to higher heat transfer during braking process. Based on the mechanical and friction test results, it could be concluded that the decision of the best formulation can be selected based on the fricton test results rather than the mechanical test results.

Figure 4 shows the friction coefficient characteristics during second recovery cycle. The first fade characteristic is not discussed because the temperature generated during this braking operation is below 288°C which is below the decomposition temperatures. It can be seen from Figure 3 that all samples developed almost recover to their respective base line friction coefficient readings when the brake drum and the sample are cooled to room temperature.

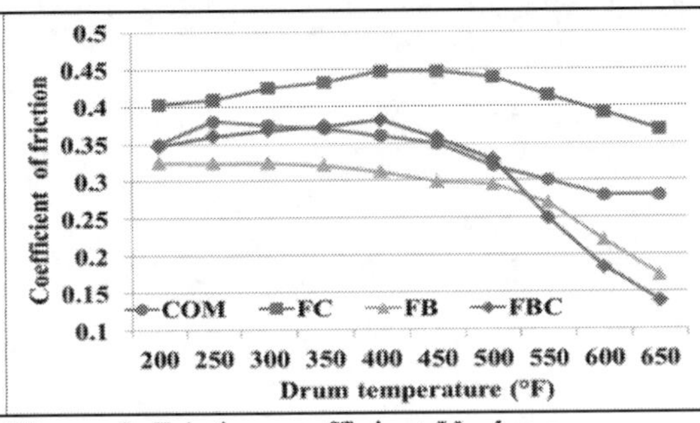

Figure 3. Friction coefficient Vs drum temperature during second fade cycle

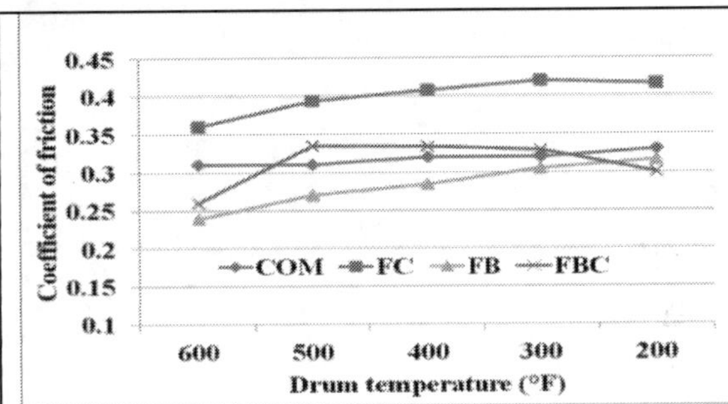

Figure 4. Friction coefficient Vs drum temperature during second fade recovery

Figure 5 shows the effect of porosity and hardness on hot friction coefficient and thickness loss of the developed samples. For homogeneous materials, the lower the porosity and the higher the hardness, the higher will be friction coefficient. However, for friction materials this phenomena does not apply. Sample FC is the highest hot friction coefficent evenhough it porosity and hardness is not the highest among the developed samples (Figure 5(a), Table 3). Thus, it could be concluded that the physical and mechanical properties of friction materials are not simply correlated with hot friction properties. The physcial and mechanical properties of friction are use for quality control in manufacturing friction materials that using the same formulation.

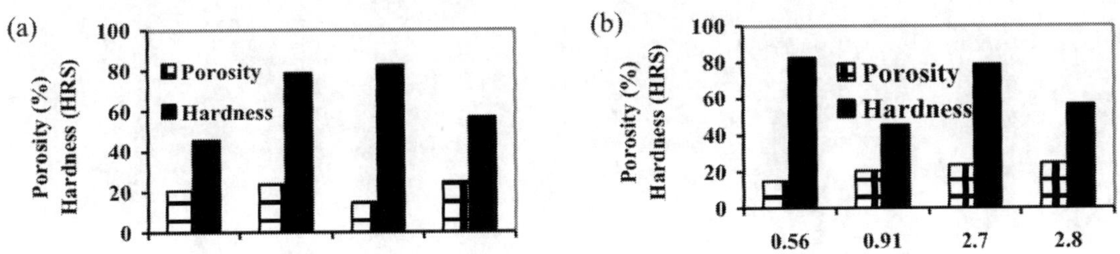

Figure 5. Effect of porosity and hardness on, (a) hot friction coefficient, and (b) thickness losss

3.3 Wear

The lowest thickness loss was observed for sample FB which composed 10% wt percentage of Barium (Table 3). Addition of Barium has improved the wear resistance of the developed friction material even though has the lowest hardness. It can be seen in Figure 5(b) that there is no simple correlation between hardness with thickness loss. However, it was also noticed that the thickness loss increased with increasing porosity. Higher porosity indicates higher pore content in the formulation which result in less contact area, subsequently increased the thickness loss.

3.3 Wear Mechanism

It can be seen in Figure 6 that the harder peak asperities on the brake disc ploughed into the friction material surface during braking process. This process was a manifestation of abrasion wear mechanism. On the other hand, Figure 7 shows the symptom of adhesion wear mechanism. Figure 7(a) shows the formation of transfer patches while Figure 7(b) shows the formation of transfer layer where the transfer patches and wear fragment have been flattened, compacted and smeared on the worn surface. Thus, it could be postulated that both the abrasion and adhesion wear mechanisms are in operating during braking process.

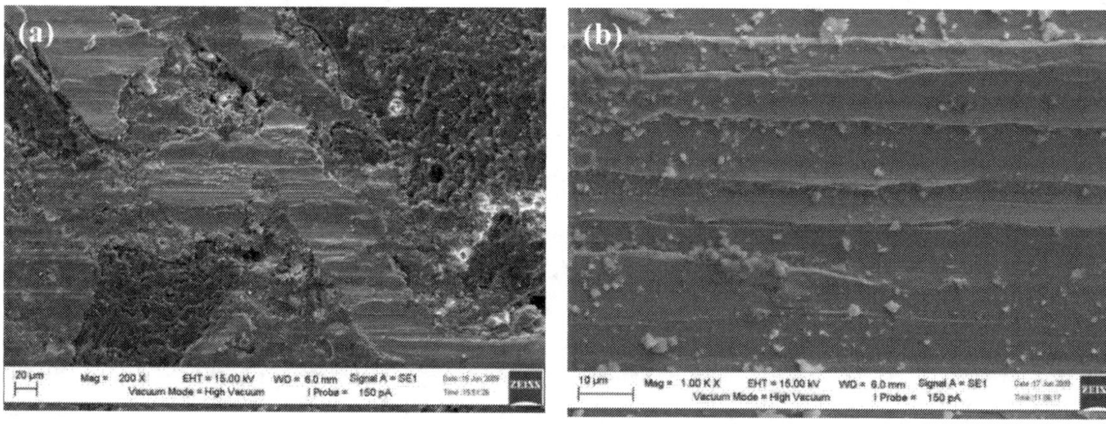

Figure 6. Abrasion wear mechanism; (a) harder peak asperities ploughed on the sample surface, (b) close-up view

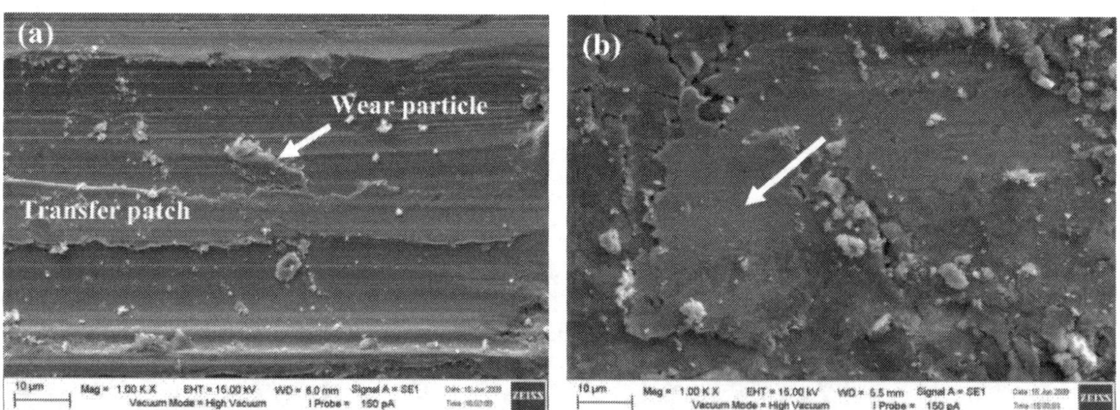

Figure 7. Adhesion wear mechanism; (a) formation of transfer patches, (b) formation of transfer layer and wear fragment to be detached

4 Conclusions

Three newly developed brake friction material with different bamboo fiber compositions have been successfully developed using powder metallurgy process. The samples had been subjected to mechanical and CHASE dynamometer friction tests. Based on test results obtained, the following phenomena could be postulated as follows; (i) the bamboo fiber could be used as a reinforcing fiber with composition of 10 wt. % and 10 wt. % of Copper (ii) there are no simple correlations between the material composition with mechanical properties with tribological characteristics, (ii) there are no simple correlations between the mechanical properties with tribological characteristics, (iii) CHASE dynamometer friction results could be used in screening the formulations, and (iv) addition of copper improve COF due higher heat transfer during friction test while addition of barium improve the thickness loss.

Acknowledgements

This research was supported by Ministry of Science, Technology and Innovation, Malaysia with sciencefund grant No. 06-01-01-SF0809 and Univertsiti Teknologi MARA. The authors are grateful to SIRIM Berhad for providing research facilities.

References

1. M. Y. Cho, S. J Kim, D. Kim, H. Jang, (2005). Wear 258, 1682–1687
2. Ling Han, Li Huang, Jinsheng Zhang, Yafei Lu, (2006). Composites Science and Technology 66, 2895–2906
3. A. M. Zaharudin, M. N. Berhan, and R. J. Talib. (2011). American Institute of Physics Conf. Proc. 1400, 274
4. K. Tanaka, S. Ueda, & N. Noguchi, (1973). Wear 23: 349-365
5. S. M. Hsu, M. C. Shen, A. W. Ruff, (1997). Tribology International 30, 377
6. Talib Ria Jaafar, Mohmad Soib Selamat, Ramlan Kasiran, (2012). Selection of Best Formulation for Semi-metallic Brake Pads Development in Powder Metallurgy, In Katsuyoshi Kondoh (ed), Powder Metallurgy, INTEC Crotia, 1-30
7. JIS D 4418 (1996), Japanese Industrial Standard. "Test procedure of porosity for brake linings and pads of automobiles",
8. MS 474. Part 2 (2003), Malaysia Standarad. "Methods of Test for Automotive Friction materials (Brake Linings, Disc Pads and Bonded Shoe): Rockwell Hardness Test" (First Revision),
9. SAE J866 (Jan 96), Friction coefficient identification system for brake lining, Society of American Engineers, Warrendale, USA
10. SAE J661 (Feb 97), Brake lining quality test procedure, Society of American Engineers, Warrendale, USA
11. Vehicle equipment safety commission Regulation V3, minimum requirements and uniform test procedures for motor vehicle brake lining. Automotive Manufacturer Equipment Companies Agency Virginia, USA

MATEC Web of Conferences **82**, 02004 (2016) DOI: 10.1051/matecconf/20168202004
D2ME 2016

3D Fractal Investigation on Ultrasonic Vibration Assisted End Grinding Surface of SiCp/Al Composites

Wei Zheng[1, a], Ming Zhou[2] and Li Zhou[3]
School of Mechanical and Electrical Engineering, Harbin Institute of Technology, Harbin, 150001, PR China
[1]zhengwhit@gmail.com, [2]zhouming@hit.edu.cn, [3]zhoulihit@163.com

Abstract. This paper presents a 3D fractal investigation on ultrasonic vibration assisted end grinding surface of SiCp/Al composites. According to variational method, 3D surface fractal dimension is calculated with MATLAB software based on profile data, and evaluates the influence of process parameters on the surface quality. The experimental results indicate that 3D surface fractal dimension increases with spindle speed and vibration amplitude, and decreases with feedrate and cutting depth. A large value of 3D surface fractal dimension corresponds to a better surface finish.

1 Introduction

Superior properties such as high specific strength and specific stiffness make SiCp/Al composites highly competitive against conventional materials and are widely used in aerospace, automobile and other fields. However, the addition of SiC particles makes SiCp/Al composites hard to be manufactured because of its extreme abrasive properties. This has limited the practical application of SiCp/Al composites, and a fair assessment of the machinability of MMCs is still a difficult issue [1]. Therefore, it is necessary to develop non-traditional machining process to meet the demand of SiCp/Al composites machining. Ultrasonic vibration assisted end grinding is a combination of two ordinary material remove processes, i.e. grinding and ultrasonic machining during which ultrasonic vibration is applied to the cutting tool, as shown in Fig. 1 [2]. Compared with other traditional machining methods, this machining method has many advantages, such as lower grinding force, higher surface quality, higher machining efficiency and longer tool life. Fig. 2 shows the motion trajectory of a single grain in end surface of the grinding tool.

Surface topography can be characterized using an excessive number of statistical parameters depicting various aspects of the surface lay, roughness, waviness and the form. Unfortunately, many of them strongly depend on how they are actually measured, including for instance the sampling and the scan lengths, and the instrumental resolution. The machining of SiCp/Al composites has been extensively studied experimentally and numerically to assess surface integrity. In general, based on available references, most of them focus on turning and the analysis of surface integrity is using 2D or 3D surface roughness [3-11].

Several research results show that the surface of SiCp/Al composites has fractal characteristics, such as self-similarity and self-affinity [12]. There are few publications focus on fractal investigation

[a] Corresponding author : zhengwhit@gmail.com

on surface of SiCp/Al composites. So this paper presents a 3D fractal investigation on ultrasonic vibration assisted end grinding surface of SiCp/Al composites to evaluate the effect pf process parameters, including cutting depth, feedrate, spindle speed and vibration amplitude, on the surface quality.

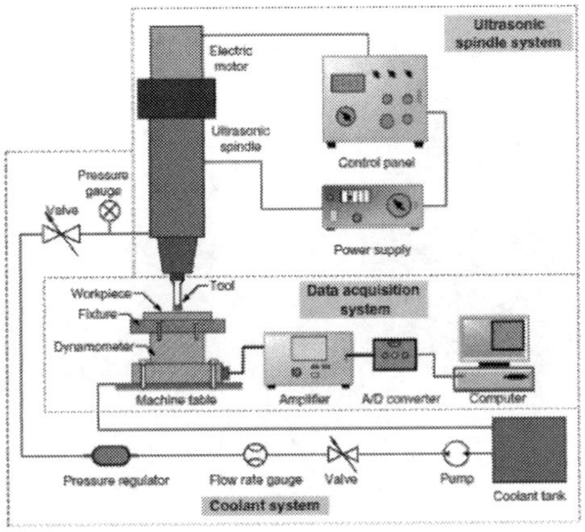

Figure. 1. Schematic diagram of ultrasonic vibration assisted end grinding.

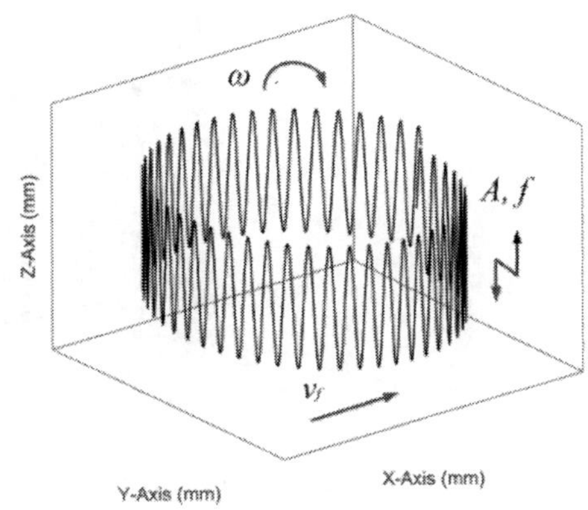

Figure. 2. The motion trajectory of grains in end surface of the grinding tool.

2 Experimental setup and procedure

2.1 Experimental equipment

Ultrasonic vibration assisted end grinding experiments were conducted on an ultrasonic vibration machine center (DMG Ultrasonic70-5 linear), which consisted of a numerical control system, an ultrasonic spindle system, a data acquisition system and a coolant system. The maximum spindle speed was 18000rpm. The frequency and amplitude supplied to the tool in the z direction were about 30 kHz and 5μm respectively. The grinding tool with metal-bonded diamond abrasives was provided by SCHOTT Company, and its properties were listed in Table 1.

Table 1. Properties of the grinding tool used in experiments.

Abrasive size s (μm)	Abrasive concentration Ca	Outer radius R_o (mm)	Inner radius R_i (mm)
75	100	30	20

2.2 Workpiece material

Figure. 3. The microstructure of the SiCp/Al composites.

The workpiece material was 45 % SiCp/Al2024 composites. The average diameter of SiC particles was 5μm. The microstructure and properties of the SiCp/Al composites were shown in Fig. 3 and Table 2 respectively.

Table 2. The properties of the SiCp/Al composites.

Parameter	Value
Modulus of elasticity E (GPa)	165
Shear modulus G (MPa)	13.2
Limit of yielding σ_s (MPa)	520
Thermal conductivity β (W/m.K)	130
Density ρ (kg/m^3)	2900
Poisson ratio γ	0.26

2.3 Experiment

Surface integrity was influenced by process parameters including spindle speed, feedrate, cutting depth and vibration amplitude. So single factor experiments based on four factors with five levels were conducted, as listed in Table 3. According to variational method, 3D surface fractal dimension was calculated with MATLAB software based on Profile data which was obtained by LCSM (laser confocal scanning microscope).

Table 3. Experimental parameters.

Levels	Spindle speed n (rpm)	Feedrate v_f(mm/min)	Cutting depth a_p (μm)	Vibration amplitude A (μm)
1	3000	5	5	0
2	6000	10	10	1.25
3	9000	15	15	2.5
4	12000	20	20	3.75
5	15000	25	25	5

3 Results and discussion

Fig. 4 shows the influence of process parameters on 3D surface fractal dimension. It can be seen that with the variation of process parameters, 3D surface fractal dimension D is almost in the range of 2.6-2.75, and nonlinearly increases with the spindle speed and vibration amplitude, and decreases with feedrate and cutting depth. Cutting depth and spindle speed have a large effect on the 3D surface fractal dimension compared with feed rate and vibration amplitude.

Fig. 4(a) shows that when spindle speed is in the range of 3000-6000rpm, 3D surface fractal dimension increases drastically, however, when spindle speed is in the range of 6000-15000rpm, the variation rate of 3D surface fractal dimension is slowly. The typical 3D topographies of SiCp/Al composite with two different spindle speed is shown in Fig. 5 (a) and (b), it can be seen that the surface quality generated with a large value of spindle speed (15000rpm) is superior to that generated with a small value of spindle speed (3000rpm).

Fig. 4(b) shows that 3D surface fractal dimension nearly linearly decreases with feedrate in the range of 5-25 mm/min. The typical 3D topographies of SiCp/Al composite with two different feedrate is shown in Fig. 6 (a) and (b), it can be seen that the surface quality generated with a small value of feedrate (5mm/min) is superior to that generated with a large value of feedrate (25mm/min).

Fig. 4(c) shows that 3D surface fractal dimension nearly linearly decreases with cutting depth in the range of 5-20μm, and decreases drastically with cutting depth in the range of 20-25μm. The typical 3D topographies of SiCp/Al composite with two different feedrate is shown in Fig. 7 (a) and (b), it can be seen that the surface quality generated with a small value of cutting depth (5μm) is superior to that generated with a large value of cutting depth (25mm/min).

Fig. 4(d) shows that 3D surface fractal dimension slowly increases with vibration amplitude in the range of 0-3.75μm, and increases drastically in the range of 3.75-5μm. The typical 3D topographies of SiCp/Al composite with two different vibration amplitude is shown in Fig. 8 (a) and (b), it can be seen that the surface quality generated with a large value of vibration amplitude (5μm) is superior to that generated with a large value of vibration amplitude (0).

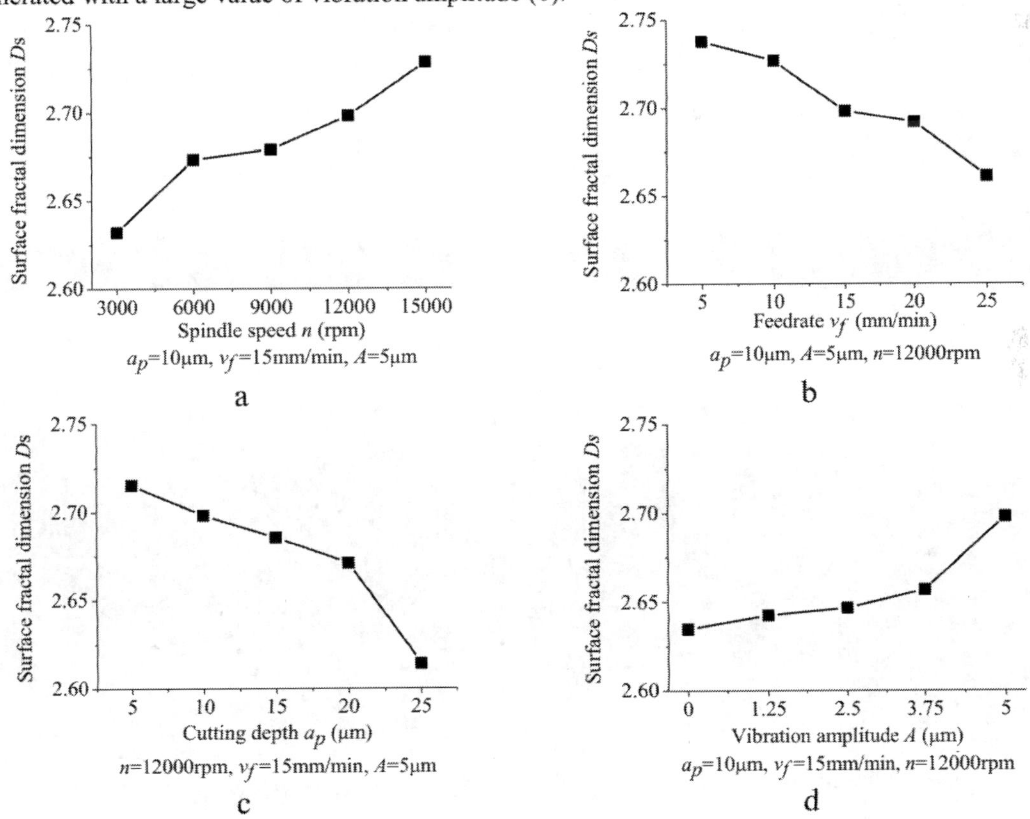

Figure. 4. The influence of process parameters on 3D surface fractal dimension.

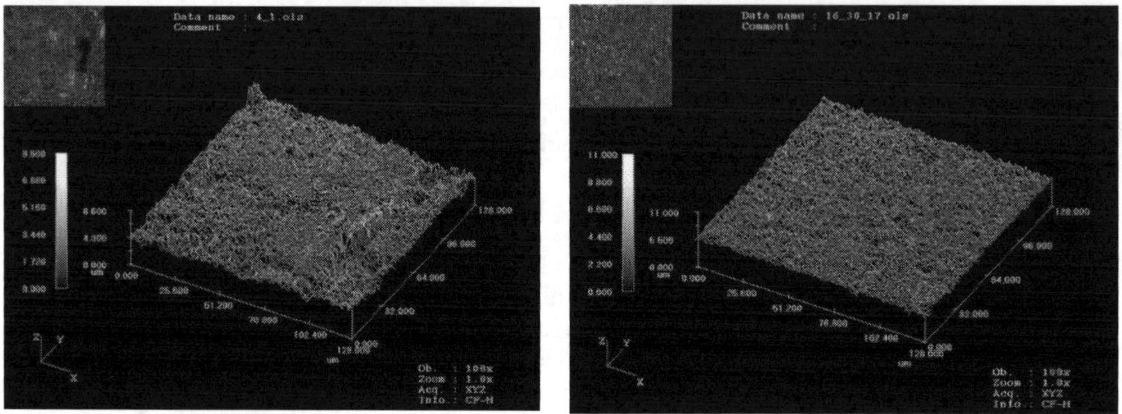

a) n=3000rpm b) n=15000rpm

Figure. 5. 3D topographies of SiCp/Al composites with different spindle speeds.

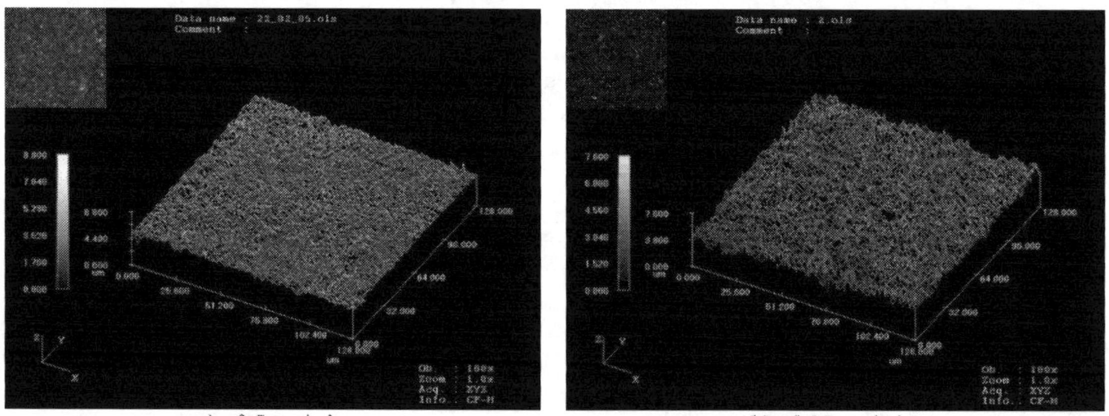

a) vf=5mm/min b) vf=25mm/min

Figure. 6. 3D topographies of SiCp/Al composites with different feedrates.

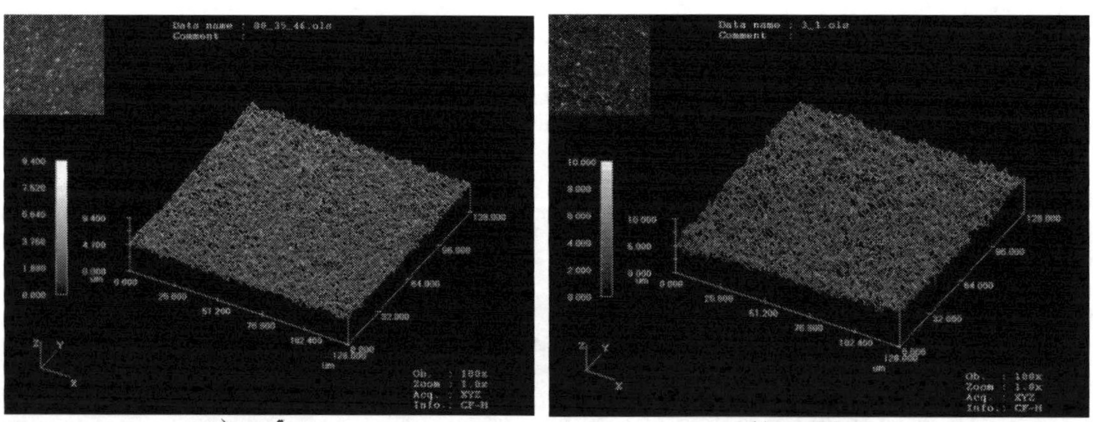

a) ap=5μm b) ap=25μm

Figure. 7. 3D topographies of SiCp/Al composites with different cutting depths.

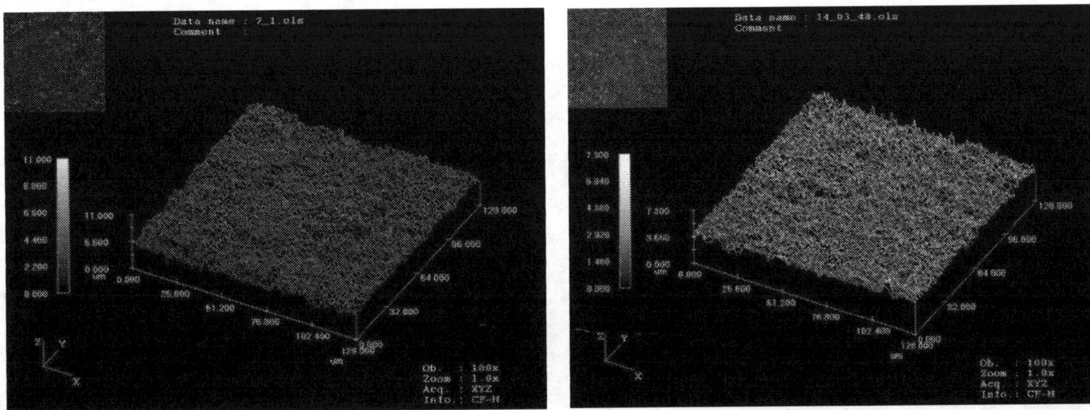

a) A=0 b) A=5μm

Figure. 8. 3D topographies of SiCp/Al composites with different vibration amplitudes.

4 Conclusion

This paper presents a 3D fractal investigation on ultrasonic vibration assisted end grinding surface of SiCp/Al composites. Several conclusions are obtained:

1. With the variation of process parameters, 3D surface fractal dimension D_s is almost in the range of 2.6~2.75.

2. 3D surface fractal dimension D_s nonlinearly increases with an increase of the spindle speed and vibration amplitude, and decreases with an increase of feedrate and cutting depth.

3. Compared with feed rate and vibration amplitude, cutting depth and spindle speed have a large effect on the 3D surface fractal dimension D_s.

4. A large value of 3D surface fractal dimension D_s corresponds to a better surface finish.

This work is supported by the National Natural Science Foundation of China (NSFC) (Grant No. 51375119).

References

1. A. Pramanik, L.C. Zhang, J.A. Arsecularatne,, Int. J. Mach. Tools Manuf. **48**, 13 (2008)
2. W.L. Cong, Z.J. Pei, X. Sun, C.L. Zhang, Ultrasonics 54, **13** (2014)
3. S. Kannan, H.A. Kishawy, Int. J. Mach. Tools Manuf. **46** (2006)
4. U.A. Dabade, S.S. Joshi, R. Balasubramaniam, V.V. Bhanuprasad, J. Mater. Process. Technol. **192**, 9 (2007)
5. K. Palanikumar, R. Karthikeyan, Mater. Des. **28**, 9 (2007)
6. Y.F. Ge, J.H. Xu, H. Yang, S.B. Luo, Y. Fu, J. Mater. Process. Technol. **203**, 10 (2008)
7. N.S.K. Reddy, S. Kwang-Sup, M.Y. Yang, J. Mater. Process. Technol. **201**, 6 (2008)
8. R.K. Bhushan, S. Kumar, S. Das, Int. J. Adv. Manuf. Technol. **50**, 11 (2010)
9. X.H. Shen, J.H. Zhang, D.L.X. Xing, Y.F. Zhao, Int. J. Adv. Manuf. Technol. **58**, 9 (2012)
10. J.L. Jiang, P.Q. Ge, W.B. Bi, L. Zhang, D.X. Wang, Y. Zhang, Int. J. Mach. Tools Manuf. 74, 12 (2013)
11. T. Wang, L.J. Xie, X.B. Wang, T.Y. Shang, Def Technol **11**, 6 (2015)
12. J.N. Muguthu, D. Gao, Mater. Manuf. Process. **28**, 6 (2013)

MATEC Web of Conferences **82**, 02005 (2016) DOI: 10.1051/matecconf/20168202005
D2ME 2016

Stable Grinding Micro-wall of Flexible Joints with Controllable Force

Da Qu[1], Fei Ding[1] and Bo Wang[1,a]

[1]*CPE, Harbin Institute of Technology, Harbin, 150001, P. R. China*

Abstract. The micro-wall between two adjacent pores on the flexible joint is the core structure affecting the performance of gyro. The thickness of the micro-wall is only about $30\mu m$ and decreases to $20\mu m$ after being machined. It is a low-stiffness structure easily affected by grinding force. Larger force can break the micro-wall. Hence, appropriate grinding force must be controlled to make balance between machining efficiency and surface quality. Besides, different thickness before being machined affects the machining stability as well. The present paper firstly introduces the developed machine tool used for grinding the micro-walled. With this machine tool, removal of the process can be measured as well. Detailed measuring theory and grinding processes are presented. The force can be adjusted and monitored in the micro-grinding process. To achieve stable removal, the related thickness is analysed by experiments. With this machine tool, the process of machining-measuring micro-wall is efficient.

1 Introduction

The flexible joint is the core part of dynamically tuned gyro (DTG) which is widely used in avigraph. Varying structures and principles of gyro started to appear in 1942. There into, DTG is highly required in the manufacturing field of aircrafts and missiles due to its small size, high precision and high reliability [1]. The working part of flexible joints, the micro-wall, is shown in figure 1.

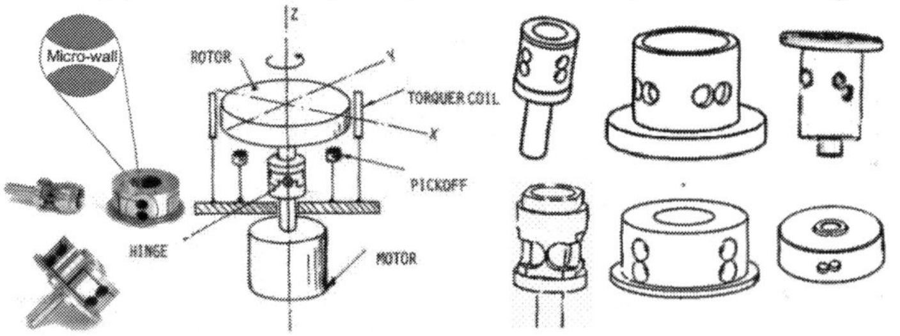

Figure 1. Components and operating principle of DTG

In service process of the flexible joint, the micro-wall is always broken firstly. Hence, it is very significant to enhance its dimensional accuracy and processed quality. As its importance, many researchers carried out relevant studies on micro machining. Wang et al. proposed a composite

---

[a] Corresponding author : bradywang@hit.edu.cn

processing technology which combines electrical discharge machining with ultrasonic vibrating grinding to process micro-pores [2]. A spiral polishing method applied in finishing the micro-pores is proposed by Yan et al. [3]. This method makes good processing accuracy but only can be used in processing a through hole. Ultrasonic assisted grinding technology was carried out by Peng et al.[4-6]. Their experiments show good performance on obtaining high quality surface. Yi et al.[7] developed a pen shaped controlling polishing device for polishing optical mold. Brinksmeier et al. [8,9] advanced thorough research of polishing microstructure array employing multiaxial finishing equipment which is suitable for studying removal mechanism. Guo et al. [10-12] and Suzuki et al. [13,14] developed a real-time force feedback controlling system to process ultra-precision optical surface. However, for stable removal, little researches were studied.

In this paper, a machine tool with two-dimensional force feedback is developed. With the two-directional force feedback, it is easy to measurement and process the micro-wall. A series of experiments were carried out and the relationship between the thickness and removal amount is illustrated. Detailed processing theory and measuring method were presented as well.

2 Setup

Figure 2 shows the structure of the machine tool which is applied to grind the micro-pore. They consist of force sensor, moving components, and control system.

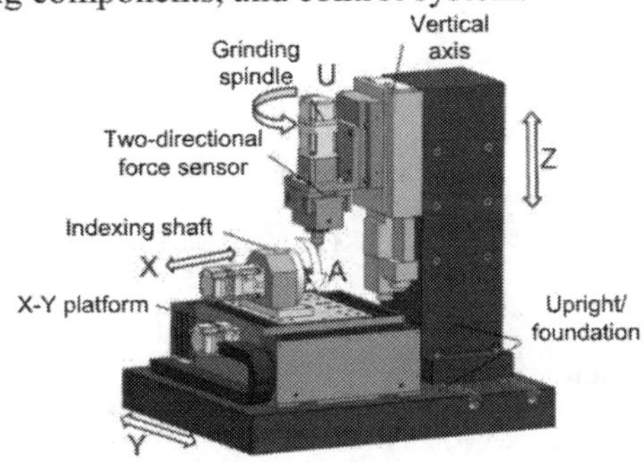

Figure 2. Structural distribution of the machine tool

Forece sensor: The force sensor was made of magnalium. Four strain gages with high sensitive coefficient were attached on each sensitive spot. They consist of a half bridge arms form of Wheatstone bridge. The force resolution of the sensor is $6mN$ which occupies 0.06% of the whole measuring range after its compensation since the force sensor is anisotropy. The direction of the force can be measured by two directional force components as well.

Moving components: Five axes are used in the machine tool. X, Y and Z axis is used for linear movement. Axis A is applied to adjust the cocoaxiality between the micro-pore and Z axis. The directional moving accuracy reaches about $50nm$. The repeated accuracy in Z direction is $300nm$. Axis U makes the grinding bar rotate. The angular displacement resolution of axis U is 1.24 second of arc. Rotation precision of axis A is less than $1\mu m$.

Control system: The hardware components of the control system are elaborated as figure 3a. Dynamic strain meter and high-speed data acquisition system are employed. Moving control module is composed of motion control card, actuators, feedback grating, main power source and control power supply circuit.

Software: The software used for controlling the movement of each axes, data output and input, and monitoring the grinding process is shown in figure 3b. Functions of different areas in the software are pointed out. The display area shows the captured force amplitude and its direction.

DOI: 10.1051/matecconf/20168202005

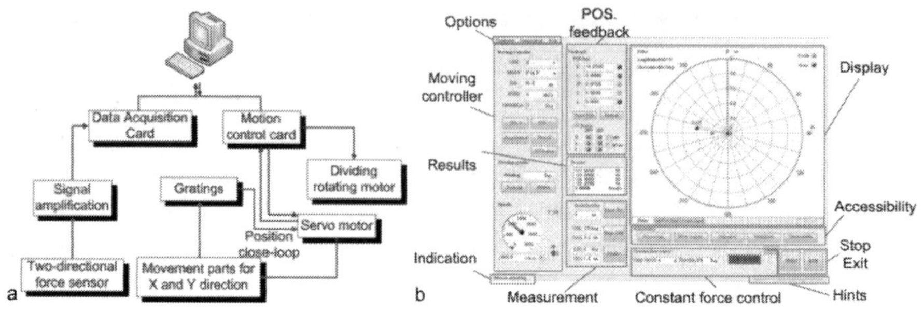

Figure 3. Schematic diagram: a. Components of control system b. The software and its functions

3 Processing steps

This part illustrates some key steps of micro-grinding flexible joints, including adjustement of coaxiality, choice of grinding bar and measurement of removal. All these steps affect the accuracy of measuring and machining.

3.1 Adjustment of coaxiality

Before micro grinding, coaxiality between the micro-pore and Z direction should be adjusted. The difference of the coaxiality is elaborated in figure 4.

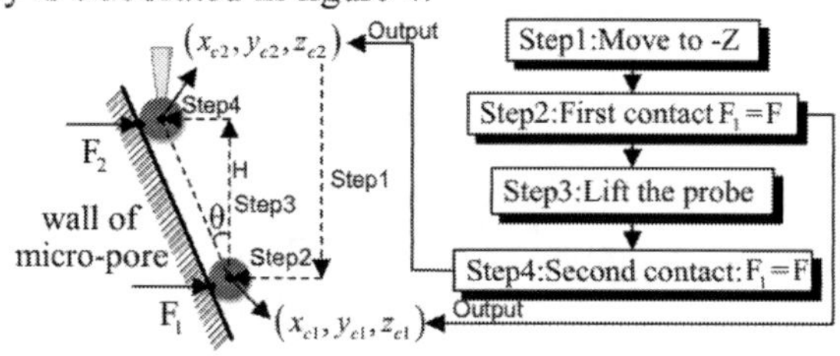

Figure 4. Flow diagram of coaxiality measurement

In the coaxiality measurement, the micro-wall is moved to make contact with the probe until $F_1 = F$. The coordinate after being compensated is put out. Where the compensation of the deformation of probe and force can be done by measuring a fixed gauge block. (x_{c1}, y_{c1}, z_{c1}) and (x_{c2}, y_{c2}, z_{c2}) are the calculated coordinates in the first and second contact after compensation, respectively. F_1 is the reactive force on the probe from the wall in the first contact. F is the contacting force which can be set to make the judgement of contact. H is the height of the lifting distance which needs setting before the measurement. The coaxiality can be calculated as

$$\theta = \arctan \frac{\sqrt{(x_{c2} - x_{c1})^2 + (y_{c2} - y_{c1})^2}}{H} \tag{1}$$

Rotating direction of axis A can be judged with sign symbol of θ. The origin of coordinates can be set as any position in the moving range of axes.

3.2 Selection of grinding bar

In the grinding process, the material of the grinding tool significantly impacts the machined dimensional accuracy, efficiency and surface quality. The grinding paste is applied between the micro-wall and the grinding bar. Grinding bar made of cast iron is used in this experiment. Besides, the size of the grinding bar needs choosing since directional grinding generates unrounded profile which affects the performance of the micro-wall.

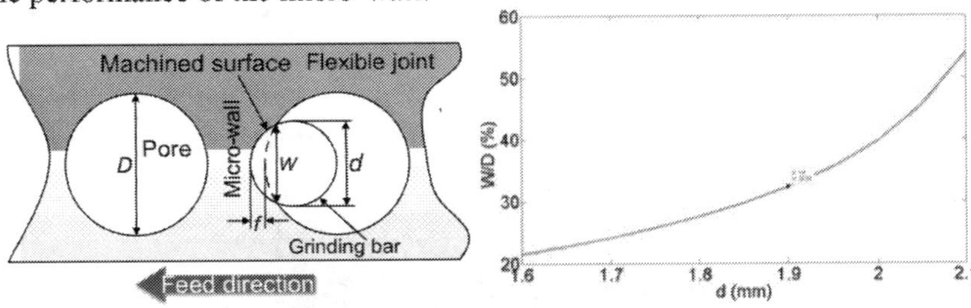

Figure 5. Schematic diagram of geometrical relationship in grinding process

The relationship between diameter of grinding bar and machined width is shown as figure 5 and is presented as

$$D-d+2f+\sqrt{d^2-W^2}-\sqrt{D^2-W^2}=0 \quad \left(2f\leq\sqrt{D^2-W^2}+d-D\right) \tag{2}$$

Where D denotes the diameter of pore, d denotes the diameter of grinding bar, W is the machined width and f is the feed of grinding bar. In this experiment, D is 2.2mm. f is 0.01mm and the appropriate d can be chosen from the relationship between W/D and d. W/D should be larger where possible. However, the chosen d should be smaller than D since the grinding bar can insert the pore easily. Consider its balance, diameter of the grinding bar used in the experiment is 1.9mm which makes W/D reaches over 30%.

3.3 Measurement of removal

The thickness measuring theory is elaborated in figure 6. In step 2, move the platform to make the micro-wall contact with the probe and precise adjusting until the captured reactive force F_1 reaches the set contacting force F. Then, the contacing position coordinate (x_{c1}, y_{c1}) calculated after deforming compensation of the sensor and probe is put out. After the first contact is fullfiled, repeating the first contact to fullfil a second contact with the other side of the micro-wall in the same depth D_c with the first-time contact to obtain the contacting position coordinate (x_{c2}, y_{c2}). Where D_c is the depth from the upper surface of the wall. Then thickness T_w can be calculated as

$$T_w=\sqrt{(x_{c1}-x_{c2})^2+(y_{c1}-y_{c2})^2}-2R_p \tag{3}$$

Where R_p denotes the radius of the probe head. The repeated accuracy of measuring thickness in X direction is $\pm 42nm$.

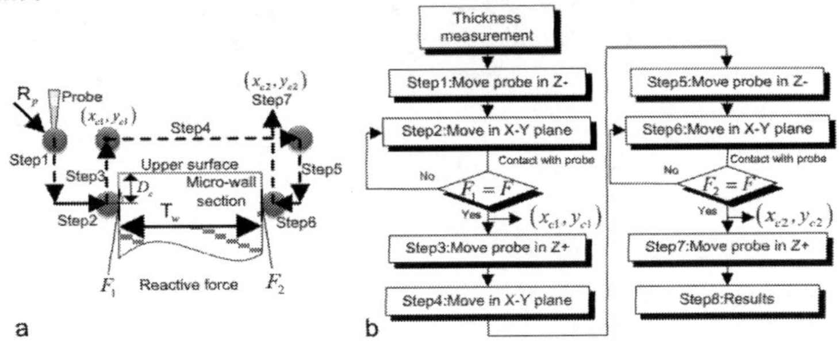

Figure 6. Thickness measuring theory

4 Experimental results

4.1 Experimental setting

In this experiment, the applied machine tool is shown in figure 7. Some setting details are shown in table 1. Diamond grinding paste with W2.5 was applied in this experiment. The material of flexible joint is a sort of high elastic alloy whose brand is 3J33A. In addition, the reciprocating movement of axis Z was taken in the experiment to complicate the machining track and decrease the surface roughness.

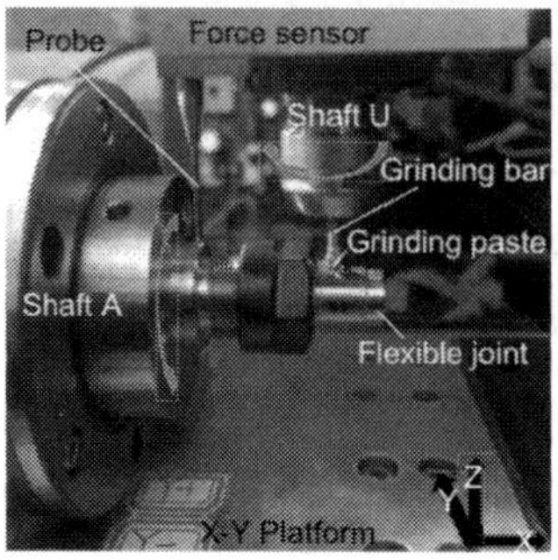

Figure 7. The machine tool used for micro grinding

Table 1. The set parameters in micro grinding process

F (N)	R_p (mm)	H (mm)	D_c (mm)
0.1	0.5	1.5	1.5

4.2 Stably grinding

In this grinding experiment, the used parameters are listed in table.2. The grinding time is 60s. The original thickness was measured before the micro-grinding was taken and the removal amount can be calculated by the difference between the original thickness and the thickness after machined. The relationship between the original thickness and the removal amount is shown in figure 8. Three-times measurements were taken in a same grinding experiment.

Table 2 The set parameters in micro grinding process

F (N)	Grinding paste	Reciprocating DIS.(mm)/speed(mm/s)	Spindle speed(rpm)
1.6	W2.5	0.2	1000

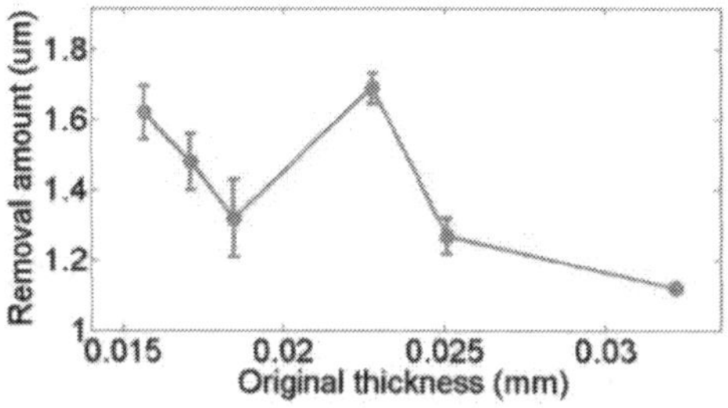

Figure 8. The relationship between the original thickness and removal amount

In figure 8, the removal amount decreases with the increase of original thickness. As the smaller thickness has larger elastic deformation when the same grinding force is applied, more grinding paste can flow between the grinding bar and the micro-wall. This causes the larger removal amount. Besides, the removal amount changes in range of 480nm when the thickness changes in range of 4.5um.

5 Conclusions

The present developed machine tool performs well on stably grinding the internal micro-wall. With the force feedback, grinding efficiency can be enhanced with high-stability. In addition, appropriate grinding parameters can be chosen aiming at varying thickness. Some conclusions can be expounded as

(1) Appropriate grinding force can be chosen with the force feedback in case of breaking the micro-wall.

(2) Fast measurement of the micro-wall can be carried out without clamping again which takes clamping error. The coaxiality of the flexible joint can be measured and adjusted to decrease grinding error.

(3) The measured thickness has measuring error since when contacting with the probe, elastic deformation appears to decrease the measuring results of the thickness. The removal amount is affected by the viscosity of the grinding paste as well. This induces the experimental results.

(4) With the feedback of grinding force, more applications can be achieved. For further study, the grinding time can be taken as an index of removal amount under the condition of certain processing parameters.

References

1. M.B. Bafghi, A. Vahedi, J. Soleimani. Optimized Design of PM Torquer for Dynamically Tuned Gyroscope. IEEE. T. Ind. Appl. **48** (2012)
2. A.C. Wang, B.H. Yan, X.T. Li, F.Y. Huang. Use of micro ultrasonic vibration lapping to enhance the precision of microholes drilled by micro electro-discharge machining. Int. J. Mach. Tool. Manu. **42** (2002)
3. B.H. Yan, H.J. Tzeng, H.F. Yuan, Y.C. Lin, H.M. Chow. Finishing effects of spiral polishing method on micro lapping surface. Int. J. Mach. Tool. Manu. **47** (2007)
4. Y. Peng, Z. Liang, Y. Wu, Y. Guo, C. Wang. Characteristics of chip generation by vertical elliptic ultrasonic vibration-assisted grinding of brittle materials. Int. J. Adv. Manuf. Tech. **62** (2011)
5. Y. Peng, Z. Liang, Y. Wu, Y. Guo, C. Wang. Effect of vibration on surface and tool wear in ultrasonic vibration-assisted scratching of brittle materials. Int. J. Adv. Manuf. Tech. **59** (2011)
6. Y. Peng, Y.B. Wu, Z.Q. Liang, Y.B. Guo, X. Lin. An experimental study of ultrasonic vibration-assisted grinding of polysilicon using two-dimensional vertical workpiece vibration. Int. J. Adv. Manuf. Tech. **54** (2010)
7. A.Y. Yi, M. Hezlep, T. Pol. A computer controlled optical pin polishing machine. J. Mater. Process. Tech. **146** (2004)
8. E. Brinksmeier, O. Riemer, A Gessenharter. Finishing of structured surfaces by abrasive polishing. Precis. Eng. **30** (2006)
9. E. Brinksmeier, O. Riemer, A. Gessenharter, L. Autschbach. Polishing of Structured Molds. CIRP. Ann-Manuf. Techn. **53** (2004)
10. J. Guo, S-y. Morita, M. Hara, Y. Yamagata, T. Higuchi. Ultra-precision finishing of micro-aspheric mold using a magnetostrictive vibrating polisher. CIRP. Ann-Manuf. Techn. **61** (2012)
11. J. Guo, H. Suzuki, T. Higuchi. Development of micro polishing system using a magnetostrictive vibrating polisher. Precis. Eng. **37** (2013)
12. J. Guo, H. Suzuki, S-y. Morita, Y. Yamagata, T. Higuchi. A real-time polishing force control system for ultraprecision finishing of micro-optics. Precis. Eng. **37** (2013)
13. H. Suzuki, S. Hamada, T. Okino, M. Kondo, Y. Yamagata,et al. Ultraprecision finishing of micro-aspheric surface by ultrasonic two-axis vibration assisted polishing. CIRP. Ann-Manuf. Techn. **59** (2010)
14. H. Suzuki, T. Moriwaki, T. Okino, Y. Ando. Development of Ultrasonic Vibration Assisted Polishing Machine for Micro Aspheric Die and Mold. CIRP. Ann-Manuf. Techn. **55** (2006)

Improvement of Wearable Wrist Rehabilitation Device Using Flexible Pneumatic Cylinders

Naoki Kato, Shujiro Dohta, Tetsuya Akagi [a], Wataru Kobayashi and Mohd Aliff

Department of Intelligent Mechanical Engineering, Okayama University of Science, 1-1, Ridai-cho, Kita-ku, Okayama, 700-0005, Japan

Abstract. In the condition of the insufficient number of PT, the novel wearable rehabilitation devices using a soft pneumatic actuator for the elderly are actively proposed and developed. In the previous study, a wearable wrist rehabilitation device based on a flexible robot arm using three flexible pneumatic cylinders was proposed and tested. In this study, the improvement of the device is carried out so as to use it without limitation of attitude of patient's arm. The attitude control of the improved device based on the analytical model is also carried out. As a result, it is confirmed that the device can give the motion to patients with any attitude of their arm according to the desired position.

1 Introduction

Based on the situation of aging society and decreasing birth rate[1],it becomes serious concern that the number of welfare workers and physical therapists ("PT" for short) is not enough. The development of rehabilitation devices using soft pneumatic actuators for the disabled and elderly has been carried out actively [2-5]. In the previous study, the wrist rehabilitation device based on the flexible robot arm using three flexible pneumatic cylinders [6] was proposed and tested. The attitude control of the robot arm while the patient is holding the upper stage in the robot arm was executed [7] [8]. The robot arm that has a function of simple back-drivability was also proposed and tested [9]. In the therapy using the device, the patient's hands must be put on the robot on the table. Therefore, the wrist rehabilitation device that a patient worn it was also proposed and tested [10]. In the next step, in order to use it while the device is not fixed on the table, the improvement of attitude control system of the device is required. In this study, the improvement of attitude control system of the device using an embedded controller and two accelerometers is investigated. The flexible pneumatic cylinder is a kind of new soft actuator. This actuator has an advantage of flexibility based on the air compressibility and the usage of flexible tube. We believe that the flexible cylinder will contribute to extend the robotics application fields as proposed in this paper.

2 Flexible pneumatic cylinder and wearable wrist rehabilitation device

a Corresponding author : akagi@are.ous.ac.jp

The construction and movement of the flexible pneumatic cylinder which is a basic component in the proposed rehabilitation device is shown in Fig.1. The cylinder consists of these main parts; a flexible tube as a cylinder, the steel ball as a cylinder head, and a slide stage. The slide stage has two rollers set on the inner bore of the stage to press and deform the tube. The steel ball is held by two slide stages from both sides of the ball. When the pressure is applied to one side of the cylinder, the tube is moved forward or backward while holding the slide stage. This is a kind of rodless-type cylinder. The slide stage can move smoothly on the curved cylinder tube.

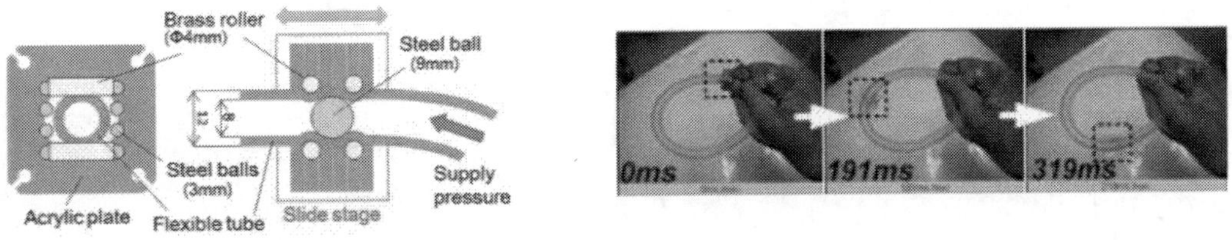

Figure 1. Construction and movement of flexible pneumatic cylinder.

Figure 2 shows the construction of the tested wearable type wrist rehabilitation device using flexible pneumatic cylinders that we had developed [11]. The device consists of three flexible pneumatic cylinders and two round stages. Each flexible pneumatic cylinder is arranged on the round stage with radius of 87.5 mm every 120 deg. from the center. An end of each flexible cylinder is fixed in the base stage. The end stage has a hole with the inner diameter of 100 mm so that a human arm can insert to it. On the end stage that has a handle, the slide stage of each cylinder is fixed. In rehabilitation, the patient arm passes through the holes of the base stage. The patient also holds the handle while working. In order to realize a wearable and portable device, the attitude control system that consists of an embedded controller (Renesas Co. Ltd., SH7125) and six quasi-servo valves[12] are mounted on the device. The quasi-servo valve consists of two on/off type valves. One is a switching valve for supply or exhaust and the other is PWM controlled valve for adjusting flow rate. The embedded controller and three quasi-servo valves are mounted on the base stage. On the end stage, there are three quasi-servo valves and an accelerometer for measuring the inclined angle of the stage. In order to supply the compressed air from the base stage to the end stage, a coil type tube that covers on a flexible pneumatic cylinder was used. By this method, the device requires only an electric power cable and an air supply pipe to drive it. The device has the outer diameter of 200 mm and the length of 420 mm. The total mass of the device including a controller and six quasi-servo valves is 1.14 kg.

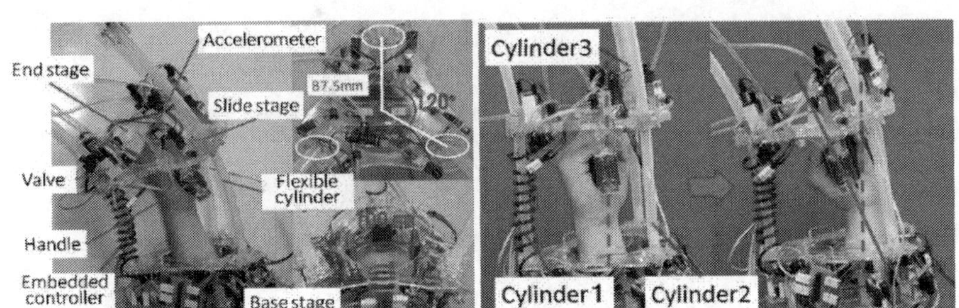

Figure 2. Construction of wearable wrist rehabilitation device using flexible pneumatic cylinders.

3 Attitude control system of tested device

Figure 3 shows the schematic diagram of the control system of the previous wrist rehabilitation device. The system consists of the flexible robot arm using three flexible pneumatic cylinders, an accelerometer, the embedded controller (Renesas Co. Ltd., SH7125), and six quasi-servo valves. In order to execute the attitude control of the device, it is necessary to measure the displacement of the flexible pneumatic cylinder. However, it is difficult to measure the displacement of the cylinder directly because of its flexibility. Therefore, indirect measurement based on the theoretical model of the arm was carried out.

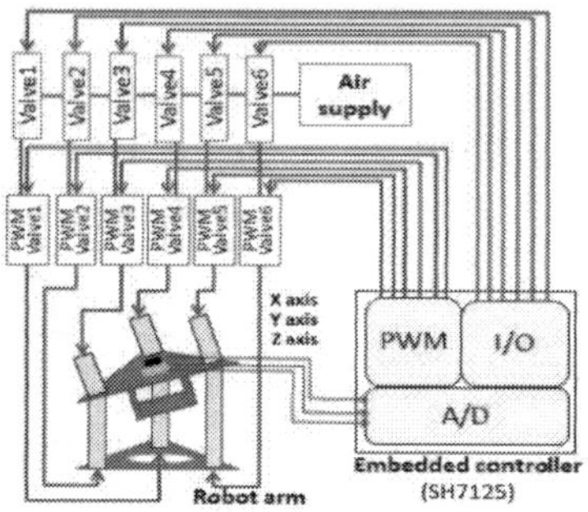

Figure 3. Schematic diagram of control system.

Figure 4 shows the definition of the cylinder length and the analytical model of the wearable wrist rehabilitation device for the control. In this model, the shape of the flexible pneumatic cylinder is assumed to be a circular arc when the device is bent. From the center of the rehabilitation device, the bending angle from X axis is called a bending direction angle α while the bending angle β is defined as the angle between the normal vectors from the center of the end stage of the Z axis of the device. From the geometric relationship, the following equations about cylinder length L_1, L_2, L_3 and the radius of curvature R can be obtained.

$$L_1 = (R - r \cdot \cos\alpha) \cdot \beta \tag{1}$$

$$L_2 = \{R - r \cdot \cos(\tfrac{2\pi}{3} - \alpha)\} \cdot \beta \tag{2}$$

$$L_3 = \{R - r \cdot \cos(\tfrac{4\pi}{3} - \alpha)\} \cdot \beta \tag{3}$$

$$R = L/\beta \tag{4}$$

where, L means the length of human arm, and r = 87.5 mm is the distance from the center to the cylinder. Angles α and β can be obtained from the output voltage of the accelerometer.

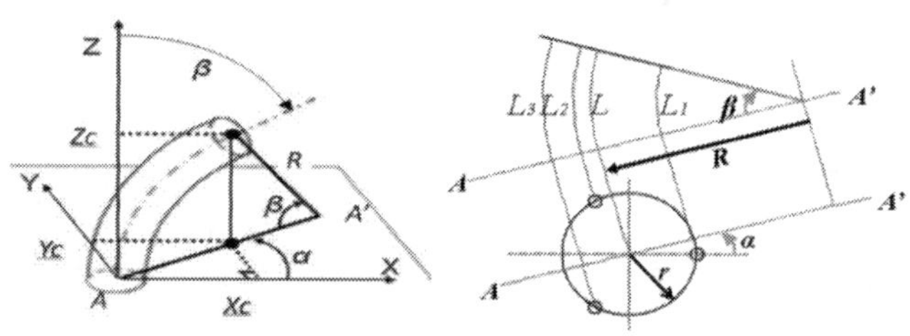

Figure 4. Analytical model of wearable device (flexible robot arm).

The attitude control of the device using the analytical model is done as follows. The embedded controller gets the output voltages from the accelerometer through A/D converter and angles α and β are calculated. Then, each length of the flexible pneumatic cylinder is calculated by Eqs.(1) to (4).

And, the embedded controller also calculates the deviation from the desired position for each cylinder. Finally, the quasi-servo valves are driven according to the control scheme. The desired position is set on the embedded controller in advance.

DOI: 10.1051/matecconf/20168202006

4 Attitude control result of the device

Figure 5 shows the transient response of each cylinder length in multi-position control. In the control, the following PID control scheme with sampling period of 7 ms was used.

$$u_{i(j)} = Kp \cdot d_{i(j)} + Kd \cdot \left(d_{i(j)} - d_{i(j-1)} \right) + Ki \cdot \sum d_{i(j)} \quad (5)$$

$$\begin{array}{ccc} & \text{Switching valves} & \\ \text{End stage : OFF,} & \text{Base stage : ON} & (u_{i(j)} > 0) \\ \text{End stage : ON,} & \text{Base stage : OFF} & (u_{i(j)} < 0) \\ \text{End stage : OFF,} & \text{Base stage : OFF} & (u_{i(j)} = 0) \end{array} \quad (6)$$

where, $d_{i(j)}$ and $u_{i(j)}$mean the current deviation of cylinder displacement and the differential input duty ratio for PWM valves, respectively. As an input duty ratio, 22.5 % is added to compensate the dead zone of PWM valves[11]. i shows the cylinder number 1, 2, and 3. The control parameters of Kp = 3.5 %/mm, Kd = 2.5 %/mm, and Ki = 0.05 %/mm were decided by trial and error.

In the experiment, the initial length L was set to be 150 mm. In Fig.5, the broken and solid lines show the desired and controlled length of each cylinder, respectively. It can be seen that the cylinder can trace the desired length well. For changing desired position, it reaches at the desired position within approximately one second even if the force from the human wrist is applied to the device as a load.

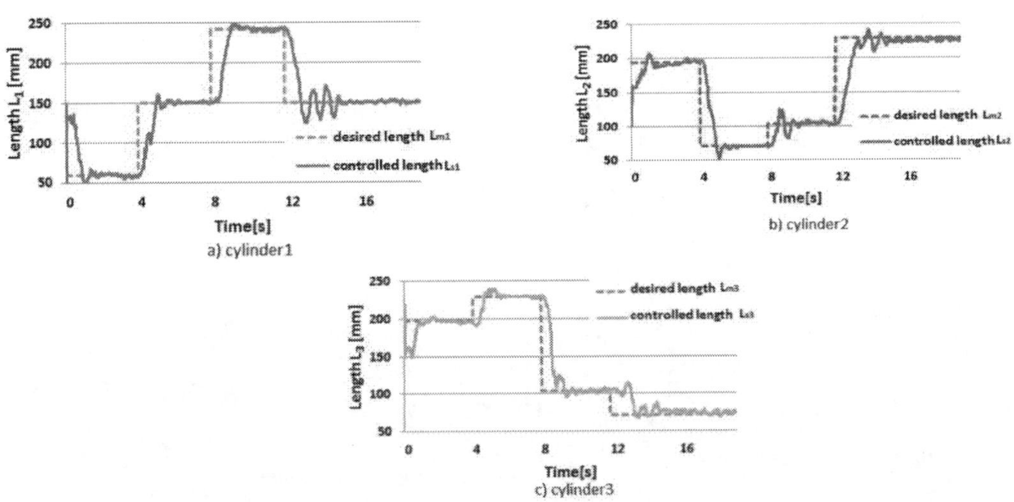

Figure 5. Transient response of cylinder length in multi-position control.

5 Improvement of wearable wrist rehabilitation device

Figure 6 shows the schematic diagram of the control system of the improved wearable wrist rehabilitation device. The fundamental construction of control system is similar to the device. Compared with the previous system, an accelerometer is added and set on the base stage. The attitude control of the device is carried out as follows. The embedded controller gets the output voltages from two accelerometers through A/D converter. From A/D value of x, y, and z direction of each accelerometer, a bending direction angle α and the bending angle β of each stage can be calculated. Differential bending angles α and β are given by Eq.(7) and Eq.(8), respectively.

$$\alpha = \alpha_e - \alpha_b \quad (7)$$

$$\beta = \cos^{-1}(\cos\beta_e \cdot \cos\beta_b - \sin\beta_e \cdot \sin\beta_b \cdot \cos\alpha) \quad (8)$$

where, subscripts e and b mean the end and base stages, respectively. Eq.(8) can be obtained by the geometical relationship between the base stage and the end stage. When $\alpha_e = \alpha_b$, that is $\alpha = 0$, the bending angle β can be expressed by the difference $\beta_e - \beta_b$. This is obviously obtained by Eq.(8). By using the calculated α and β, each length of the flexible pneumatic cylinder is calculated by Eqs.(1) to (4). The embedded controller also calculates the deviation from the desired position for each cylinder. The quasi-servo valves are driven according to the P control scheme.

Figure 7 shows the transient response of each cylinder length in multi-position control. In the experiment, the base stage has the inclined angle of approximately 20 deg. from the horizontal plane as an initial condition. In the control, PID control scheme with sampling period of 7 ms and the control parameters gain of $Kp = 3.5$ %/mm, $Kd = 3.5$ %/mm, and $Ki = 0.3$ %/mm was used. The same desired positions as the previous experiment were given. In Fig.7, the broken and solid lines show the desired and controlled length of each cylinder, respectively. It can be seen that the cylinder can trace the desired position. As a result, it can be confirmed that the device can give the motion to patient's wrist even if the patient's arm is not set parallel to the vertical plane. It can be observed that there are oscillation around the desired position. This seems to be caused by adding the accelerometer to the control system. The controlled performance will be improved by applying superior control scheme.

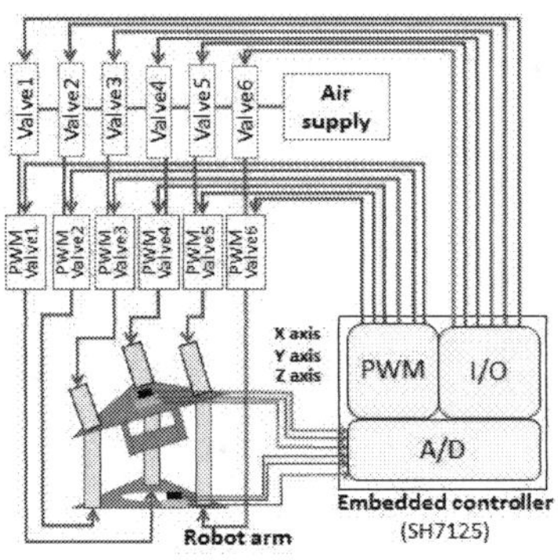

Figure 6. Schematic diagram of improved control system.

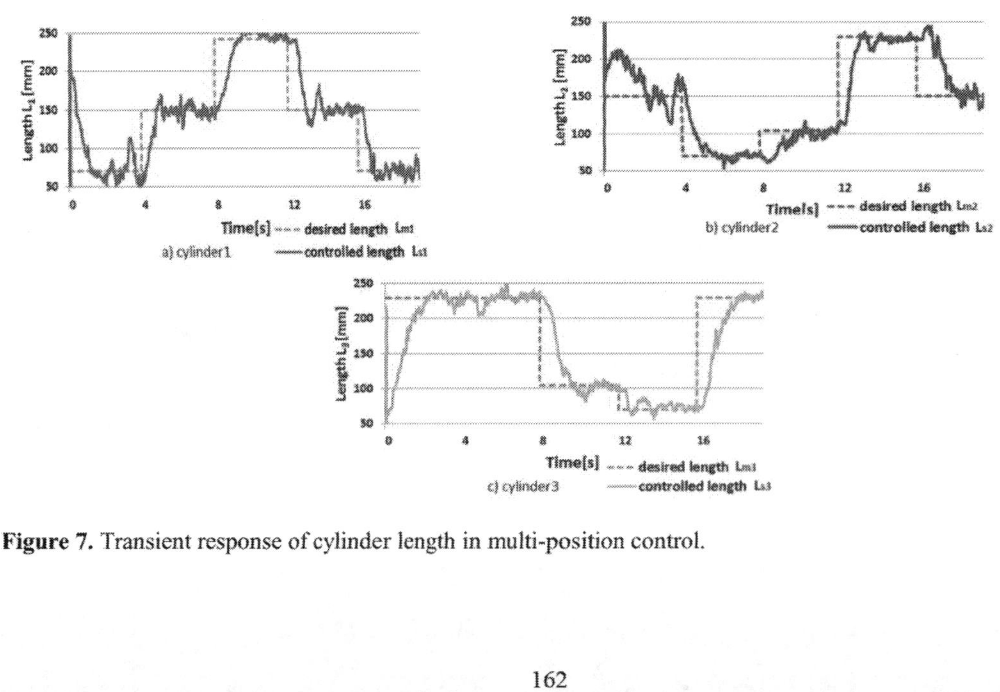

Figure 7. Transient response of cylinder length in multi-position control.

6 Conclusions

The improved wearable wrist rehabilitation device using three flexible pneumatic cylinders, an embedded controller, six quasi-servo valves, and two accelerometers is proposed and tested. The measuring system of differential bending angle between the base and the end stage of the device is also constructed in the built-in embedded controller. By using this measuring system, the attitude control of the improved device can be realized. As a result, it can be confirmed that the device can give the motion to patient's wrist even if the patient's arm is not set parallel to the vertical plane.

Acknowledgements

This research was supported by the Ministry of Education, Culture, Sports, Science and Technology of Japan through a Financial Assistance Program for QOL Innovative Research (2012-2016).

References

1. Ministry of Internal Affairs and Communications, Statistics Bureau, Statistics, Population Estimates, Result of the Population Estimates, Monthly Report. Result of the Population Estimates (online), Available: http://www.stat.go.jp/english/data/jinsui/tsuki/ (2015)
2. M. Ishii, K. Yamamoto, K. Hyodo. Stand-Alone Wearable Power Assist Suit -Development and Availability-, *Journal of Robotics and Mechatronics*, Vol.**17**, No.5, 575–583 (2005)
3. J. Piquion, A. Nayar, A. Ghazaryan, R. Papann, W. Klimek, and R. Laroia, Robot-assisted gynecological surgery in a community setting, *Journal of Robotics and Surgery*, Vol.**3**, Issue 2, 61–64 (2009)
4. T. Noritsugu, M. Takaiwa, D. Sasaki, Development of Power Assist Wear Using Pneumatic Rubber Artificial Muscles, *Journal of Robotics and Mechatronics*, Vol. **21**, No. 5, 607-613 (2009)
5. H. Kobayashi, T. Shiban, Y. Ishida, Realization of all 7 motions for the upper limb by a muscle suit, *Journal of Robotics and Mechatronics*, Vol. **16**, 504-512 (2004)
6. T. Akagi, S. Dohta, Development of a Rodless Type Flexible Pneumatic Cylinder and Its Application, *Transactions on Robotics and Automation of the JSME (C)*, Vol. **73**, No. 731, 2108–2114 (2007) (in Japanese)
7. T. Akagi, S. Dohta, Feifei Zhao, T. Fujikawa, Development and Attitude Control of Flexible Robot Arm Using Flexible Pneumatic Cylinder with Simple Structure, *International Journal of Automation Technology*, Vol.**5**, No.4, 523-530 (2011)
8. M. Aiff, T. Akagi, S. Dohta, Control and Analysis of Robot Arm Using Flexible Pneumatic Cylinder, *Mechanical Engineering Journal*, Vol.**1**, No.5, 1-13 (2014)
9. T. Morimoto, T. Akagi, S. Dohta, Development of Flexible Haptic Robot Arm Using Flexible Pneumatic Cylinders with Backdrivability for Bilateral Control, *Lecture Notes in Electrical Engineering 345,* Springer, 231-237 (2016)
10. T. Morimoto, M. Aiff, T. Akagi, S. Dohta, Development of Flexible Pneumatic Cylinder with Backdrivability and Its Application, *International Journal of Materials Science and Engineering*, Vol.**3**, No.1, 7-11 (2015)
11. N. Kato, S. Dohta, T. Akagia, T. Morimoto, Y. Soga, Development of Wearable Wrist Rehabilitation Device Using Flexible Pneumatic Cylinders, *MATEC Web of Conferences*, Vol.**51**, No.02003, 1-5 (2016)
12. F. Zhao, T. Akagi, S. Dohta, Development of Small-Sized Quasi-Servo Valve for Flexible Bending Actuator, *Proc. 2009 CACS Int. Automatic Control Conference*, 558-562 (2009)

MATEC Web of Conferences **82**, 02007 (2016) DOI: 10.1051/matecconf/20168202007
D2ME 2016

Microstructure and Tribological Behaviour of CrCuFeNiTi High Entropy Alloys

Kehinde Sobiyi[1,3,a], Michael Bodunrin[2] , Esther Akinlabi[3] and Babatunde Obadele[4]

[1]*School of Chemical and Metallurgical Engineering, University of the Witwatersrand, South Africa*
[2]*Department of Metallurgical and Materials Engineering, Federal University of Technology, Akure, PMB 704, Nigeria*
[3] *Department of Mechanical Engineering Science, University of Johannesburg, South Africa*
[4] *Department of Chemical Engineering, University of Johannesburg, South Africa*

Abstract. The microstructure, hardness and wear performance of CrCuFeTiNi high entropy alloy were evaluated. The alloy was produced by mixing constituent elements, consolidated and melted using vacuum arc furnace. The X-ray diffraction analysis confirmed that the alloy is composed of FCC and BCC simple phases. The microstructure contains three regions that were rich in Cu, Cr and Fe which resulted from segregation of constituent elements. The Vickers micro hardness of the alloy was determined as $HV_{0.5} = 510 \pm 7$ MPa. The wear performance of the alloy was also studied using WC balls under two load conditions. The volume loss was evaluated, accompanied by analysis of the wear tracks and debris using SEM images and EDS. The main wear mechanisms were ploughing, adhesion and oxidation-dominated wear.

1. Introduction

New design concept in alloy development was established independently by Yeh [1] and Cantor [2] about eleven years ago. The concept involved the production of solid solution metallic systems containing a minimum of five elements in equiatomic or equimolar ratios. These metallic systems are referred to as high entropy alloys (HEAs). Unlike in traditional metallic systems, where one element serves as the matrix and other alloying elements are carefully added in specified proportions, to avoid the formation of unwanted phases, HEAs are known to have simple structures such as face centered cubic (FCC), body centered cubic (BCC) or a mixture of both (FCC+BCC) structures [3]. This unique attribute of HEAs has been ascribed to high entropy of mixing which often suppresses the formation of metastable and complex phases [4].

In previous studies, it was reported that HEAs are becoming increasingly attractive for use in several engineering fields due to superior mechanical [5], corrosion [6] and tribological properties [7] over the conventional alloys. In addition, the ability of HEAs to retain some of its unique properties at high temperatures has made the alloys a potential substitute for nickel based super alloys in high temperature environments as well as for making machine tools where creep and wear resistant materials are desirable.

In this research, an attempt has been made to develop new CrCuFeTiNi high entropy alloy. The microstructural features and wear performance were examined.

[a] Corresponding author : kk_sob@yahoo.com

2. Experimental Procedure

Elemental powders of Cr, Cu, Fe, Ni and Ti (purity 99.5 wt %) were mixed in a planetary ball mill for 6 hrs at 250 rpm. The mixed powder compositions was compacted and arc-melted in a high-purity argon atmosphere. The melting process was repeated three times to improve the homogeneity of the alloys.

The microstructure of the as-cast samples were observed using a scanning electron microscopy (SEM, TESCAN), equipped with Oxford instrument energy dispersive X-ray spectrometer (EDS). The alloys phase constitutions were analysed with an X-ray diffractometer (X-bunker) with Cu Kα radiation operated at 40kV, 30mA. Vickers testing machine (EMCOTEST®) was used to determine the microhardness of the sectioned and polished samples under a test load of 500 g/f and dwell time of 10 seconds. Fifteen indentations were applied per sample and the average values were recorded.

Dry sliding wear test was performed on the samples using the ball on disk tribometer (CETR UMT-2, now Bruker Nano Inc. Campbell, CA). Tungsten carbide ball of diameter 9.5 mm was applied at normal loads of 25 N and 35 N for a period of 1000 seconds at a frequency of 5 Hz. The wear volume loss was plotted. The sample worn surfaces were analysed under the SEM. The schematic diagram of the experimental procedure is shown in Figure 1.
Wear volume loss of material on the substrate is calculated from Equation 1.

$$V_{total} = V_A + V_B = L\left\{r^2 \sin^{-1}\left(\frac{W}{2r}\right) - \frac{W}{2}\left(r^2 - \frac{W^2}{4}\right)^{1/2}\right\} + \frac{\pi}{3}\left\{2r^3 - 2r^2\left(r^2 - \frac{W^2}{4}\right)^{1/2} - \frac{W^2}{4}\left(r^2 - \frac{W^2}{4}\right)^{1/2}\right\} \quad (1)$$

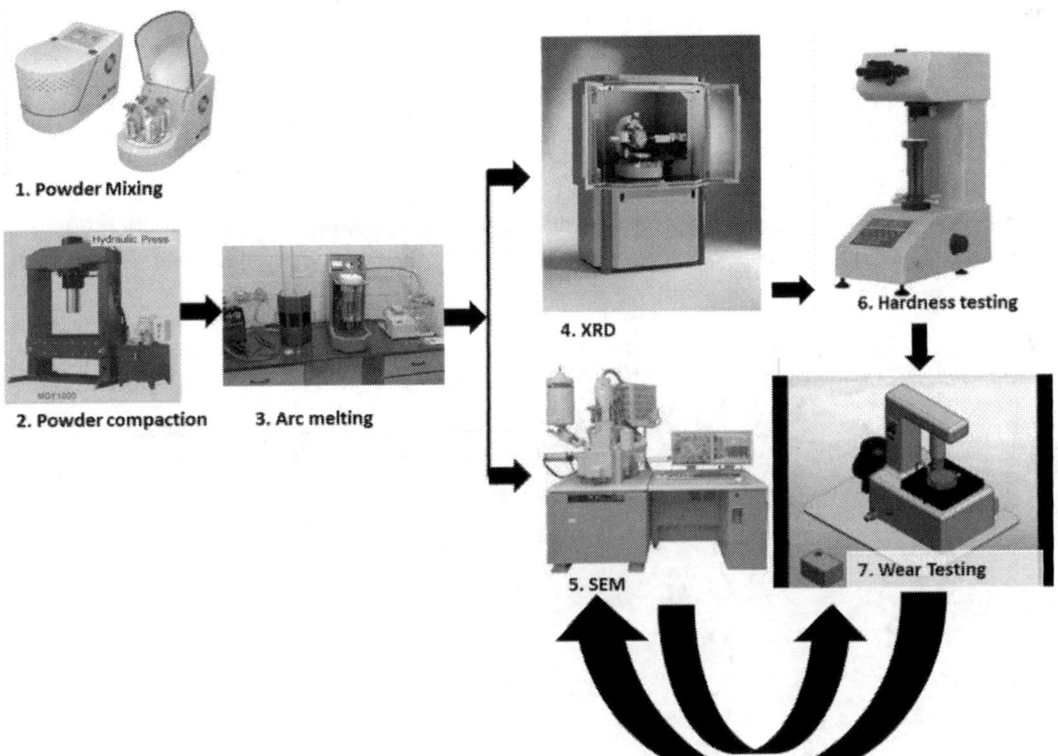

Figure 1. Schematic diagram of the experiments

3 Results and discussions

3.1 Phase analysis and microstructure of the alloys

The XRD pattern of the CrCuFeNiTi high entropy alloy is shown in Figure 2. It can be observed that the CrCuFeNiTi alloy composed mainly of simple phases of face centered cubic (FCC), and body centered cubic (BCC). The presence of Ti in the alloy helps to strengthen the alloy due to its large negative enthalpy of mixing with the other elements in the composition. The phase formation is determined by the competition between the enthalpy and entropy [7].

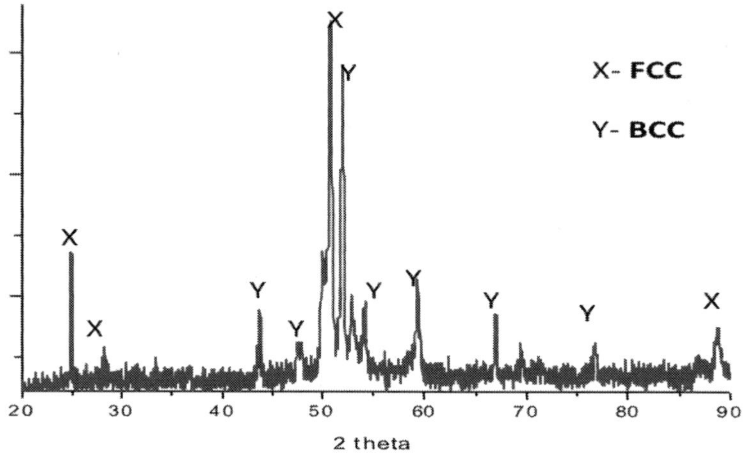

Figure 2. XRD pattern of the CrCuFeNiTi high entropy alloys

The SEM/BSE image of the CrCuFeNiTi alloy is presented in Figure 3. It is observed that the microstructure contains three majors regions (region I-light, region II greyish dark and region III-grey) with an agglomeration of black dot-like structure dispersed mostly within the dark-greyish and grey regions. EDS spectra for each region is shown in Fig. 2/ Table 1, region I which is the light phase is rich in Cu, Cr is/are dominant in region II (greyish dark) while region III (grey) is predominantly Fe. The high amount of different elements contained in different regions confirmed segregation of elements during the production of the alloy. Cu segregation in $Al_xCoCrCuFeNi$ HEA has been reported by [8].

The microhardness of the alloy is 510 $\pm$7 $HV_{0.5}$. The hardness is as a result of the presence of titanium in the alloy, thus enhancing expansion lattice [9].

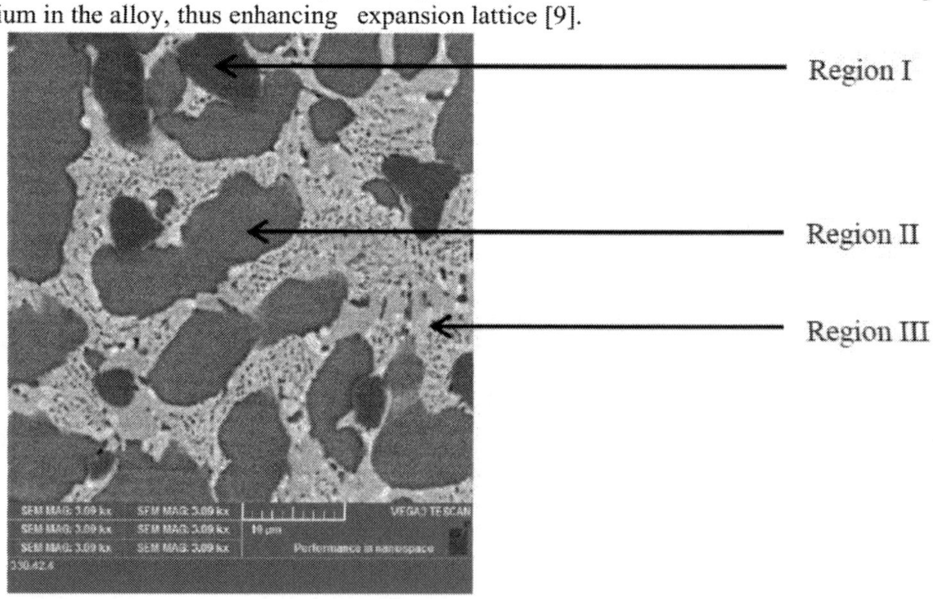

Figure 3. SEM micrograph of the CrCuFeNiTi high entropy alloy

3.2 Wear resistance behaviour

To understand the wear mechanism on the CuCrFeNiTi alloy against WC balls during reciprocating sliding condition, the wear tracks on the surfaces of the alloy under two different loads of 25 N and 35 N were observed by SEM as shown in Figure 3. Wear tracks comprised abraded regions with continuous microgrooves parallel to the direction of sliding. The presence of plastic deformation in the sliding contact area is responsible for the surface morphology [10]. There were many accumulated fine and irregular wear debris particles, delaminated scars on the worn tracks and micropatches within the worn zone. From the EDS point analysis of the debris, elemental oxygen, iron and titanium were detected which showed that the alloy experienced considerable oxidation (presence of oxygen) as evidenced by the formation of materials accumulation regions as shown in Fig. 4d. The main wear mechanism is in the form of oxidation-dominated wear as evident by the presence of oxygen in the debris and also adhesion wear with ploughing.

The wear volume loss plot against the applied load is shown in Figure 5. There was a marginal increase in the material volume loss from about 0. 35 mm^3 to 0. 36 mm^3 as the applied load is increased from 25N to 35N.

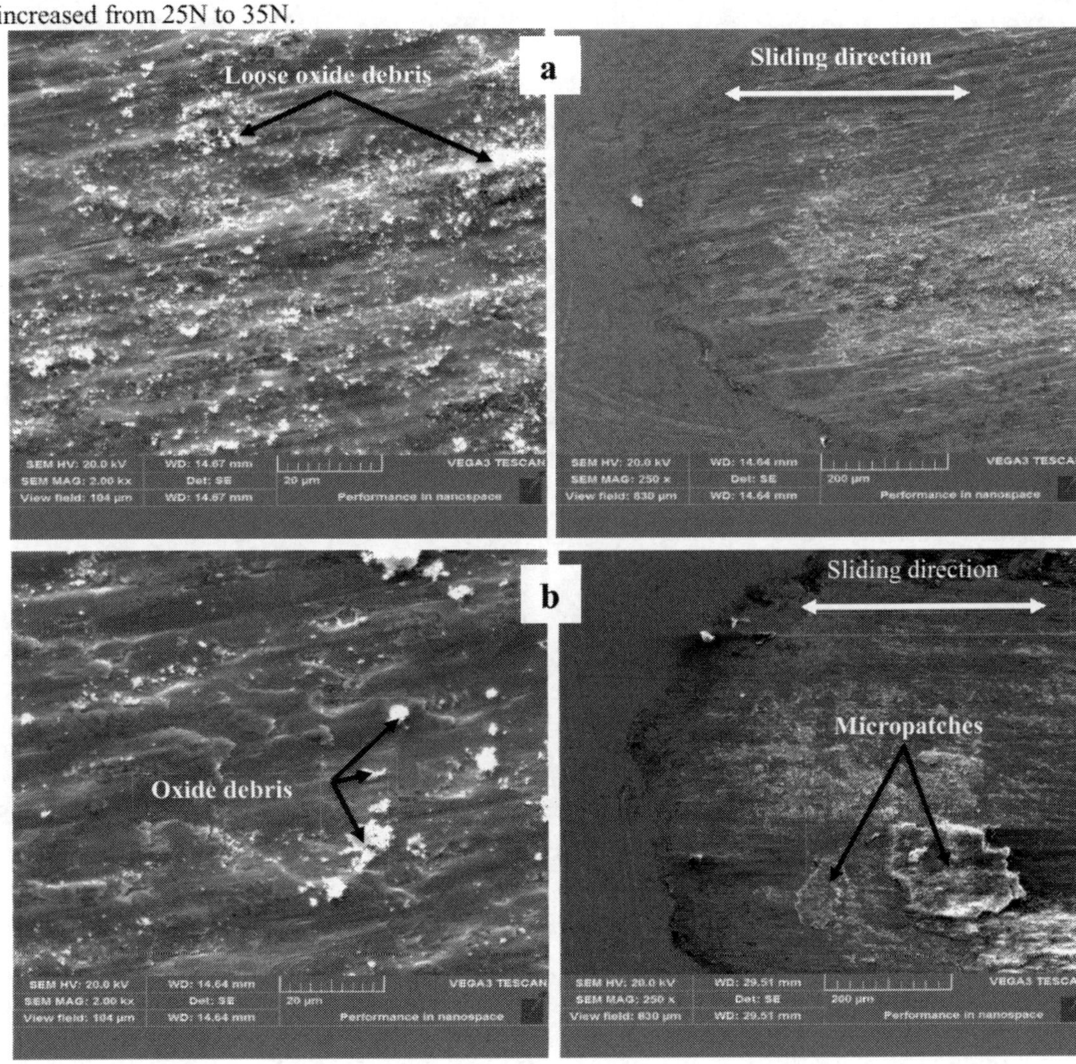

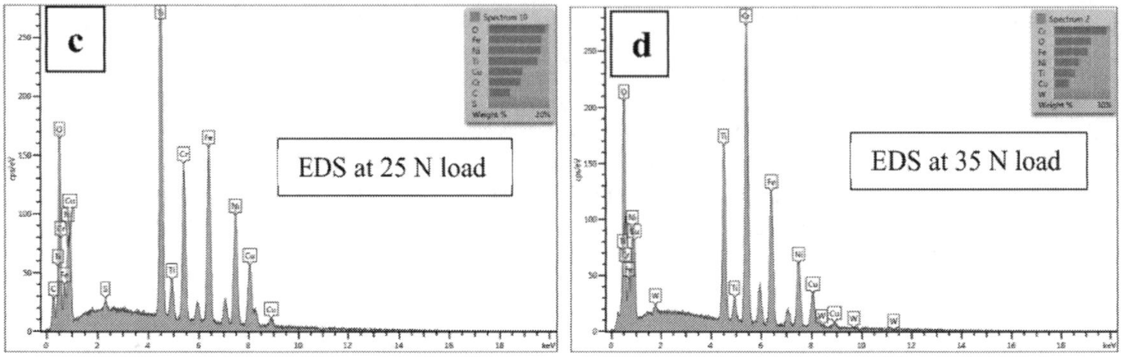

Figure 4. SEM images of the worn surface and the corresponding EDS spectra of CuCrFeNiTi alloy under applied load of (a,c) 25 N and (b,d) 35 N.

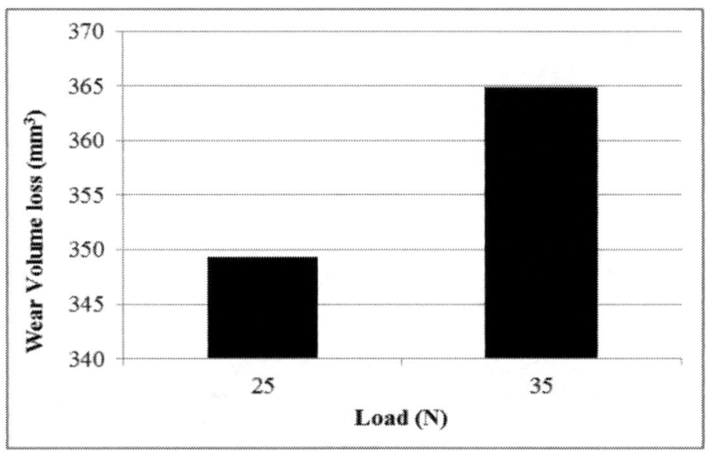

Figure 5. Wear volume loss of the alloy under different loads

4 Conclusions

The study investigated the microstructure and the wear performance of CrCuFeTiNi high entropy alloy produced by arc melting. The following results were obtained:

- The alloy consisted of simple BCC and FCC phases. It is also observed that the microstructure contains three majors regions; Fe rich, Cr rich and Cu rich regions.
- Vickers microhardness of the alloy is determined as 510 ± 7 $HV_{0.5}$. The hardness is as a result of the presence of titanium in the alloy.
- Wear tracks comprised abraded regions with continuous microgrooves parallel to the direction of sliding. The main wear mechanism is in the form of oxidation-dominated wear as evident by the presence of oxygen in the debris and also adhesion wear with ploughing.

References

1. J.W. Yeh, S.K. Chen, S.J. Lin, J.Y. Gan, T.S. Chin, T.T. Shun, C.H. Tsau, S.Y. Chang, Adv. Eng. Mater., **6**, 299–303 (2004)
2. B. Cantor, I. Chang, P. Knight, A. Vincent, Mater. Sci. Eng. A, 375–377 (2004), pp. 213–218
3. H. Zhang, Y. Pan, Y. He, H. Jiao, Appl. Surf. Sci, **257**, 2259–2263 (2011)
4. T.T. Zuo, X. Yang, P.K. Liaw, Y. Zhang, Intermetallics, **67,** 171–176 (2015)
5. F.J. Wang, Y. Zhang, G.L. Chen, H.A. Davies, Int. J. Mod. Phys. B, **2**, 1254–12593 (2009)

6. X.W. Qiu, Y.P. Zhang, Mater. Sci. Eng. Power Metall., 17, 377–382 (2012)
7. J. M. Wu, S. J Lin, J W Yeh, S. Chen, Y haung, H Chen, Wear, **261**, 513–519 (2006)
8. M.Tsai and J. Yeh, Mater. Res. Lett., **2,** 107–123 (2014)
9. W. Wang , W. Wang b , J. Yeh, J. Alloys Comp., **589**, 143–152 (2014)
10. S. Ge, S. Wang, N. Gitis, M. Vinogradov and J. Xiao, Wear, **264,** 571 – 578 (2008)

MATEC Web of Conferences **82**, 02008 (2016) DOI: 10.1051/matecconf/20168202008
D2ME 2016

Axial Crushing and Energy Absorption of Empty and Foam Filled Jute-glass/ Epoxy Bi-tubes

Asad A. Khalid[1,a] and Maziri Morsidi[1]

[1]*Department of Mechanical Engineering, Faculty of Engineering, Universiti Teknologi Brunei, Brunei*

Abstract. Experimental work on the axial crushing of empty and polyurethane foam filled bi-tubular composite cone-tube has been carried out. Hand lay-up method was used to fabricate the bi-tubes using woven roving glass, jute and hybrid jute-glass/epoxy materials. The tubes were of 56 mm diameter, and the cones top diameters were 65 mm. Cone semi-apical angles of 5°, 10°, 15°, 20° and 25° were examined. Height of 120 mm was maintained for all the fabricated specimens. Effects of material used, cone semi apical angle and foam filler on the load-displacement relation, maximum load, crush force efficiency, and the specific energy absorption and failure mode were investigated. Results show that the foam filler improved the progressive crushing process, increased the maximum load and the absorbed energy of the bi-tubes. The maximum crushing load and the specific energy absorption increased with increasing the cone semi apical angle up to 20° for the empty bi-tubes and up to 25° for the foam filled bi-tubes. Progressive failure mode with fiber and matrix cracking was observed at the top narrow side of the fractured bi-tubes as well as at the bottom surface of 20° and 25° cone semi-apical angle bi-tubes.

1 Introduction

Crashworthiness analysis is an important factor to design structures and to determine the capability of the structure to absorb energy and to protect vehicle occupants during collision event [1]. Fiber-reinforced composite materials are used now in automotive industry to provide better crashworthiness with using lightweight materials [2]. The crush behaviour of composite structures offers distinct advantages for automotive applications and has been the subject of numerous investigations [3]. Several studies performed on composite structures like tubes [3–10] and cones [11–14] have been carried out to investigate the energy absorption capability and failure mechanism of these structures at different testing conditions. Researchers found that, high energy absorption can be obtained during the progressive crushing of composite tubes and cones under axial loading. In addition, researchers found that the stability, mode of failure and energy absorption of thin-walled tubular components can be improved by using foam fillers [10].

The mechanism by which the structure collapses has a direct influence on the energy absorption capability [11]. In this paper, a new foam filled bi-tubular arrangement is used for the aim of improving the crashworthiness under axial crushing. The main objectives of the present research are

[a] Corresponding author : asad.khalid@utb.edu.bn

to investigate the crushing characteristics, specific energy absorption and failure mode of empty and foam filled bi-tubular cone-tube under axial compression loading. Also, the research aims to investigate the effect of cone-semi apical angle and material used on the performance of the bi-tubular arrangement.

2 Experimental Work

2.1 Materials and Fabrication

Composite cone and tube specimens were fabricated by hand lay-up method using woven roving jute and glass fibers with epoxy resin matrix. Cone and tube moulds were fabricated by using zinc coated mild steel sheets. Then the sheets were bent and welded to form the tube cone shape specimens. Plywood was inserted inside these hollow moulds for support purpose. A PVC circular cross-section sheets were cut with diameters equal to the top and bottom diameters of the cone and the tube and then fixed at the mould sides. The plywood was made in a way that can be easily extracted from these moulds after the completion of the fabrication process. Figure 1 shows the geometrical configuration of the fabricated bi-tubular cone-tube.

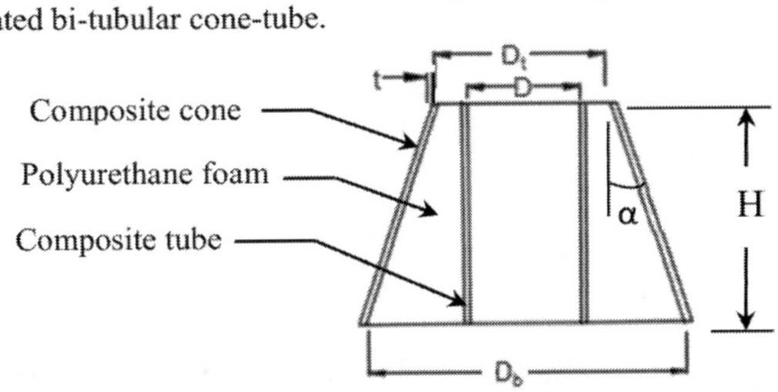

Figure 1. Geometrical configuration of the bi-tubular cone-tube

Where; t is the wall thickness, D is the tube diameter, D_t and D_b are respectively the top and bottom diameters of the cone. Table 1 shows the dimensions of the fabricated bi-tubular specimens. (α) is the cone semi-apical angle and H is the height of the specimen.

Table 1. Dimensions of the cone and tube arrangements

Specimen type		Outer cone			Tube		(H), mm
	(α), deg	(D_t), mm	(D_b), mm	Number of layers	(D), mm	Number of layers	
bi-tubular cone-tube	5	65.63	86.62	6	56	6	120
	10	65.63	107.95	6	56	6	120
	15	65.63	129.94	6	56	6	120
	20	65.63	152.98	6	56	6	120
	25	65.63	177.54	6	56	6	120

Tensile tests on composite specimens were performed to determine the mechanical properties of jute and glass/ epoxy materials. The properties of jute/epoxy specimens were E_{11}=17.68 GN/m^2, G_{12}=3.0 GN/m^2, G_{23}=2.86 GN/m^2, ν_{12}=0.29 and ρ = 1600 kg/m^3. The properties of the glass/epoxy specimens were E_{11}=52.25 GN/m^2, G_{12}=3.08 GN/m^2, G_{23}=4.20 GN/m^2, ν_{12}= 0.35 and ρ= 2100 kg/m^3. Where E_{11} is the modulus of elasticity in the longitudinal direction, G_{12} and G_{23} are the In-plane and out of plane modulus of rigidity, ν_{12} is the minor Poisson's ratio and ρ is the density. For tubes with foam filler, polyurethane foam of 20 kg/m^3 density was used to fill the gap between the inner tube and the outer cone.

2.2 Testing Procedure.

Constant slow speed compression testing was performed using a computer-controlled servo-hydraulic Instron machine type 5584 (Figure 2). The cross-head speed was adjusted at 4 mm/min. Composite bi-tubular cone tube specimens were axially crushed between two parallel steel flat platens, one is static and one is moving. The fixed platen was fitted with a load cell from which the load signal was taken directly to the computer. For each test, the crush load was plotted on the Y-axis and the crosshead displacement on the X-axis.

Figure 2. Instron machine type 5584 and test fixture.

3 Results and Discussion

3.1 Crushing Characteristics

Load-displacement graphs for the tested empty and foam filled bi-tubes are shown in Figure 3. Cone semi-apical angle of 5° was used. The terms G, JG and J refer to the glass, jute-glass and jute materials respectively. The terms E and F refer to the empty and foam filled bi-tubes. It can be seen from this Figure that the load increased linearly till the initial failure load and then fluctuated with a slight increase till failure. The glass/ epoxy empty bi-tubes shown in Figure 3a supported load higher 31.28% and 48.78% than the hybrid jute-glass and jute/epoxy empty bi-tubes respectively. The foam filled glass/ epoxy bi-tubes shown in Figure 3b supported load higher 27.48% and 45.88% than the hybrid jute-glass and jute/epoxy foam filled bi-tubes respectively. The foam filled glass, jute-glass and jute/ epoxy bi-tubes supported load higher 16.69%, 21.05% and 21.15% than the empty glass, jute-glass and jute epoxy tubes respectively.

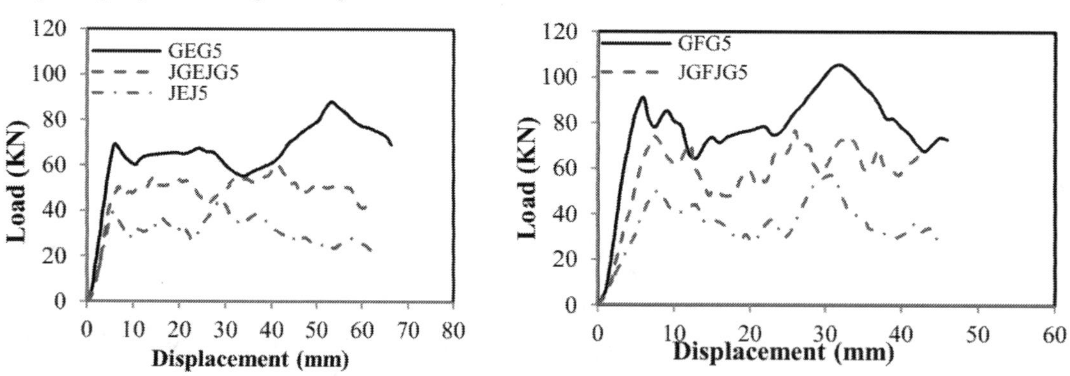

(a) (b)

Figure 3. Load-displacement graphs for the (a) empty bi-tube and (b) foam filled bi-tube (Cone semi-apical angle = 5°)

Figure 4 shows the load-displacement graphs of the empty and foam filled bi-tubes with semi-apical cone angle of 25°. As shown, the foam filled glass, jute-glass and jute/ epoxy bi-tubes

 DOI: 10.1051/matecconf/20168202008

supported load higher respectively 37.97%, 38.98% and 46.40% than the empty glass, jute-glass and jute/ epoxy bi-tubes. The increase of cone semi apical angle from 5° to 25° increased the maximum load 25.25%, 32.20% and 33.40% respectively for the glass, jute-glass and jute/ epoxy foam filled bi-tubes. The increase in the load is attributed to the increase in the cross section area of the bi-tubes with the increase in the cone semi-apical angle and crush distance. With the increase in the cone semi apical angle from 5° to 20°, the maximum load increased 11.72%, 19.52% and 18.70% respectively for the glass, jute-glass and jute/ epoxy empty bi-tubes. Further increase in the cone semi-apical angle of the empty bi-tubes, decreased the maximum.

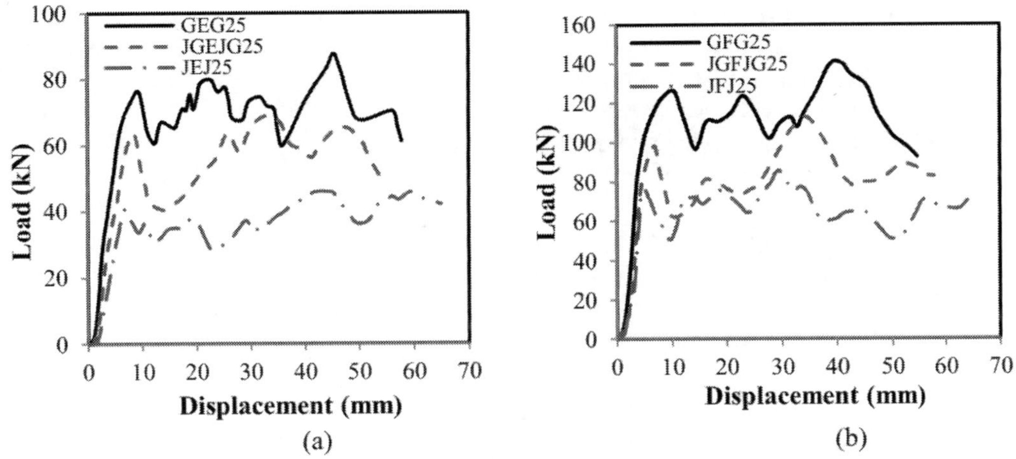

Figure 4. Load-displacement graphs for the (a) empty bi-tubes and (b) foam filled bi-tubes (cone semi-apical angle = 25°)

Figure 5 shows effect of cone semi-apical angle on the maximum axial compression load of the empty and foam filled bi-tubes. Cone semi-apical angles of 5°, 10°, 15°, 20° and 25° were examined. It can be seen that the maximum load increased with the increase in the cone semi-apical angle from 5° to 20° and from 5° to 25° for the empty and foam filled bi-tubes respectively. The foam filler provide stability to the bi-tubes to withstand the applied load. Maximum load values of 141.16 kN, 112.84 kN and 85.73 kN respectively were obtained from the glass, jute-glass and jute/ epoxy foam filled bi-tubes of 25° degree cone semi apical angle. The empty glass, jute-glass and jute/ epoxy bi-tubes supported maximum load values of 99.56 kN, 75.05 kN and 55.37 kN respectively for the 20° cone semi apical angle.

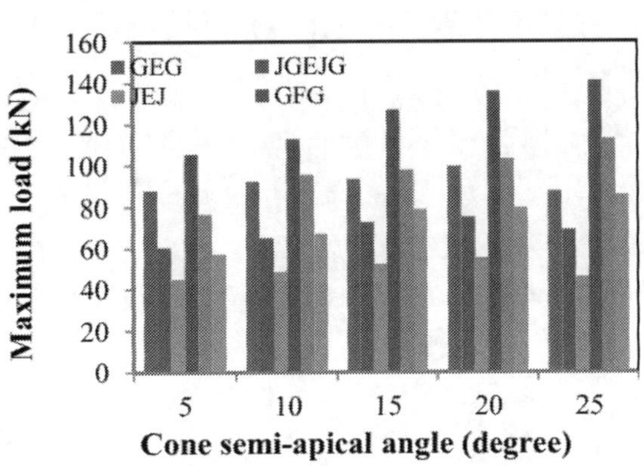

Figure 5. Maximum load versus cone semi-apical angle for the empty and foam filled bi-tubes

3.2 Crush Force Efficiency

The crush force efficiency (CFE) is important factors to measure the crush performance and to evaluate the crashworthiness of the energy absorber component. The CFE for the composite tubes under lateral loading is determined using the following equation [2,5].

$$CFE = \frac{F_{mean}}{F_{max}} \qquad (1)$$

Where F_{mean} and F_{max} are the mean and maximum crush failure loads respectively.
Maximum absorption capacity is obtained when the crush force efficiency approach to unity cartographic damage [15]. On the other hand as the crush force efficiency decreases, the absorption capacity decreased resulted in a catastrophic failure [16].
Figure 6 shows the crush force efficiency of the axially crushed bi-tubes. As shown, crush force efficiency of the foam filled bi-tubes is significantly higher than the empty bi-tubes. Maximum crush force efficiency of 0.91, 0.86 and 0.84 were obtained by the foam filled glass, jute-glass and jute/ epoxy bi-tubes respectively. Maximum crush force efficiency of 0.82, 0.79 and 0.74 were obtained from the empty glass, jute-glass and jute/ epoxy bi-tubes respectively.

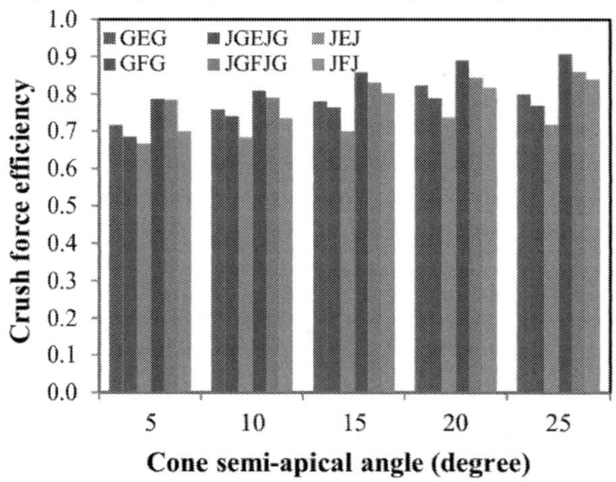

Figure 6. Crush force efficiency versus cone semi-apical angle for the empty and foam filled bi-tubes

3.3 Specific Energy Absorption

The specific energy absorption (SEA) is defined as the total absorbed energy (E_t) per unit mass (m) of the absorbed structure as in the following equation [4, 8, 13] :

$$SEA = \frac{E_t}{m} \qquad (2)$$

Where the total absorbed energy is the work done by the crushing force represented by the area under the axial force versus axial displacement curve and can be determined as follows [4, 8, 13] :

$$E_t = \int_0^{\delta_{max}} F d\delta \qquad (3)$$

Where *F* is the crush force in axial direction, and δ is the displacement in axial direction.
The specific energy absorption of the axially crushed tubes is shown in Figure 7.

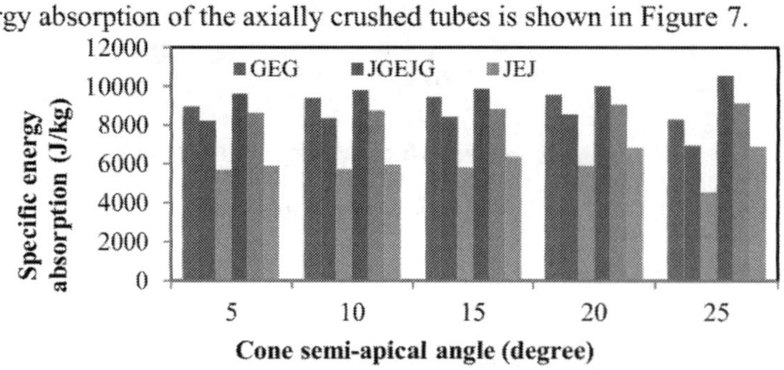

Figure 7. Specific Energy absorption of the axially crushed empty and foam filled bi-tubes

 DOI: 10.1051/matecconf/20168202008

It can be seen that, the specific energy absorption increased with the increase in the cone semi-apical angle. The energy absorbed by the foam filled bi-tubular cone-tubes is found higher than that obtained by the empty bi-tubes for all the cone semi-apical angles tested. Maximum specific energy absorption values of 1056 J/kg, 9115 J/kg and 6885 J/kg were obtained by the glass, jute-glass and jute/epoxy foam filled bi-tubes respectively. Maximum specific energy absorption values of 9562J/kg, 8553 J/kg, and 5902 J/kg were obtained by the empty bi-tubes of glass, jute-glass and jute/epoxy respectively

3.4 Failure Mode

Figure 8 shows the fractured specimens of empty and foam filled bi-tubular cone-tubes tested under axial compression load. The fracture initiated at the narrow upper end of the specimens at the region of the highest stress. It has been found that more fiber splitting obtained from the foam filled bi-tubes. The more fiber splitting provides stable crushing process with higher load to cause failure. Fiber break and matrix cracking were observed at the bottom end of the empty bi-tubular specimens of 20° and 25° cone semi apical angles. Samples of the fractured specimens are shown in Figure 9. The specimens were of 5° cone semi-apical angle. It has been found that the foam filled bi-tubular specimens fractured by progressive crushing splaying mode.

	Failure mode			
Specimen type	Empty bi-tube		Foam filled bi-tube	
α, (degree)	5	25	5	25
Jute/ epoxy				
Jute-glass/ epoxy				
Glass/ epoxy				

Figure 8. Failure mode of the bi-tubular cone-tube specimens

Empty bi-tube		Foam filled bi-tube	
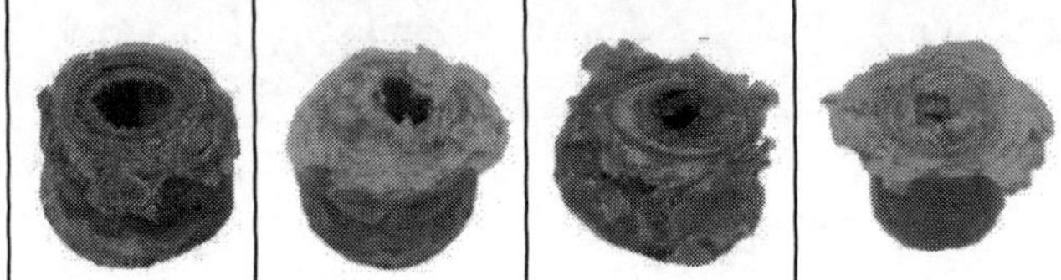			

Figure 9. Comparison of the failure mode (cone semi-apical angle= 5°)

4 Conclusions

The main conclusions that could be drawn from this investigation are:

1. The maximum load, crush force efficiency and the energy absorption increased with increasing cone semi-apical angle up to 20° for the empty bi-tubes and up to 25° for the foam filled bi-tubes.

2. Maximum axial load of 141.16 kN, 112.84 kN and 85.73 kN were obtained from the foam filled glass, Jute-glass and jute/ epoxy bi-tubes respectively with the cone semi-apical angle of 25°. The maximum loads obtained by the empty bi-tubes were 99.56 kN, 75.05 kN and 55.37 kN respectively with the cone semi-apical angle of 20°.
3. The crush force efficiency of the foam filled bi-tubes is higher than the empty bi-tubes. Maximum crush force efficiency of 0.91, 0.86 and 0.84 were obtained from the foam filled glass, jute-glass and jute/ epoxy bi-tubes. The values were 0.82, 0.79 and 0.74 for the empty bi-tubes.
4. Maximum energy absorption of 10561 J/kg, 9115 J/kg, and 6885 J/kg were obtained from the foam filled glass, jute-glass and jute/ epoxy bi-tubes respectively with the cone semi-apical angle of 25°. The maximum energy absorbed by the empty bi-tubes were; 9562 J/kg, 8553 J/kg, and 5902J/kg respectively with the cone semi-apical angle of 20°.
5. The bi-tubular specimens were fractured by progressive failure splaying mode with fiber and matrix cracking at the top narrow side of the specimens. Further cracks were noticed at the bottom surface of 20° and 25° cone semi-apical angle bi-tubes.

References

1. P.D. Bois, C.C. Chou, B.B. Fileta, T.B. Khalil,, A.I. King, Michigan: American Iron and Steel Institute (2004)
2. R. A. Eshkoor, S. A. Oshkovr, A. B. Sulong, R. Zulkifli, A. K. Ariffin, and C. H. Azhari, Mater. Des., **47**, 248–257 (2013)
3. D. Hull, Compos. Sci. Technol., **40**, 377–421 (1991)
4. R. A. Eshkoor, A. U. Ude, S. A. Oshkovr, A. B. Sulong, R. Zulkifli, A. K. Ariffin, and C. H. Azhari, Int. J. Impact Eng. **64**, 53–61 (2014)
5. S. A. Oshkovr, R. A. Eshkoor, S. T. Taher, A. K. Ariffin, and C. H. Azhari, Compos. Struct. **94**, 2337–2342 (2012)
6. A. Zuraida, A. A. Khalid, and A. F. Ismail, Mater. Des., **28**, 71–77 (2007)
7. A. A. Khalid, Sahari B.B., Y.A. Khalid, Inst. Eng., **61** (1) 1–12 (2000)
8. M. Mirzaei, M. Shakeri, M. Sadighi, and H. Akbarshahi, Compos. Struct. **94**, 1959–1966 (2012)
9. M. Haghi Kashani, H. Shahsavari Alavijeh, H. Akbarshahi, and M. Shakeri, Mater. Des., **51**, 1095–1103 (2013)
10. L. Mirfendereski, M. Salimi, and S. Ziaei-Rad, Int. J. Mech. Sci. **50**, 1042–1057 (2008)
11. A. A. Khalid, B. B. Sahari, and Y. A. Khalid, Compos. Sci. Technol. **62**, 17–27 (2002)
12. M. Kathiresan, K. Manisekar, and V. Manikandan, Compos. Struct. **94**, 3510–3519 (2012)
13. M. A. Guler, M. E. Cerit, B. Bayram, B. Gerçeker, and E. Karakaya, Int. J. Crashworthiness, **15** (4) 377–390 (2010)
14. M. Kathiresan, K. Manisekar, and V. Manikandan, Compos. Struct. **108**, 584–599 (2014)
15. F. Tarlochan and S. Ramesh, Compos. Struct. **94**(3) 904–916 (2012)
16. L. Guoxing and Y. Tongxi, England, Woodhead Publishing Limited, 1-23 (2003)

MATEC Web of Conferences **82**, 02009 (2016) DOI: 10.1051/matecconf/20168202009
D2ME 2016

Reflow Process Parameters Analysis and Reliability Prediction Considering Multiple Characteristic Values

Yu Guo[1,a], Zhili Sun[1], Minghe Liu[2] and Mingang Yin[1]

[1]*School of Mechanical Engineering &Automation, Northeastern University, Shenyang, China, 110819*
[2]*School of Mechanical Engineering, Shenyang Jianzhu University, Shenyang, China, 110168*

Abstract. As a major step surface mount technology, reflow process is the key factor affecting the quality of the final product. The setting parameters and characteristic value of temperature curve shows a nonlinear relationship. So parameter impacts on characteristic values are analyzed and the parameters adjustment process based on orthogonal experiment is proposed in the paper. First, setting parameters are determined and the orthogonal test is designed according to production conditions. Then each characteristic value for temperature profile is calculated. Further, multi-index orthogonal experiment is analyzed for acquiring the setting parameters which impacts the PCBA product quality greater. Finally, reliability prediction is carried out considering the main influencing parameters for providing a theoretical basis of parameters adjustment and product quality evaluation in engineering process.

1 Introduction

In the era of rapid development for communication and information technology, electronic production has gradually become an indispensable necessity in daily life. To quickly adapt to market demand, electronic product design and manufacturing technology advances rapidly. Surface mount technology (SMT) has become the main way of electronic product manufacturing for its good coupling performance and high production efficiency. Industrial production usually aims at meeting customer requirements for product quality. As the main process in SMT, reflow profile quality reflects the final product quality directly in reflow soldering process [1]. During the reflow process, it is difficult for reflow profile to fully meet the customer's requirements. For this reason, we need to control the curve error within a reasonable range. Currently, the main method for reflow profile prediction is try and error, which requires that the operator has a wealth of experience [2]. Since the reflow process has a certain randomness which is affected by various uncertainties and the environment, it is necessary for reflow profile setting and analysis from the perspective of reliability. Considering the complexity of reflow process [3, 4], the paper conducts the study with the production examples from a company in Kunshan.

In order to investigate the influence on characteristic values, orthogonal design for parameters is carried out first to determine the main setting parameters and its influence. The study provides the basis for the reflow process parameter adjustment. Then the evaluation index for reflow profile characteristic values is defined to fit the reliability formula with Weibull distribution. The reliability

[a] Corresponding author : guoyu@me.neu.edu.cn

of reflow profile characteristic values is also evaluated. Consequently, reliability analysis of reflow profile provides a theoretical basis for determining the characteristic value error range.

2 Reflow parameters analysis based on orthogonal experiment

2.1 Orthogonal Experimental Design

Orthogonal Experimental design is a design approach which arranges and analyzes multivariate experiment using normalized orthogonal [5]. Study of the reflow parameters impact on reflow profile and key indicators based on orthogonal experiment is able to determine the greatest impact parameter on all key indicators and their impact direction [6].

For the reflow process, there are many parameters to characterize the quality of reflow profile. Besides, each input parameter is the factors affecting the final quality of the reflow profile comprehensively. Therefore, the use of orthogonal experiment method can efficiently analyze the reflow process with multiple indicators.

2.2 Orthogonal Experimental Design

From the analysis, the setting parameters of the reflow process are the key factors to affect the reflow profile. So the setting parameters of eight oven zones((℃)) and the belt speed(m/min)marked A~J are selected as input parameters. Three levels are determined for each factor in the experiment. The characteristic values of reflow profile are peak temperature (℃), soaking time(s), climbing time(s), melting time(s), cooling rate(m/min) and infiltration temperature (℃) marked K~P. The factors and their levels are listed in table 1.

Table 1. Factor levels

Factor / Level	Zone 1 (℃)	Zone 2 (℃)	Zone 3 (℃)	Zone 4 (℃)	Zone 5 (℃)	Zone 6 (℃)	Zone 7 (℃)	Zone 8 (℃)	Belt speed (m/min)
1	150	155	175	180	200	230	255	225	0.8
2	160	165	180	185	205	240	260	230	0.85
3	170	175	185	190	210	250	265	235	0.9

$L^{27}(3^{13})$ orthogonal table is used for arranging 27 sets orthogonal experiments. Then the experiments are carried out in an electronics production company in Kunshan. The experimental results are recorded in table 2.

Table 2. Orthogonal design& experimental results

No.	A	B	C	D	E	F	G	H	J	K	L	M	N	O	P
1	1	1	1	1	1	1	1	1	1	237.03	107.36	25.49	59.16	2.30	190.22
2	1	1	1	1	2	2	2	2	2	238.30	105.70	20.96	57.57	2.37	188.04
3	1	1	1	1	3	3	3	3	3	240.05	103.41	14.92	57.52	2.48	187.30
4	1	2	2	2	1	1	1	2	2	238.59	105.46	24.87	58.33	2.27	185.92
…	…	…	…	…	…	…	…	…	…	…	…	…	…	…	…
25	3	3	2	1	1	3	2	3	2	237.53	104.86	14.32	56.07	2.27	185.77

26	3	3	2	1	2	1	3	1	3	239.90	104.51	22.12	54.84	2.35	188.86
27	3	3	2	1	3	2	1	2	1	238.47	105.41	22.95	57.61	2.27	185.54

2.3 Experimental analysis of multiple indicators

Range analysis aims at determining progression of factors based on range value calculated with mathematical statistics calculation method.

There are *m* factors in the experiment according to orthogonal table. Experimental results are recorded by $x1$, $x_2, \ldots, x_N$. Each factor has *t* levels, and *n* experiments are carried out for each level. Therefore, the total experiment time *N* is calculate by $N=tn$. It is known that

$$K_{ij} = \frac{T_{ij}}{n} \tag{1}$$

$$R_j = K_{ij(\max)} - K_{ij(\min)} \tag{2}$$

K_{ij} represents the mean value of experimental result for the *j*th column and the *i*th level. T_{ij} is the sum of experimental results for the *j*th column and the *i*th levels. $K_{ij\,(\max)}$ is the maximum K_{ij} value in the *j*th column. $K_{ij(\min)}$ is the minimum K_{ij} value in the *j*th column. *c* is range value of the *j*th column.

Range analysis is capable of studying the degree of setting parameters influence on the characteristic values. Thus, range analysis is implemented on orthogonal test results in table 2. *K* value and *R* value of each characteristic value is computed by formula (1) and formula (2). According to range analysis, the main factors affecting the peak temperature is temperature of zone 7, temperature of zone 6, belt speed and temperature of zone 4 in succession; the main factors affecting the soaking time is belt speed, temperature of zone 6, temperature of zone 3, and temperature of zone 4 in succession; the main factors affecting the climbing time is temperature of zone 6, belt speed, temperature of zone 5 and temperature of zone 4 in succession; the main factors affecting the melting time is belt speed, temperature of zone 6, temperature of zone 8, and temperature of zone 7 in succession;the main factors affecting the cooling rate is temperature of belt speed, zone 8, temperature of zone 7 and temperature of zone 2 in succession; the main factors affecting the infiltration temperature is temperature of zone 6, temperature of zone 5, temperature of zone 7 and temperature of zone 8 in succession. Analysis of *K* value is shown in figure 1.

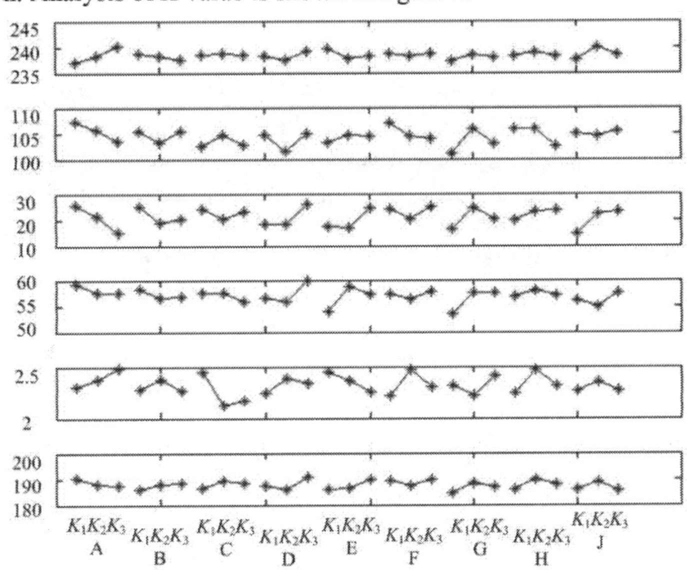

Figure 1. Analysis of *K*-value

It is known from figure 1 that the output distribution shows a range of differences with different factor influence degree. The factor with a relatively large difference is corresponding to the important one, which is consistent with the characteristic values in the oven zone accordingly during the reflow process. From this point of view, we conclude that the analyzed result meets the actual production process. To sum up, belt speed, zone 7 and zone 8 are the major factors to influence the reflow profile, which needs to monitor particularly in the production process. During the production process, the above findings may provide a theoretical reference for the production of parameter adjustment.

3 Reliability of characteristic value index

The sample No. 2Q012D000-600-GB/T-A-HF is used in the experiment for analysis. The setting parameters are listed in table 3. In the experiment, four major parameters are used as random variables for reliability analysis. From the above analysis, the major parameters are temperature of zone 4, zone 6, zone 7 and belt speed. Due to differences in the production conditions, errors are produced for various parameters inevitably, which affects product reliability significantly. For analyzing the randomness of parameters, the above major parameters are supposed as random variables which obey normal distribution as table 4.

Table 3. Temperature setting of reflow soldering oven

Oven zone	#1	#2	#3	#4	#5	#6	#7	#8	#9	#10
Setting temperature/℃	160	165	180	185	205	240	260	230	60	20

Table 4. Mean and standard deviation of all variation

Varible	T_4(℃)	T_6(℃)	T_7(℃)	v (m/min)
Mean value	185	240	260	0.85
Standard deviation	2	2	2	0.01

The probability level is chose as $p_1 = 0.01$、$p_2 = 0.5$、$p_3 = 0.99$. Then the sample value is calculated according to Box-Behnken matrix sampling method [7]. Furthermore, characteristic value is calculated with BP-Neural Network. For evaluating the fitting reflow profile comprehensively, reflow profile quality evaluation index is defined as equation (4). According to the definition of evaluation index, it is known that the smaller $f(x)$ represents the product meeting customer requirements better. The calculated results are listed in table 5. xi represents the reliability of evaluation index.

$$f(x_i)=\sum_{k=1}^{P}L(y_k) \tag{3}$$

Table 5. Sample points of Box-Behnken method and data of reliability tests

Sample	T_4(℃)		T_6(℃)		T_7(℃)		v(m/min)		x_i
	Level	Value	Level	Value	Level	Value	Level	Value	
1	p_2	185	p_2	240	p_2	260	p_2	0.85	0.0353
2	p_1	180.34	p_1	235.34	p_2	260	p_2	0.85	0.0307
3	p_3	189.66	p_1	235.34	p_2	260	p_2	0.85	0.0121
4	p_1	180.34	p_3	244.66	p_2	260	p_2	0.85	0.0720
5	p_3	189.66	p_3	244.66	p_2	260	p_2	0.85	0.0405
6	p_2	185	p_2	240	p_1	255.34	p_1	0.827	0.0235

7	p_2	185	p_2	240	p_3	264.66	p_1	0.827	0.0417
8	p_2	185	p_2	240	p_1	255.34	p_3	0.873	0.0333
9	p_2	185	p_2	240	p_3	264.66	p_3	0.873	0.0531
10	p_1	180.34	p_2	240	p_2	260	p_1	0.827	0.0416
11	p_3	189.66	p_2	240	p_2	260	p_1	0.827	0.0195
12	p_1	180.34	p_2	240	p_2	260	p_3	0.873	0.0565
13	p_3	189.66	p_2	240	p_2	260	p_3	0.873	0.0285
14	p_2	185	p_1	235.34	p_1	255.34	p_2	0.85	0.0304
15	p_2	185	p_3	244.66	p_1	255.34	p_2	0.85	0.0523
16	p_2	185	p_1	235.34	p_3	264.66	p_2	0.85	0.0264
17	p_2	185	p_3	244.66	p_3	264.66	p_2	0.85	0.0688
18	p_1	180.34	p_2	240	p_1	255.34	p_2	0.85	0.0403
19	p_3	189.66	p_2	240	p_1	255.34	p_2	0.85	0.0175
20	p_1	180.34	p_2	240	p_3	264.66	p_2	0.85	0.0896
21	p_3	189.66	p_2	240	p_3	264.66	p_2	0.85	0.0305
22	p_2	185	p_1	235.34	p_2	260	p_1	0.827	0.0169
23	p_2	185	p_3	244.66	p_2	260	p_1	0.827	0.0477
24	p_2	185	p_1	235.34	p_2	260	p_3	0.873	0.0248
25	p_2	185	p_3	244.66	p_2	260	p_3	0.873	0.0644

MATLAB toolbox is applied for parameters estimation. The estimative values are

$$\beta = 20256,\ \eta = 0.045$$

K-S test is applied for testing the fitness of parameters estimation. The observed value $\tilde{D}_n$ is

$$\tilde{D}_n = \max\left\{\tilde{d}_i\right\} = 0.1108$$

Significance level is taken as α=0.10, so $D_{n,\alpha}$=0.2377. Since it is known that $\tilde{D}_n < D_{n,\alpha}$, the hypothesis H_0 is acceptable. Therefore, the reliability of evaluation index obeys the Weibull distribution, which is shown as follows.

$$R(x) = \exp\left[-\left(\frac{x}{\eta}\right)^{\beta}\right] = \exp\left[-\left(\frac{x}{0.045}\right)^{2.256}\right]$$

The fitting curve of reliability is shown in figure 2. It is obvious that the smaller evaluation index is, the higher the reliability is.

The fitting equation of reflow profile quality evaluation index reliability has significant practical value. On one hand, when the evaluation index is calculated according the characteristic value of the reflow profile, the reliability of the evaluation index can be calculated. Furthermore, reliability of product can be acquired quantitatively as well. On the other hand, precision condition of characteristic value acquired by custom can also be determined with the index reliability analysis.

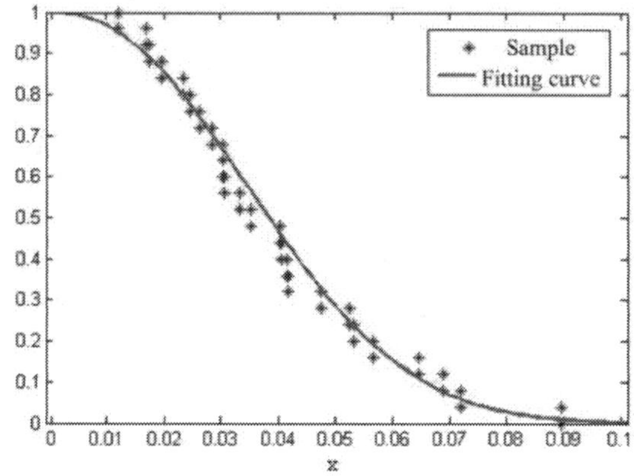

Figure 2. Fitting curve of reliability

4 Conclusions

(1) Considering the effect of characteristic values of reflow profile comprehensively, characteristic values and their influence degrees under different factor levels are calculated based on orthogonal design. The study result shows that the major influences on the reflow profile are belt speed, temperature of zone 6 and zone 7, which need to be monitored seriously.

(2) Reflow profile quality evaluation index is defined to calculated the characteristic value. Weibull distribution is applied for fitting the equation of index reliability. The reliability equation is capable of acquiring the product reliability with required evaluation index.

Acknowledgment

This project is supported by the National Natural Science Foundation of China(No. 51375083) and the Fundamental Research Funds for the Central Universities of China (Grant No. N140303011).

References

1. Y.Y. Su, K. Srihari, C.R. Emerson, Computers and industrial engineering, *A profile identification system for surface mount printed circuit board assembly*, **33**, 377-380(1997)
2. Y. Guo, Z.L. Sun, E.S. Pan, Q. Yang, Journal of Northeastern University(Natural Science), *Forecast of SMT Reflow Soldering Profile Based on Improved Artificial Neural Network*, **32(12)**, 1749-1752 (2011)
3. T.N. Tsai, Expert Systems with Applications, *Development of a soldering quality classifier system using a hybrid data mining approach*, **39**, 5727-5738 (2012)
4. L.G. Barajas, M.B. Egerstedt, E.W. Kamen, A. Goldstein, IEEE Transactions on Electronics Packaging Manufacturing, *Stencil printing process modeling and control using statistical neural networks*, **31(1)**, 9-18 (2008)
5. L.J.Pan, J Q.Chen. *Experimental Design and Data Processing*. Nanjing: Southeast University Press (2008).
6. Z.G. Feng, D.W. Yu, Y.H. Zhu, Electronics Process Technology, *Effect of parameters of reflow oven on the Reflow Profile*, **25(6)**, 243-246 (2004)
7. M. Yan, Z.L. Sun, Q. Yang, Chinese Journal of Mechanical Engineering, *Analysis method of reliability sensitivity based on response surface methods*, **43(10)**, 67-71 (2007)

MATEC Web of Conferences **82**, 02010 (2016)　　DOI: 10.1051/matecconf/20168202010
D2ME 2016

Effect of Interface on the Deformation of Aluminium Bicrystal: Atomistic Simulation Study

Shaohua Yan, Bobin Xing and Qing-Hua Qin [a]

Research School of Engineering, The Australia National University, Acton, ACT 2601, Australia

Abstract. Molecular dynamic (MD) simulation has been conducted to study the effect of interface structure on the mechanical response of eight <110> symmetric tilt grain boundaries in high stacking-fault Al. It is found that the grain boundaries with E structure unit (SU) have higher energy, but the grain boundary energy alone cannot be used as a parameter to determine the mechanical properties of grain boundary. The SUs, especially E units, do have an influence on the mechanical response of grain boundaries. Our results show that the dislocation imitates from E units preferably, but this depends on the dissociation at grain boundary.

1 Introduction

In the last decade, experiment of single crystals at micron/submicron scale length has been extensively studied. One of the well-known conclusions from these microscale tests is the relation between the sample size and flow stress, which can be described as the below power-law:

$$\sigma \propto D^{-n} \tag{1}$$

Where σ is the flow stress, D presents the diameter of the sample, and n is the fitting value. Unlike the deformation in the metallic foil [1-4], the deformation of single crystal at micon/submicron scale is featured with [5]: (1) a dramatic increase and stochastic variation of flow stress with decreasing sample size, (2) intermittency in plastic flow and strain burst (avalanches), (3) strength and avalanche properties that depend on the initial dislocation density. Up to date, most of the research has focused on the deformation of single crystals. However, most of real engineering materials are polycrystalline or multiphase. It is reported that the interface has had an influence on the strength of materials, which depends on the type of the interface. Allison Kunz et al. [6] reported that the single vertical grain boundary of pure aluminium reduce the degree of strain hardening and causes lower flow stress, while Ng and Ngan [7] reported the presence of grain boundary leads to increasing both the strain-hardening rate and flow stress. The difference is probably due to the structure of grain boundaries. As suggested by Ref. [8], the deformation/dislocation emission is largely depended on the atomistic structure of grain boundaries. For experiment, it is unrealistic to explore the relationship between interface structure and deformation mechanism systematically. Simulation may give us possible path to get insight into the plasticity of nanomaterials at submicron/submicron length.

In this paper, molecular dynamics is used to study the effect of interface structure on the deformation of eight <110> symmetric tilt grain boundaries in high stacking-fault aluminium. First,

[a] Corresponding author: qinghua.qin@anu.edu.au

the grain boundary (GB) structure at equilibrium condition is identified using structure unit theory. Then uniaxial tensile deformation is applied to GBs with minimum energy. Finally, the dislocation nucleation is then discussed to get insight into the GB structure's influence.

2 Methodology

The computational model used in this study is a bicrystal with a grain boundary located in the centre, as shown in Fig.1, all Al bicrystals used in this research have a symmetric tilt misorientation about [110] axis, which is z-axis in this research. The interface misorientation angle is measured used the [001] direction as the reference (0°). As shown in table 1, eight different GBs with different misorientation angles have been simulated.

In order to get GB structures with minimum energy, different initial GB structures have been achieved in two ways: (1) changing the initial distance between grain A and grain B, in which nine distances (-4Å,-3 Å, -2 Å, -1 Å, 0 Å, 1 Å, 2 Å, 3 Å, 4 Å) have been used, the negative distance means that there is an overlap between these two grains; (2) altering the cut-off regime, which arranges from $0.25a_0$ to $0.7a_0$ with an increment of $0.05a_0$, where a_0 is the lattice constant. Periodic boundary conditions are used for all directions when finding the minimum energy.

For this work, all the simulation is conducted using LAMMPS [9]. The embedded-atom method (EAM) potential is developed by Mishin [10]. After the minimum energy configuration is attained, the simulation model is equilibrated using MD in the isobaric-isothermal (NPT) ensemble at a pressure of 0 bar and a temperature of 10 K for 20 ps. A constant rate of 1×10^9 s^{-1} is applied perpendicular to the boundary plane (along Y direction) at a temperature of 10 K. The visualization tools Ovito [11] are used to illustrate of the bicrystal models. The common neighbour analysis (CNA) [12] technique is used to identify the defect structure and its evolution during the simulations. The dislocation extraction algorithm (DXA) [13] is used to convert identified dislocations into continuous lines and compute their Burgers vectors.

Table 1. The misorientation angles, grain boundary plane normal, dimensions, and number of atoms for the ten symmetric boundaries in this study.

GB plane normal and CSL designation	Misorientation angle θ (°)	SU period	Calculated GB energy (mJ/m^2)	Dimensions X×Y×Z (nm)	Atoms
Σ 33(554)	121.0	|DDE.DDE|	319	16.29×32.9×4	128870
Σ 11(332)	129.5	|DE.DE|	411	17.46×34.2×4	143962
Σ 9(221)	141.1	|E.E|	451	17.18×34×4	140832
Σ 19(331)	153.5	|EA.EA|	430	17.48×35.3×4	148568
Σ 33(441)	160.0	|EAA.EAA|	417	16.45×32.58×4	128520
Σ73(661)	166.6	|E'A'A'.E'A'A'|	363	14.68×34.6×4	121716
Σ 9(114)	38.9	|BC.BC|	330	17.01×34.36×4	140532
Σ 11(113)	50.5	|C.C|	150	17.10×34.92×4	143640

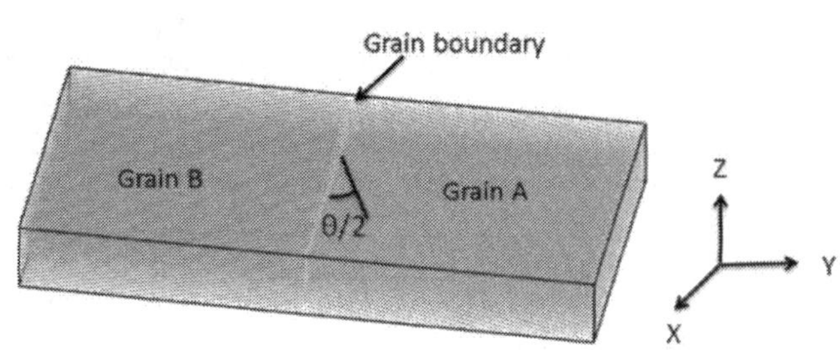

Figure 1. Schematic of Al bicrystal interface model. The interface is generated by a symmetric tilt rotation of grain A and grain B around [110] axis.

3 Results and discussion

3.1 GB structure of the Al bicrystal and energy

The GB structures with minimum energy after calculating are shown in Fig.2 (a) ~ (h). The centrosymmetric parameter was used to identify the structure units (SUs) that were used to identify the atoms associated with the GB. In this paper, the definition of SUs is similar to Ref. [14]. For the misorientation angle ranging from 121° to 141.1°, as shown in Fig.2 (a)~(c), there are two primary SUs found in these GB, which are D SU and E SU. Besides, in this misorientation angle range (121.0°<θ<141.1°), the E SU approaches each other when θ ranges from 121° to 141.1°, there is only E SU for Σ 9(221). The GB structure with θ> 141.1° are also characterized with two primary SUs, A (A') unit and E(E') unit, and the distance between adjacent units increases as well. For the equilibrium GB structure of Σ 9 (114) and Σ 11 (113), C and B units are the two SUs. From Table 1, it can be seen that the GB energy is various with respect to misorientation angle. However, GBs with E/E' SU, such as Σ 9(221) and Σ 19(331), are found to have higher energy because this SU leads to pronounced GB expansion to accommodate large free volumes [15]. It is also reported that the presence of E SU would reduce the strength of the interface [15, 16]. The influence of these GB structures on the mechanical response will be discussed later in this paper.

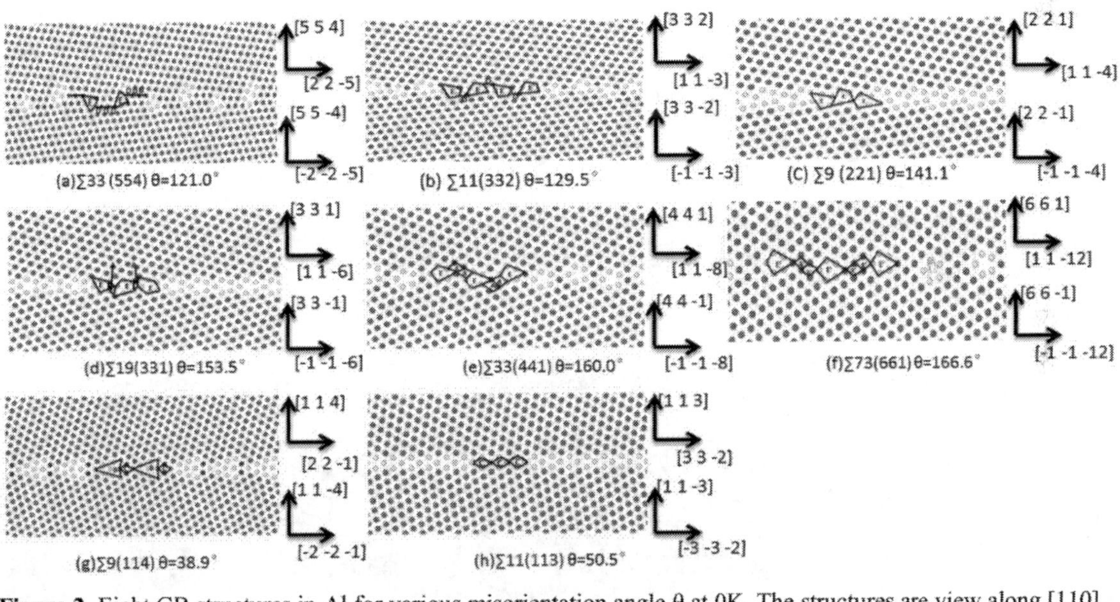

Figure 2. Eight GB structures in Al for various misorientation angle θ at 0K. The structures are view along [110] tilt axis. Images are coloured according to CNA parameters. Atoms with red colour are perfect fcc structures, while yellow atoms stand for GB and dislocation, and blue atoms represent the stacking fault.

3.2 Stress-Strain behaviour of the Al bicrystal

The uniaxial tensile stress-strain behaviour of Al bicrystals is shown in Fig.3. The results of GB without E SU are represented in Fig.3 (a), while the results of GB with E SU are displayed in Fig.3 (b). Compared with GB without E SU, the maximum tensile stress and the corresponding strain for GBs with E SU are relatively smaller. Fig.4 shows the stress required for the dislocation nucleation for the eight <110> tilt grain boundaries. It can be seen the stress for GBs with E unit is lower than those without E unit. Recall that GBs with E SU have higher energy than those without E SU, it seems the GB energy can be used as a parameter to determine the mechanical properties of GB. However, for the GB with E SU, the lowest stress is located in ∑33 (554) θ=121°, which has lowest GB energy among GBs with E unit. The highest energy grain boundary, ∑ 9(221), represents higher stress than these GBs with lower energy, which are ∑33 (554), ∑11 (332), and ∑73(661). Thus, grain boundary energy alone cannot be used to determine the mechanical properties of GB, the GB structure should also be taken into account because our results imply that the GB structure can have an influence on the mechanical properties of the interface.

Tucker et al. [16] also reported that the E SU has an effect on the properties of grain boundaries of Cu bicrystal. They found that the more ratio of E SU existing in the GB structure, the smaller stress required for the dislocation nucleation. Thus, for their case, the smallest stress was found in ∑ 9(221) ,in which only E SU is found at the GB. Same results can also be found in Ref.[15], where ∑ 9(221) of Al and Cu displays smallest shear strength, but the SUs at ∑ 9(221) in Ref. [15]are C unit and B unit. Nonetheless, for our case, the smallest one is the ∑ 33(554)/ θ=121°. This discrepancy can be interpreted by the following two facts: (1) Al has higher stacking fault than Cu, which may differentiate the mechanical behaviour, (2) different EAM potentials used for Al, which cause completely different SUs for the same ∑ tilt grain boundary, thus the mechanical response is changed accordingly.

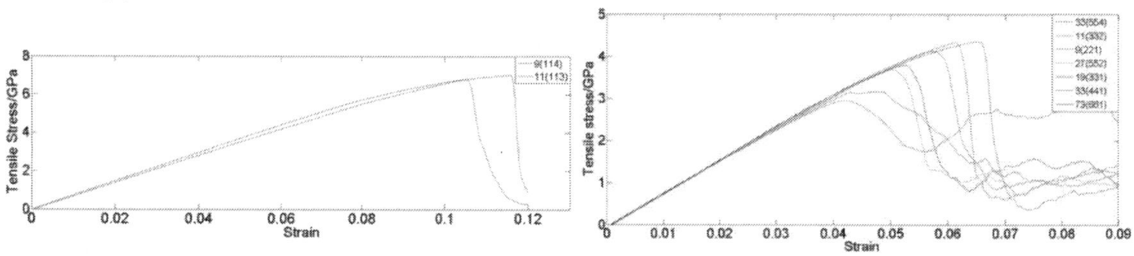

Figure 3. Stress-Strain response of the Al bicrystal at 10K , (a) without E unit, (b) with E unit

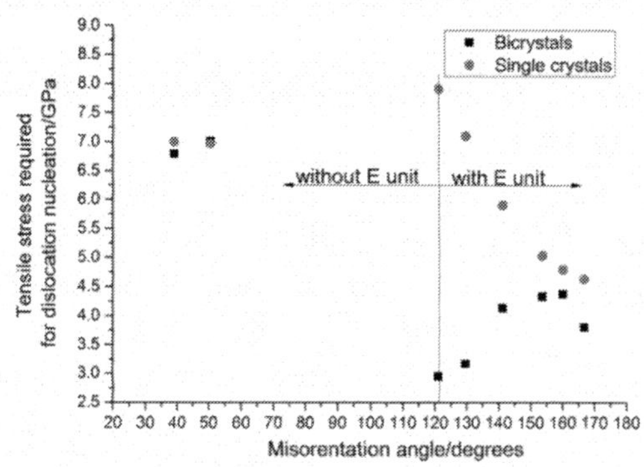

Figure 4. Tensile stress required for dislocation nucleation for the eight symmetric tilt grain boundaries.

3.3 Dislocation nucleation of the Al bicrystal

From the section 3.2, it can be found that the SU plays a key role on mechanical behaviour of the Al bicrystal. To get insight into the mechanism of the dislocation, visual inspection of the MD simulation is conducted using Ovito, the results are shown in Fig. 5 and Fig.6. Fig.5 shows the dislocation nucleation and propagation of two GB with E SU. For ∑9(221), dislocation nucleates from the upper and lower grain boundaries, and the free surface. Until ε=0.062, there are some partial dislocations growing from the middle grain boundary. But for ∑33 (554), the dislocation starts from the grain boundary, including the middle grain boundary. More specifically, the dislocation initiates from E units. Comparing the SU period of these two grain boundaries, SU period of ∑ 9(221) consists of |E.E|, while ∑ 33(554) of |DDDE.DDDE|. It seems the dissociation at GB can trigger the emission of dislocation from E SU. Fig.6 illustrates the dislocation behaviour of the Al bicrystal without E SU. For ∑ 11(113) with the SU period |C.C|, as shown in Fig.6(a), the dislocation initiates from the free surface, and dislocation can be found on the grain boundary only after the strain reaches 0.117. As for ∑ 9 (114) with the SU period |BC.BC|, the dislocation nucleates from the grain boundary at smaller strain (ε=0.106). Similar to the GB with E SU, the dissociation at GB without E SU seems to trigger the emission of dislocation as well.

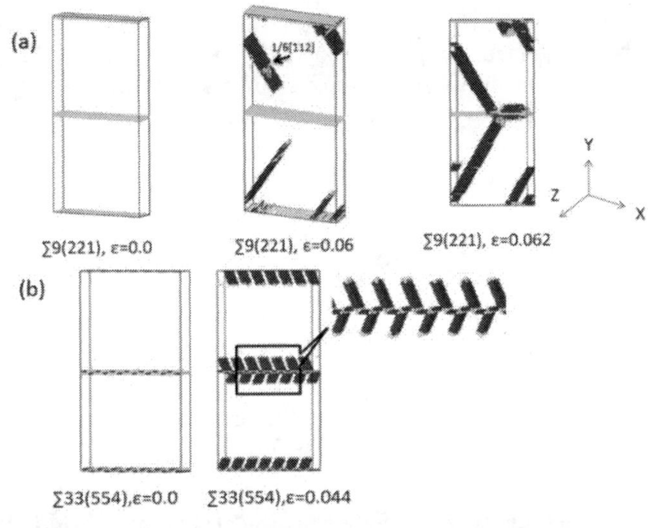

Figure 5. Snapshot of Al bicrystal with E SU GB at different deformation stage. (a) dislocation nucleation and propagation of ∑ 9(221) with the SU period |E.E|, (b) dislocation nucleation and propagation of ∑ 33(554) with the SU period |DDDE.DDDE|. Images are coloured according to CNA parameters. Atoms with red colour are perfect fcc structures, while yellow atoms stand for GB and dislocation, and blue atoms represent the stacking fault.

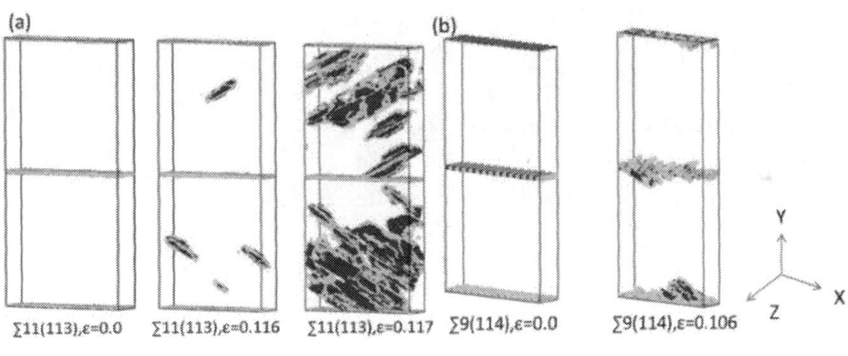

Figure 6. Snapshots of Al bicrystal without E SU GB at different deformation stages. (a) Dislocation nucleation and propagation of ∑ 11(113) with the SU period |C.C|, (b) Dislocation nucleation and propagation of ∑ 9 (114) with the SU period |BC.BC|. Images are coloured according to CNA parameters. Atoms with red colour are perfect fcc structures, while yellow atoms stand for GB and dislocation, and blue atoms represent the stacking fault.

4 Conclusions

In this paper, MD simulation has been performed to study the influence of interface structure on the mechanical properties of <110> symmetric tilt grain boundaries in Al. Our results show that GBs with E SU have higher energy, but the GB energy alone cannot be used as a parameter to determine the mechanical properties of GB. The dissociation at GB also has an effect on the deformation behaviour of grain boundaries. When there is dissociation at GB structure, the dislocation prefers to initiate from E SU, while dislocation nucleates from surface atoms when there is no dissociation at GB structure.

ACKNOWLEDGEMENT

The MD simulation was conducted using NCI National Facility of Australia. The author, Shaohua Yan, would like to acknowledge the financial support from China Scholarship Council (CSC). This work was financially supported from Australian Research Council under Linkage project (grant no. LP130101001).

References

[1] Kang YL, Zhang ZF, Wang HW, Qin QH. Mater. Sci. Eng., A **394**, 312 (2005).
[2] Kang YL, Wang HW, Zhang ZF, Qin QH. Microscopic Experimental Analysis of Fracture Toughness in Fracture of Copper Foils. Proc Of 11th International Conference on Fracture (in CD-ROM, ISBN: 9788890318818), (2005).
[3] Kang YL, Zhang ZF, Wang HW, Qin QH. Experimental investigations of size effect on fracture toughness of metallic foils, in: ECF15, Stockolm (2004).
[4] Wang HW, Kang YL, Zhang ZF, Qin QH. Int. J. Frac. **123**, 177 (2003).
[5] Uchic MD, Shade PA, Dimiduk DM. Annu. Rev. Mater. Sci. **39**, 361 (2009).
[6] Kunz A, Pathak S, Greer JR. Acta Mater. **59**, 4416 (2011).
[7] Ng KS, Ngan AHW. Philos. Mag. A **89**, 3013 (2009).
[8] Spearot D, Tschopp M, Jacob K, McDowell D. Acta Mater. **55**, 705 (2007).
[9] Plimpton S. J. Comput. Phys. **117**, 1 (1995).
[10] Mishin DF. Phys. Rev. B **59**, 3393 (1999).
[11] Stukowski A. Modell. Simul. Mater. Sci. Eng. **18**, 015012 (2010).
[12] Schiøtz J, Di Tolla FD, Jacobsen KW. Nature **391**, 561 (1998).
[13] Stukowski A, Albe K. Modell. Simul. Mater. Sci. Eng. **18**, 085001 (2010).
[14] Rittner J, Seidman D. Phys. Rev. B **54**, 6999 (1996).
[15] Sansoz F, Molinari JF. Acta Mater. **53**, 1931 (2005).
[16] Tucker GJ, Tschopp MA, McDowell DL. Acta Mater. **58**, 6464 (2010).

Chaotic Characteristics and the Application of SVM in the Tool Wear State Recognition

Guan Shan [a], Pang Hongyang and Kang Zhenxing
Northeast Dianli University school of mechanical engineering, Jilin, China,132012

Abstract. Metal cutting process is a nonlinear system to obtain the tool wear state and chaos theory are introduced tool wear and feature extraction of acoustic emission signal analysis and classification of tool wear state and wear prediction based on support vector machine (SVM). First, optimal embedding dimension of the time delay of phase space reconstruction of nonlinear dynamic system, the chaotic attractor; secondly, three characteristics: correlation dimension, the largest Lyapunov exponent and the Kolmogorov is extracted from the AE signal denoising feature vector and construct the different wear conditions. Finally, the feature vector is fed into the support vector machine (SVM), and the tool wear condition is classified. Research shows that: the cutting tool wear acoustic emission signal possesses the characteristics of chaos, chaotic characteristic parameters and tool wear status has intrinsic relationship; combined with chaos theory and support vector machine (SVM), can be very good to achieve the tool wear state recognition and prediction.

1 Introduction

As for automated processing, monitoring the tool wear state can give system error level in real-time, realize cutting compensation automatically, reduce the processing cost effectively and reduce scrap rate [1]. Li Xiaoli [2] used the wavelet analysis method to extract the vibration and acoustic emission tool wear characteristics, set up the neural network model based on fuzzy reasoning and better recognition results were obtained. Tae Jo KO et al [3] used time series AR model to extract characteristics of cutting force, classify and cluster analysis, and monitor single crystal diamond tool wear state based on the fuzzy recognition technology. Wafaa Rmili et al [4] used a three-axis accelerometer to measure vibration signal during the process of turning, to extract the feature parameters reflect the vibration response, and input the feature parameters to the automatic monitoring system to realize online monitoring of tool wear. The process of tool wear can produce abundant acoustic emission signal. Through tool wear acoustic emission signal sampling, this paper introduced chaos theory analysis, extracted three chaotic characteristic parameters: the correlation dimension and Lyapunov index and Kolmogorov entropy [5], constructed feature vector under different wear conditions, and the characteristic vector were put into the support vector machine (SVM) to identify the tool wear state and forecast tool wear loss.

[a] Corresponding author : guanshan1970@163.com

2 Test data collection

Tool wear is nonlinear, and the result of many factors, so in this paper, by using the principle of the orthogonal experiment to simplify the factor levels, extract the key factors (cutting speed v, feeding f, by turning the ap) to do experiment. Test was carried out in CKA6136i CNC lathe, using Kennametal KC9125 carbide coating blade, cutting material is GH625 nickel-based superalloy. Acoustic emission sensor is the NI Company R15, preamplifier is PXPA Ⅱ broadband acoustic emission amplifier, 1.5 ~ 2 MHZ bandwidth, 40 db gain selection, acquisition card is USB6366 NI Company, and sampling rate is 2 MHZ.

In order to eliminate the installation error caused by tool change and measuring tool wear loss, when completing a specified cutting time, a cutting edge is no longer used. Take a new knife to cut, after cutting 10 s, parking, record the 6 ~ 10 s of data, and tool microscope is used to measure the flank wear loss VB. Replace a new tool, cutting 20 s in the case of cutting condition unchanged. Two pieces of data were recorded, a period of 6 ~ 10 s, a period of 15 ~ 20 s. In theory, the first piece of data and the 6 ~ 10 s of cutting records for the first time signal, their degree of tool wear is the same. Measure the tool wear after parking. A new knife is used for cutting for a longer time repeatedly, until the cutter grinding is blunt. To illustrate the results of this article, with the cutting speed of 560 r/min, feed speed of 0.3 mm/r, cutting depth of 0.4 mm, the acquisition of the acoustic emission signal, for example, is described. Figure 1 is the timing signals whose tool wear loss are: 0.10 mm, 0.13 mm, 0.16 mm and 0.23 mm, 0.27 mm and 0.30 mm.

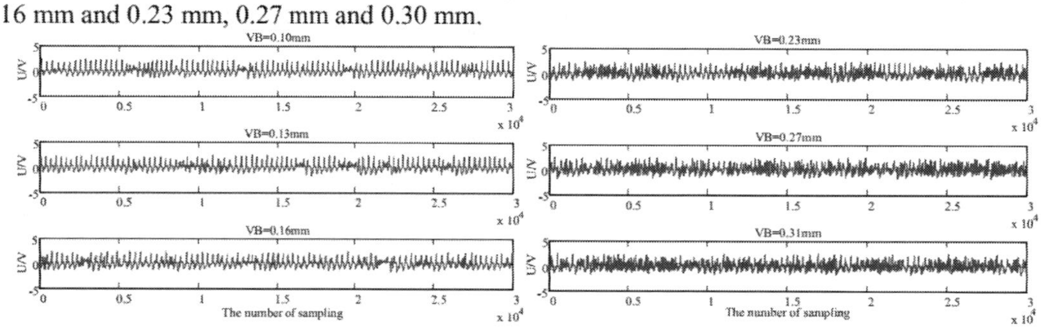

Figure 1. Timing signals of wear loss

3 Feature extraction based on chaos theory

The chaos [6] was produced by a deterministic nonlinear dynamic system, it seems to be no rules, similar to the random phenomenon. For nonlinear dynamic system, to fully excavate internal information, we need to expand one-dimensional temporal signal dimension, reconstruct the phase space of dynamic system, and then extract the characteristic parameters of correlation dimension and Lyapunov index and Kolmogorov entropy, which are used for describing system.

3.1 The phase space reconstruction

Phase space reconstruction refers to the fact that actual system in the given time series is extended to 3 d or more higher dimensional space, the key lies in the determination of delay time and embedding dimension. This paper mainly used the autocorrelation method to evaluate delay time τ, CAO method for embedding dimension.

(1)The determination of delay time

Autocorrelation method is to determine the time delay, it is very mature, the basic idea is to solve the linear correlation between the time series. For a given a time series, write the related function, with the increase of time, when autocorrelation function descends to the initial value of 1 to 1/e, the best time delay is to reconstruct phase space, it is best delay time τ to reconstruct phase space. The selection of different wear stages delay time is shown in figure 2. It can be seen from the diagram, τ values showed a trend of overall decrease with the increase intent of tool wear. With the increase of tool wear, acoustic emission phenomenon grows with the increase of friction between flank and artifacts, the increase number of ringing per unit time makes maximal time reduced. Therefore, τ values can indirectly reflect the characteristics of acoustic emission phenomenon parameters - the size of the ringing frequency.

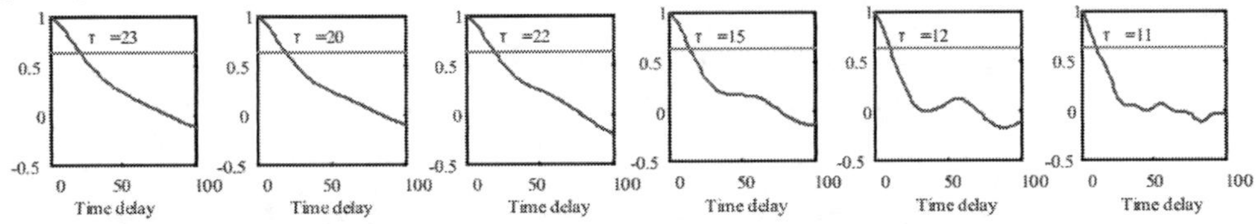

Figure 2. Autocorrelation method corresponding to different wear quantity signal delay time

(2)The CAO method for the embedding dimension

CAO solved minimum embedding dimension on the basis of FNN (False on his Neighbors).Table 1 is the delay time and embedding dimension based on the acoustic emission signal of different wear loss. The table shows that with the increase of wear and tear, the delay time and embedding dimension shows a trend of decrease.

Table 1. Corresponding to the calculation results of different wear loss

Wear loss/mm	Delay time τ Range	Delay time τ Mean	Embedding dimension m Range	Embedding dimension m Mean
0.10	20~25	23	10~12	10
0.13	19~22	20	9~13	11
0.16	20~23	22	10~11	11
0.23	12~17	15	9~12	12
0.27	10~13	12	8~11	8
0.31	10~12	11	8~10	9

3.2 Attractor phase diagram.

Each dimension of reconstructed phase space vector contains information of system evolution, two adjacent phase space vector can map the phase space trajectory, this is the attractor. Attractor has self-similar hierarchical structure, for a dissipative system, as the growth of the time, phase volume shrinks. Orbital folding, twisted and stretching, results unlimited nested self-similar geometric structure. It makes the trajectory of convergence to the attractor steadily. Figure 3 is attractor corresponding to the three different wear, respectively structured track structure from the 2 d and 3 d. Attractor has obvious folding nested structure, and on the direction y=x, and tool wear acoustic emission signal has chaos characteristics.

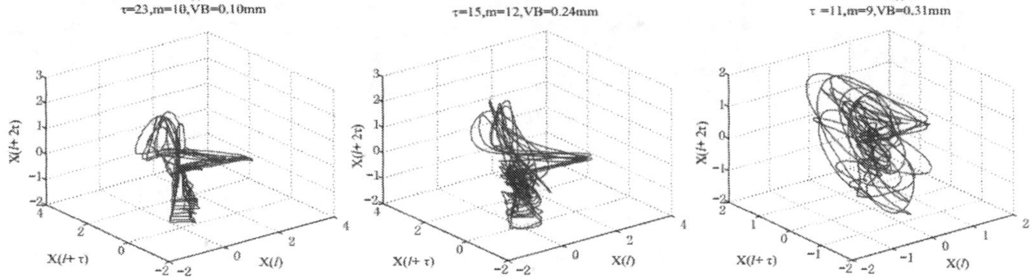

Figure 3. Attractor figures of different wear loss

3.3 Correlation dimension.

Correlation dimension is one of the types of extraction dimension, extracted from the chaos attractor in a series of points, and reflects the system complexity. This paper mainly calculated correlation dimension by GP algorithm. The main idea of the GP algorithm is to give two points X_i, X_j in phase space, the distance between them is r_{ij}. The distance between the point and the point is closer, their correlation degree is higher. A time series in the M dimensional phase space generate N phase points, X_1, X_2,...,X_N. Giving a positive number, the distance of points of logarithm is $r_{ij} < \varepsilon$, $C(\varepsilon)$ is the proportion of the total points of logarithm, there is:

$$C(\varepsilon)=\frac{1}{N^2}\sum_{i,j=1,i\neq j}^{N}\theta(\varepsilon-|X_i-X_j|) \tag{1}$$

$\theta(x)$ is the unit step function, correlation dimension is showed as follows:

$$D=\lim_{\varepsilon\to 0}\frac{\ln C(\varepsilon)}{\ln\varepsilon} \tag{2}$$

The type 5 shows that positive scale ε determines the accuracy of the correlation dimension. In practical applications, there is no specific scale, by giving a m, get a curve of lnC(ε) and lnε, calculate a group of logarithmic curve corresponding to m. Curve of linear scale-free in linear slope changes no longer and is saturated, and the straight slope is the correlation dimension. As it is shown in figure 4, the curve and the gradient of correlation dimensions lnC(ε) – lnε with the wear quantity is 0.24mm.

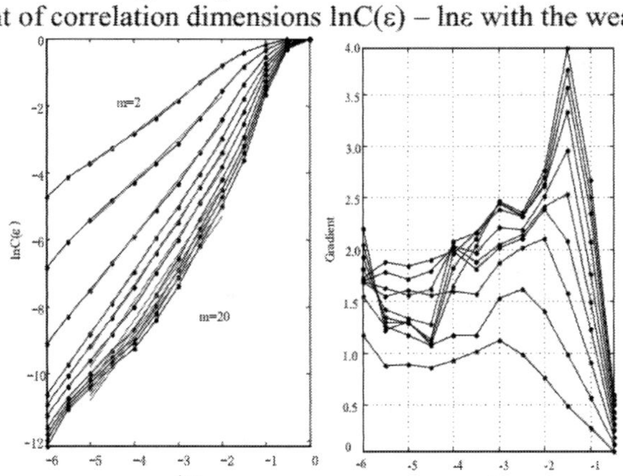

Figure 4. ln C(ε) and the gradient with lnε diagram

Linear scale-free region straight slope is correlation dimension, and graphic correlation dimension is equal to 2.2052. The relations between wear loss and correlation dimension is shown in figure 5.

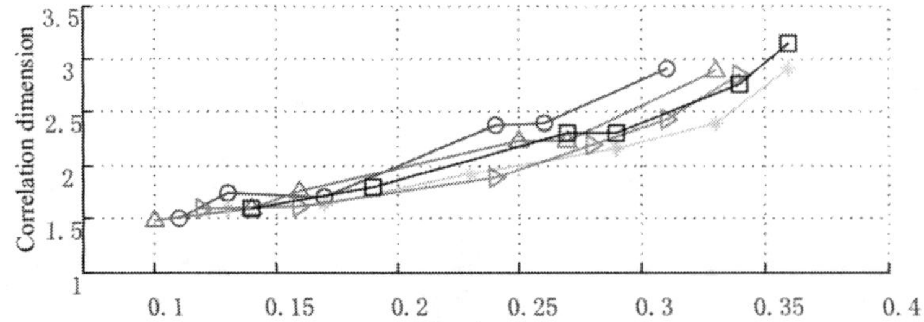

Figure 5. Different tool loss corresponds to the correlation dimension

As can be seen from the figure 5, as erosion rate increases, the increase of correlation dimension and nonlinear dynamic system is more complex, the adjacent sequence correlation decreases.

3.4 Lyapunov index

Lyapunov index is a kind of strange attractor nature of the measure and statistics, and maximum Lyapunov index reflects the system is in chaos. For one dimensional nonlinear dynamic system, Lyapunov index is expressed as:

$$\lambda = \lim_{n\to 0} \frac{1}{n} \sum_{i=0}^{n-1} \ln \left| \frac{df(x)}{dx} \right|_{x=x_i} \tag{3}$$

If it is extended to the n power system, can get n Lyapunov index, $\lambda_1 \geq \lambda_2 \cdots \geq \lambda_n$.The smallest Lyapunov index determines the dynamic system orbit contraction speed, and maximum Lyapunov index reflects the speed of covering the track attractor. In this paper, calculate Lyapunov index according to the small data method [7].

Figure 6 shows the curve evolution of distance with the wear loss is 0.16 mm, the basic orbit jth pair adjacent points, after i discrete time step. The red part of the figure is to use the least squares fitting to get maximum slope of the straight line segment, calculated maximum Lyapunov index is 0.2318. It said that system has chaotic characteristics.

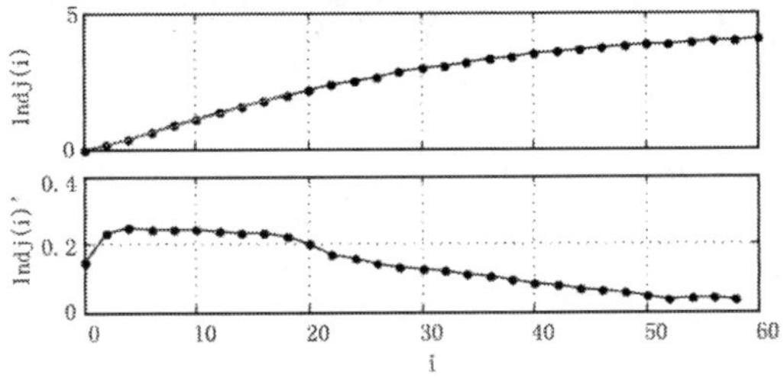

Figure 6. Linear area and the largest Lyapunov index curve

The sampling signal under different cutting condition Lyapunov index and correlation dimension showed that with the increase of wear, Lyapunov index and correlation dimension presented the same change trend, illustrates the enhancement of chaotic characteristic. It also reflects the state of tool wear.

3.5 The Kolmogorov entropy.

K entropy is an important characteristic of characterization of nonlinear system complexity, it says the average of the system information loss rate, and can represent chaos system. The mathematics is based on the definition of infinite subdivision, estimate the amount of information which is calculated by measure. Usually in order to simplify the calculation, according to thermodynamics entropy S to define Q order generalized entropy:

$$K_q = \lim_{\tau\to 0} \lim_{r\to 0} \lim_{m\to\infty} \frac{1}{m\tau} \frac{1}{q-1} \ln \sum_{i=0}^{m} P^q(i_0, i_1, \cdots, i_m) \tag{4}$$

When q = 0, it is the topological entropy, and when q = 1, K entropy.
GP algorithm gives the calculation method of K2. Correlation integral is shown in 10, combined with the correlation integral to get the following formula:

$$K_2 = \frac{1}{m\tau} \ln \frac{\varepsilon^D}{C(\varepsilon)} (\varepsilon \to 0) \tag{5}$$

Figure 7 is K entropy's change curve with the embedding dimension.

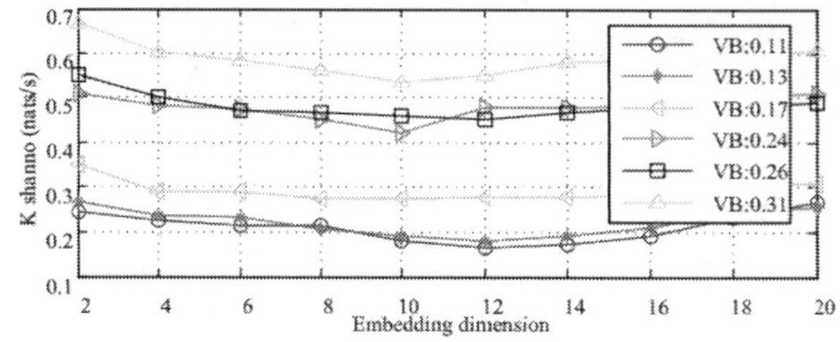

Figure 7. K entropy change with embedding dimension curve

As can be seen from the figure 7, the embedding dimension has less effect on the K entropy. With wear loss change, entropy K showed a trend of increase, wear was positively correlated with the chaos of the system.

4 The application of SVM in the tool wear state recognition

Support vector machine (SVM) is the latest in statistical learning theory and practical content, has a wide application in engineering field. Support vector machine classification, the main parameters are the kernel function is t, loss function, penalty coefficient c and kernel function parameter g, they are need to be set up. Parameters' selection directly influences the final identification accuracy.

Choose features samples, they are under different conditions, different wear conditions. Take 60 samples into support vector machine (SVM) for training and recognition. Keep other parameters constant, change the kernel function type, get the recognition accuracy as shown in table 2.

Table 2. Different kernel functions corresponding to the recognition accuracy under roughly selected parameters

Kernel function type	Parameters' selection	Recognition accuracy
Linear	'-c 1.2 –g 2.8 –t 0'	85.0000%
Polynomial	'-c 1.2 –g 2.8 –t 1'	96.6667%
RBF	'-c 1.2 –g 2.8 –t 2'	91.6667%
Sigmoid	'-c 1.2 –g 2.8 –t 3'	33.3333%

Table 2 is the recognition accuracy of different kernel functions, under the condition of roughing parameters. When the kernel function is polynomial kernel function, the recognition accuracy is higher. LIBSVM [8] can use Cross test Valid (Cross) to realize parameter optimization, referred to as CV method. The increase of the classification accuracy is more obvious after parameters optimized by CV method. Interactive inspection essence is to seek the optimal parameters under the condition of the best classification accuracy. Within a certain range, the two parameters is divided into a grid, finding the best parameters based on the given step value. The calculated best punishment coefficient c is 5.2780, the kernel function parameter g is 48.5029. As can be seen from table 3, after the parameters were optimized, identification accuracy was obviously improved.

Table 3. Different kernel functions corresponding to the recognition accuracy after parameter optimization

Kernel function type	Parameters' selection	Recognition accuracy
Linear	'-c 5.2780 –g 48.5029 –t 0'	91.6667%
Polynomial	'-c 5.2780 –g 48.5029 –t 1'	100%
RBF	'-c 5.2780 –g 48.5029 –t 2'	96.6667%
Sigmoid	'-c 5.2780 –g 48.5029 –t 3'	33.3333%

5 Summary

The chaotic characteristics of acoustic emission signals are determined by the tool wear condition, by reconstructing the phase space of the time sequence signals, the phase trajectory structure of the nonlinear dynamic system can be recovered. The complexity of the system can be characterized by chaos: correlation dimension, Lyapunov exponent, Kolmogorov entropy to measure. The constructed feature space of tool wear acoustic emission signal chaos features , combined with the SVM, can well realize classification of tool wear state. The selection of SVM kernel function has a great influence on the identification results, and the parameter optimization selection can improve the recognition accuracy of the model significantly.

References

1. Zhang Kai-feng, Yuan Hui-qun, Nie Peng. Tool wear condition monitoring based on generalized fractal dimensions[J]. Journal of Vibration and Shock, 2014, **33**(1): 162-164.
2. Li Xiaoli. Fuzzy Pattern Recognition of the Wear of Metal Cutting Tools [J]. Journal of KunMing Institute of Technology, 1995, **20** （5） :26-32.
3. Tae Jo KO，Dong Woo Cho，et al. Fuzzy Pattern Recognition for Tool Wear Monitoring in Diamond Turning[J].CIRP Annals-Manufacturing Technology，1992， **41**(1)：125-128.
4. Wafaa Rmili，Abdeljalil Ouahabi，et al. An automatic system based on vibratory analysis for cutting tool wear monitoring[J]. Measurement ，2016，**77**(3)：117-123.
5. Li Gang-hu，Li Ya-an，JIA Xue-song. Better feature extraction and classification of chip-radiated noise with three chaos feature parameters[J].Journal of Northwestern Polytechnical University，2006，**24**(2)：170-173. (in Chinese with English abstract)
6. Volodymyr Lynnyk， Noboru Sakamoto，Sergej Čelikovský. Pseudo random number generator based on the generalized Lorenz chaotic system[J].IFAC-Papers OnLine，2015，**48**(18)：257-261.
7. Yang Yong-feng，Wu Min-juan，Gao Zhe，et al. Parameters selection for calculating largest lyapunov exponent from small datasets[J].Journal of Vibration，Measurement＆Diagnosis，2012，**32**(3)：371-374.
8. Wu Hao，Li Qun-zhan，Liu Wei. A new algorithm of wide-area backup protection based on PSO-LIBSVM[J]. Power System Protection and Control，2013，**41**(15)：49-58.

MATEC Web of Conferences **82**, 02012 (2016) DOI: 10.1051/matecconf/20168202012
D2ME 2016

Anti-oxidation and Infrared Characteristics of ZrSiO4 Coating

Yang Wang, Zhaofeng Chen [a] and ShengJie Yu

College of Materials Science and Technology, Nanjing University of Aeronautics and Astronautics, Nanjing 210016, P.R. China

Abstract. In this paper, preparation and properties of ZrSiO4 coating prepared by pressureless sintering coated carbon/carbon (C/C) composites and the effect of different temperature on coating in 3-5μm and 8-12μm infrared emissivity were studied. The composition and structure of the ZrSiO4 coatings were examined by XRD and SEM that showed the main phase was ZrSiO4. With the addition of mineralizer, ZrSiO4 synthesis rate increased from 68.58% to 83.47%. The anti-oxidant performance of the prepared SiC layer and SiC-ZrSiO4 coating at 1500℃ in air for 24h with a mass loss was 1.5% and 0.4%, respectively. In addition, the experimental results indicated that the emissivity of ZrSiO4 coating declined with the decrease of ZrO2, and the infrared emissivity of ZrSiO4 coating in the 3-5μm and 8-14μm wavebands was 0.55 and 0.64 at 50℃, 0.18 and 0.43 at 300℃, 0.35 and 0.74 at 500℃, respectively.

1 Introduction

In recent years, the rapid development of aerospace technology and modern needs of war, has led to the continuous improvement of Mach number of flights. Aircraft parts that undergo aerodynamic heating need to withstand more high temperature and thermal shock during flight. Hence, high temperature resistant and low emissivity materials play an important role in modern weapon system [1]. Infrared stealth includes near-infrared (1-2.7 μ m), mid-infrared (3-5 μ m), and far infrared (8-14 μ m) three atmospheric windows. Among these, 3-5 μ m and 8-14 μ m wavebands are the most important infrared detected wavebands [2]. ZrSiO4 is a tetragonal structure with the islands of silicate minerals. Due to its high melting point, low thermal conductivity, low thermal expansion coefficient, excellent chemical and phase stability and other characteristics, ZrSiO4 is widely used in the high-temperature environment. Moreover, sintered ZrSiO4 has excellent thermal shock resistance. It is therefore an important candidate for high temperature structural ceramic materials [3, 4].

ZrSiO4 has rarely been reported as a high-temperature ceramic coating, especially considering the infrared stealthy characteristic and the traditional technologies, such as chemical vapor deposition, plasma or thermal spraying, sol-gel andelectrophoretic deposition. The preparation of ZrSiO4 coatings is complex, and the production cycles are long, making it not suitable for large complex shape components. Hence, pressureless sintering ZrSiO4 coating is a good choice. This paper describes a pressureless sintering method for synthesizing ZrSiO4 ceramic coatings and a preliminary investigation was undertaken to measure the performance of the coating.

[a] Corresponding author : zhaofeng_chen@163.com

2 Experimental details

Small specimens (100×100×5mm3) as substrates were cut from bulk 2D-C/C composites (prepared in Changsha China) with a density of 1.70 g/cm2. Before the coating process, the specimens were cleaned using ultrasonic cleaner for 5 min,and dried at 100℃ for 2 h. SiC bonding coating was prepared by a method of CVD (Chemical Vapor Deposition) for 10 h. An SiC coating structure was achieved with a thickness of about 20μm.

After the preparation of the SiC buffer layer, an in-situ formation process was conducted. ZrSiO4 coating was prepared using the optimal sintering additive MgO. The commercially available powders of MgO, ZrO2 and SiO2 are of analytical grade with a particle size from 1 to 4 μm. Raw materials formula are shown in Table 1, which LiF and CaCl2 are used as mineralizer.

Table 1. Raw materials formula.

Formula	**ZrO_2**	**SiO2**	**MgO**	**LiF**	**$CaCl_2$**
ZS-1	123	72	5.85	/	/
ZS-2	123	72	5.85	1.95	3.9

The starting powders were mixed for 12 h by ball milling in water-free ethanol with an agate media, and then dried at 70℃ for 12 h. First, a slurry of powders was applied to the SiC coating surface. It was made by mixing the powders, distilled water and a little PVB (polyvinylbutyral) (0.5~1wt.%) binder.

After applying the slurry to the surface of the SiC bonding layer, the samples were dried in air for sufficient time. Finally, the pressureless sintering of bar samples was conducted in a high-temperature furnace with with 1 atm argon gas, by heating at a ramp rate of 5℃•min-1 from room temperature to 500℃ and then heating at a ramp rate of 10℃•min-1 to 1500℃, holding for 1.5h at this temperature to transfer into ZrSiO4.

After sintering, the preparation of ZrSiO4 coating was applied directly to the surface of SiC coating. Product phases were characterised by X-ray diffraction (XRD) using a D/MAX-RB diffractometer and CuKα1 radiation for the range 15° ~80° in 2θ with a step-width of 0.02°. The morphologies of the as-prepared multi-layer coatings were analyzed using JEOL-6360LV scanning electron microscope (SEM). The anti-oxidation property of the ZrSiO4 coating was studied at 1500℃ using tubular oxidation furnace (Type SK-1600℃, Tianjin, China). Using emissivity measuring instrument, the average infrared emissivity of samples was measured at different temperatures in the 3-5 μm and 8-14 μm (Test in Xi'an Institute of Applied Optics).

3 Results and discussion

3.1 Composition and Structure Analysis

Fig. 1 shows the XRD spectrum of ZS-1 and ZS-2 formulas. It was found that both of the main phases were ZrSiO4, though a small amount of ZrO2 and SiO2 were detected. But compared to the ZS-1 formula, ZS-2 formula relative intensity of ZrO2 and SiO2 is significantly lower than ZS-1 formula with higher relative intensity of ZrSiO4. This is because on the basis of ZS-1 formula, ZS-2 formula accordingly added a little mineralizer which are LiF and CaCl2, improving the conversion rate of ZrSiO4 obviously. From the chemical reaction dynamics analysis, the role of mineralizer is mainly due to the formation of active reactive groups in the synthesis process, thereby reducing the activation energy. So the synthesis rate of ZrSiO4 increase at a certain temperature [5]. R. A. Epplerm [6] first quoted famous symbol method to study the ion diffusion problems, and thus explained the mechanism of how the mineralizer to improve zirconium silicate synthesis rate. By using the following formula to calculate a relative content of ZrSiO4 phase:

$$W=Ai/\Sigma Aij*100\% \quad (1)$$

Where Ai is the relative integral area of ZrSiO4 strongest peak and Aij the sum of the relative integral area of all phases strongest peak.

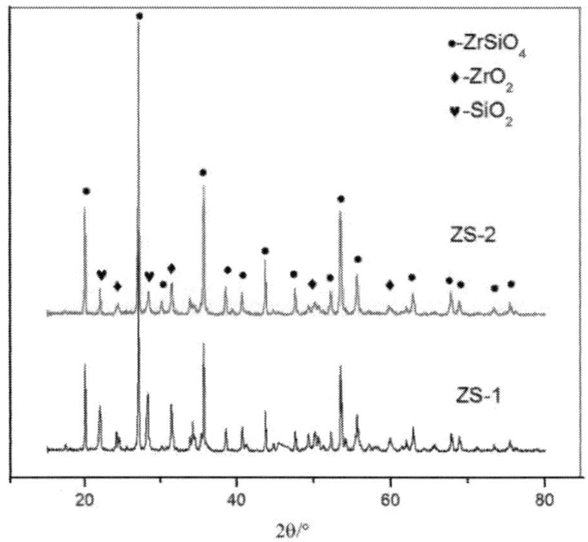

Figure 1. XRD spectrum of the ZrSiO4 coating.

From XRD analysis calculations, the peak integral area and relative content of ZrSiO4 prepared from ZS-1 formula and ZS-2 formula are about 68.58% and 83.47,respectively. So with the addition of mineralizer, ZrSiO4 synthesis rate increased obviously.

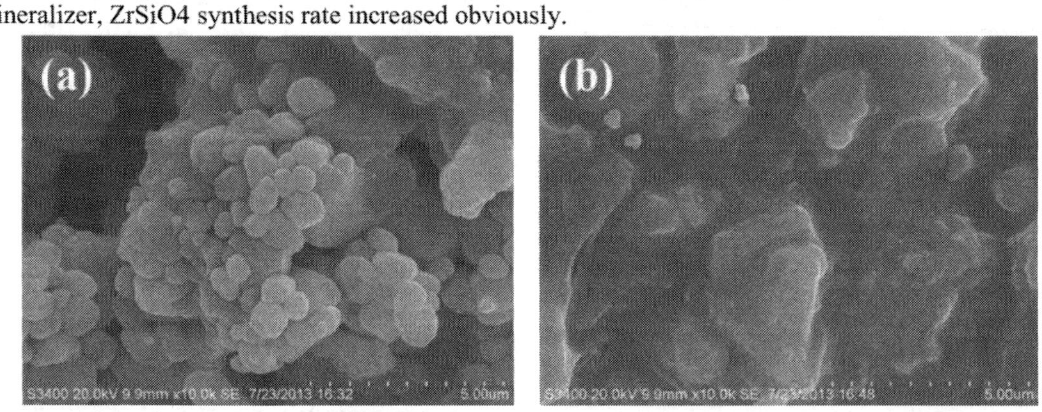

Figure 2. Surface SEM microstructures of coating: (a) ZS-1; (b) ZS-2.

Fig. 2 shows the coating surface SEM microstructures pictures of ZS-1 and ZS-2. It can be seen from the Fig. 2(a) that a lot of tiny particles covered on a large grain which is ZrSiO4 grain, and the tiny particles can be regarded as ZrO2 particles. However, in Fig. 2(b), these tiny particles covered on ZrSiO4 grains were rare. This is because as the mineralizer was added, most of the ZrO2 transform to ZrSiO4 completely. And the images displayed that the ZrSiO4 grains embedded in the glass state, which might be SiO2 melting. From Fig. 2(b), SiO2 melting wrapped ZrSiO4 grains, well filled in the gaps between the grains, which indicated SiO2 can effectively fill the cracks between the grain boundaries, prevent grain boundary diffusion of oxygen. This improves the oxidation resistance of coating.

3.2 Oxidation Resistance

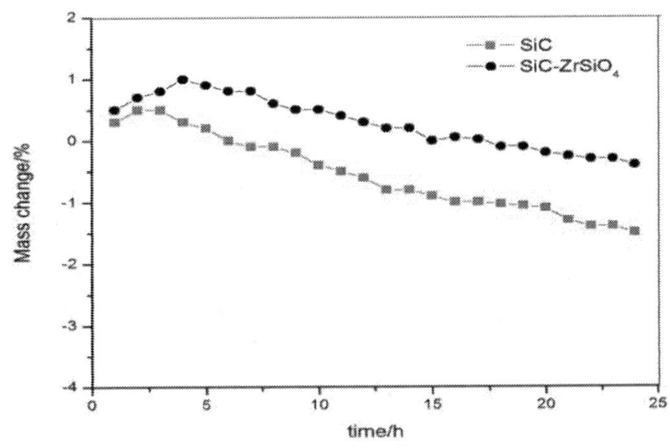

Figure 3. Isothermal oxidation curves for samples of the ZS-2 coating at 1500℃

Due to the higher synthetic rate of ZrSiO4, we chose ZS-2 for oxidation resistance test. Fig. 3 shows the isothermal oxidation curves for samples of the coating at 1500℃. The mass rate of both curves at the beginning of the oxidation was increased first, and then decreased slightly. Mass loss of SiC-coated C/C was 1.5% during 24h oxidation, which indicated that the as-prepared SiC coating could effectively protect C/C during the oxidation process. This is because in high temperature air, SiC coating surface generates a dense thin layer of SiO2, which can effectively block the diffusion of oxygen [7,8]. The formation of the SiO2 was attributed to the reactions displayed in Eqs. (2)

$$SiC+2O2 \rightarrow SiO2+ CO2 \quad (2)$$

This explains why the weight had some increase at the start of oxidation experiments. Even though SiC caoting can effectively protect C/C, it is not employed in practial application. Because at the temperature above 1400 ℃ stage, SiO2 melted into liquid, can effectively fill the cracks, which can exhibit excellent oxidation resistance, but in the medium temperature stage, the viscosity of SiO2 is very big, which can not fill the micro-cracks in the coating, and in the environment of high temperature scouring, SiO2 is easily torn away, so it is necessary to prepare a coating of high temperature oxidation resistant coating on the surface of SiC layer. From Fig. 3, ZrSiO4/SiC coating has a better oxidation resistance than SiC coating and there is almost no mass loss in the progress of oxidation. The mass loss of ZrSiO4/SiC-coated C/C was only 0.4% during 24h oxidation. As a result of the preparation of dense ZrSiO4 ceramic coating, O2- did not easily diffuse into the coating surface, to combine with the SiO2 liquid glass phase formed at high temperature. This can effectively make up for the microcracks to prevent the diffusion of oxygen. As shown in the Fig. 2(b), ZrSiO4 grains embeded in SiO2 glass state, and the coating has no microcracks, so the oxidation rate of the coating depends on the diffusion rate of oxygen in the ZrSiO4 and SiO2. At the same time, SiC used as a buffer layer, also forming a dense SiO2 layer on the surface in high temperature, can prevent oxygen diffusing, which further improves the property of oxidation resistance.

3.3 Infrared Characteristics

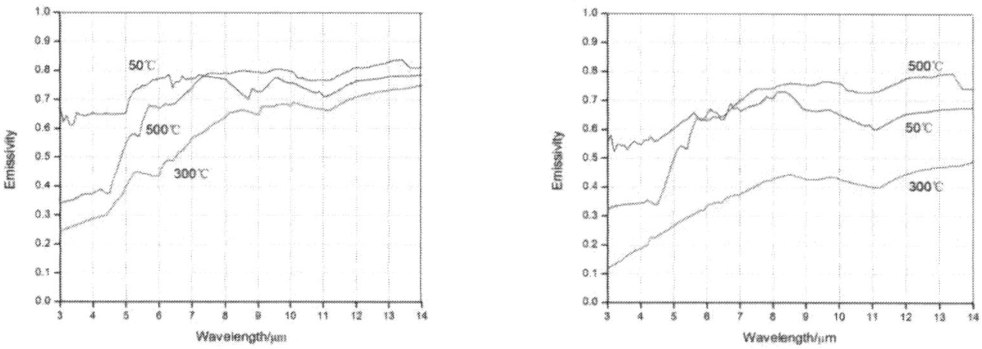

Figure 4. Spectral emissivity curves for ZrSiO4 coating at different temperature: (a) ZS-1; (b) ZS-2

Figure 4 shows the emissivity curves for ZrSiO4 coating at 50℃, 300℃ and 500℃. It can be seen from the picture that the emissivity of ZrSiO4 coating was the lowest when the temperature is at 300 ℃ and the measured value of infrared emissivity is increasing with the wavelength increase. When the temperature is above 300℃, the emissivity increases. When the wavelenght is above 7 μ m, the emissivity at 50℃ is lower than that at 500℃. Compared with Fig. 4(a) and (b), the emissivity of ZS-2 is lower than ZS-1, overall. From Fig. 4(b), the emissivity of 3-14 μ m band of ZrSiO4 coating is less than 0.8 at different temperature which expresses an excellent infrared characteristic. Kruse et al. [9] according to the experimental results thought that the emissivity decreases with temperature for non-metallic materials. However, in this study, the emissivity of ZrSiO4 coating presents a different rule. This may be because the residual ZrO2 changes the trend that the emissivity decreases with temperature increase. When the temperature is above 300℃, residual ZrO2 on the emissivity have a significant impact, the residual ZrO2 more, the temperature higher, the emissivity higher.

Table 2. Average emissivity

	The average emissivity			
	ZS-1		**ZS-2**	
	3-5μm	8-14μm	3-5μm	8-14μm
50℃	0.65	0.71	0.55	0.64
300℃	0.28	0.68	0.18	0.43
500℃	0.38	0.79	0.35	0.74

Table 2 lists average emissivity of ZS-1 and ZS-2 in the 3-5 μ m and 8-14 μ m wavebands at different temperatures. The emissivity on 8-14 μ m waveband is higher than the emissivity in 3-5 μ m waveband at any temperature. From ZS-1 formula, the emissivity of both bands are the lowest with value of 0.28 and 0.68 when the temperature is at 300℃, while the lowest value of the emissivity of ZS-2 is 0.18 and 0.43 when the temperature is at 300 ℃. Compared with ZS-1, the data of ZS-2 emissivity is lower, which indicated that the emissivity of ZrSiO4 coating declined with the decrease of ZrO2. This may be because, in ZS-1 formula, many of the remaining ZrO2 will cause vibration strongly and affect absorption of infrared radiation, thereby increasing the infrared emissivity of the coating. And the effect of ZrO2 on the emissivity is more evident in the high temperature. Before 300℃, the emissivity of ZrSiO4 is mainly determined by the temperature, so the emissivity declease with the temperature increase. Due to the fact that higher the temperature of an object, the better the crystallinity and periodicity of crystal, so emissivity declined; After 300℃, ZrO2 content on the emissivity plays a main role, so at 500℃, the emissivity has increased significantly.

4 Conclusions

$ZrSiO_4$ ceramic coatings were prepared by pressureless sintering in a argon atmosphere at 1500℃. With the addition of mineralizer, $ZrSiO_4$ synthesis rate increased from 68.58% to 83.47%. The anti-oxidant performance of the prepared SiC layer and SiC-$ZrSiO_4$ coating at 1500℃ in air for 24h with a mass loss was 1.5% and 0.4%, respectively. The average infrared emissivity of $ZrSiO_4$ coating in the 3-5 μm and 8-14 μm wavebands decreased with increase temperature. The emissivity of $ZrSiO_4$ coating declined with the decrease of ZrO_2, and the infrared emissivity of $ZrSiO_4$ coating in the 3-5 μm and 8-14 μm wavebands was 0.55 and 0.64 at 50℃, 0.18 and 0.43 at 300℃, 0.35 and 0.74 at 500℃, respectively.

Acknowledgements

The present work was supported by the Funding of Jiangsu Innovation Program for Graduate Education (the Fundamental Research Funds for the Central Universities), KYLX15_0308 and Priority Academic Program Develop-ment of Jiangsu Higher Education Institutions.

References

1. J. Zueco, F. Alhama., J. Quant. Spectrosc. Radiat. Transfer, **14**, 7 (2006)
2. Y. T. Jiang, Y. Wang. Infrared Technology, **25**, 5 (2003)
3. Y. Shi, X.X. Huang, D.S. Yan, J. Eur. Ceram. Soc., **13**, 2 (1994)
4. Y. Kubeta, S. Yamamoo, T. Mori, J. Ceram. Soc. Jpn., **102**, 1 (1994)
5. J. Wen, S. Z. Sun, Bulletin of The Chinese Ceramic Socity, **18**, 6 (1999)
6. R. A. Eppler, J. Am. Ceram. Soc, **53**, 8 (1970)
7. J. D Buekley, D. D Edie, Noyes Publications,USA (1993)
8. O. V. Mazurin, Handbook of glass data-Part A, Silica glass and binary silicat glasses, **15**, (1985).
9. W. P. Kruse, D. Laurence, McGlauchlin, Elements of Infrared Technology. New York: Wiley, (1962)

MATEC Web of Conferences **82**, 03001 (2016) DOI: 10.1051/matecconf/20168203001
D2ME 2016

Parametric Design of Outdoor Broadcasting Studio Based on Schema Theory

Li Zhu [1], Nan Zhang[1, a] and Xiaoying Qing[2]

[1]*School of Architecture and Art, Central South University, Changsha , China*
[2]*School of Art and Design, Luoyang Institute of Science and Technology, Luoyang, China*

Abstract. This paper mainly demonstrates that the schema is an important way for the architect to cognize architecture form logic. It connects schema to algorithm of parametric design in order to seek the "algorithm schema" generation in parametric design of architecture. Meanwhile, this paper discusses the generative process and methods of the "algorithm schema" in parametric design of architecture by describing a case of outdoor broadcasting studio of Hunan Economic Radio. It also reveals the importance of "algorithm schema" for the cognition and architectural form logic generation.

1 Introduction

Schema is an abstract structure extracted from cognitive experience which appears repeatedly in human life. It helps cognize the structural cognition among objective matters. The concept "schema" could be dated back to the ancient Greek philosopher Plato's basic thought: the ideal is to abstract things in a general way. Based on this idea, people form their cognitions about the world. The study of schema exerts a far-reaching influence not only in philosophy but also in psychology. F.C.Bartlett, a British psychologist, has officially put forward the concept of schema from the view of psychology.He argued that schema helps actively organize the past experience and affect the well adaptable organic reaction. According to cognitive linguistics: schema contains space structure and movement structure in the development. So it plays an important role in the process of conceptual metonymy.

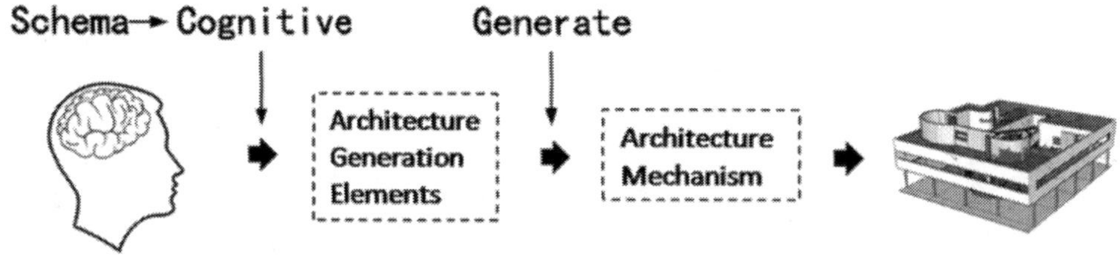

Figure 1. Construction of Architecture Mechanism Based on Schema Theory

[a] Corresponding author : 307579716@qq.com

The analysis above shows that schema can be considered as a very important mean of cognizing the abstract structure of relationship among objects. "Schema is the process of conceiving the structural image based on the conception", said by Immanuel Kant. By this token, the architecture form logic generation also has the same features as schema. Schema could help architects establish the relationship between visual form and elements that affect architecture form generation, such as functional logic of architecture, geographical environment, regional culture, social relations, etc. In this way, schema could help architects cognize the meaning implied in architecture form logic and further improve it.

2 The schema of parametric architectural design

The schema conception has been widely used in architectural design. Under the modern science environment which is dominated by Chaos theory, System theory and Control Theory, parametric design is a method that employs the powerful computing capability of computers to describe the architectural form logic. It has renewed the cognition and application of schema in architecture design. The logic of relations among different elements that could affect architecture formation is described by parametric design through an abstract algorithm of computer, thereby generating the non-linear space. Computer algorithm also has the same feature as the schema that could describe the abstract logical relations among different things, therefore we can extract the imaged cognitive schema to algorithm and provide a guide to architecture form generation.

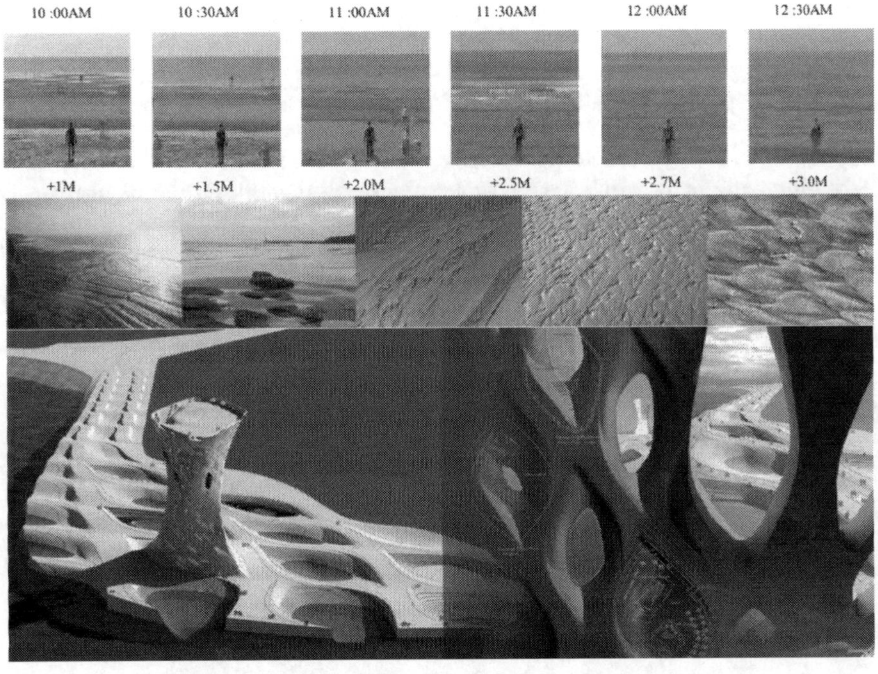

Figure 2. Perceptual schema extraction from the tides

Pure design makes perfectly explanation about the relationship between architecture's form and the logic of tidal action by using computer algorithm on the sightseeing tower project of Mersey river. This project is located at the estuary of Mersey River, with the aim to attract tourists to enjoy the beautiful natural scenery. Pure Architecture expects that the sightseeing tower is not only a place for tourists to enjoy the scenery but also a way to make it possible for tourists to cognize the relationship between architecture's form and the site. Based on the regularity of tidal action of the Mersey River, Pure Architecture designed this special "Tides Park". It is consisted of a series of interweaved paths and landscape units. These units include local plants and different heights of water landscape formed by tides. The tower indicates the way how the site was transformed from land into water.

3 The algorithm schema architecture of a parametric design for outdoor broadcasting studio

The algorithm schema architecture in parametric design has two levels: firstly, the architect forms the perceptual schema through his cognition and extraction from the generating factors of the architecture. Secondly, architects translate the perceptual schema into an algorithm schema by using computer programming language. As human's abstract cognition runs through. the whole process from the extraction of the perceptual schema to the translation of the algorithm schema, the relation among the factors that have effects on architectural form logic converts human's perceptual schema into the algorithm schema, thus forms the logic of architecture generation.

Take the outdoor studio of Hunan Economic Radio as an example, we can explore the relationship between the form logic of the architecture and its algorithm schema. The studio is located at the South Huangxing Roadpedestrian Mall, an important commercial district in Changsha, which is convenient for live broadcast and real-time interaction, besides, this area is full of fashion and modern elements, the architecture form can be easily perceived, it's easy to become a landmark of the city. In general, people perceive the live broadcast in the form of sound through radio channels, however, with the help of outdoor studio, people can not only hear but also see the scene.

3.1 The schema generation of architecture form logic

As for the study of the architecture form logic on this case, we will extract the perceptual schema from its functional image. The function of the studio is used for radio program production and broadcast. Here the program is output in the form of wave with different frequency and transmitted to the audience's receiving equipment within its signal coverage and then plays. In this process, the acoustic wave is converted into radio wave first and then transmits in the space, the receiving equipment then catches the wave and restores it back to the acoustic wave, so the audience can perceive the program. The wave becomes the vector that transmits and communicates with the program, so in our form-finding process on this case, since the form of wave has strong functional identification and can be perceived by the viewer in his intuition. The extraction form the perceptual schema of wave becomes starting point of architecture form logic generation.

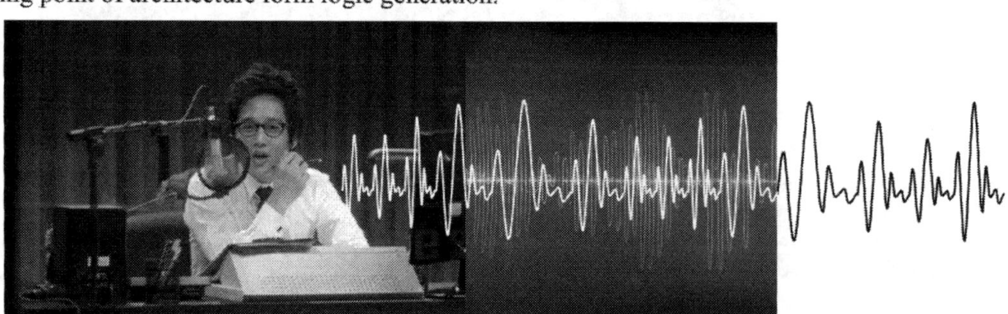

Figure 3. Acoustic wave is converted into radio wave first and then transmits in the space

For this case of the architecture form logic, we select wave image as the perceptual schema to generate the architecture prototype. Parameters such as period, amplitude, are able to affect the two-dimensional shape of the wave. After translating the two-dimensional into three-dimensional one, period and amplitude also has the possibility to shape the studio of three-dimensional architecture mass: period controls the depth dimension while amplitude controls the height of the studio and size of the bay. Afterwards, by increasing and moving the control point to adjust the modeling of the studio, we can obtain a preliminary architecture.

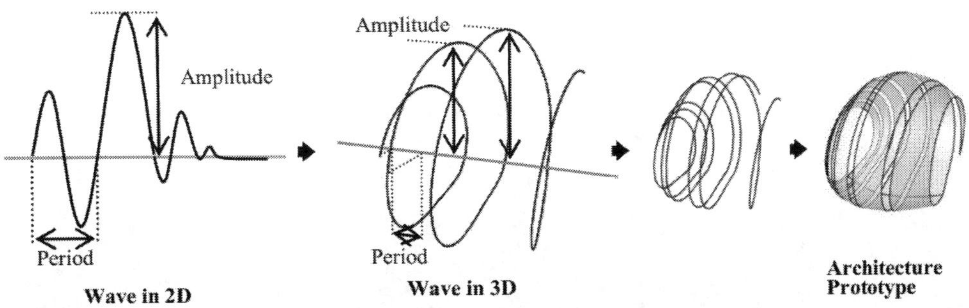

Figure 4. Period and amplitude as parameter of wave control and generate the shape of studio

The original architecture mass is confirmed by two parameters of wave, namely period and amplitude while the frequency affects the form of wave segmenting the evaluation of the architecture. After a comprehensive consideration of the function, form, aesthetic appreciation and mass of the studio, a section of frequency with 6 periods of wave was confirmed to be the prototype of the studio.

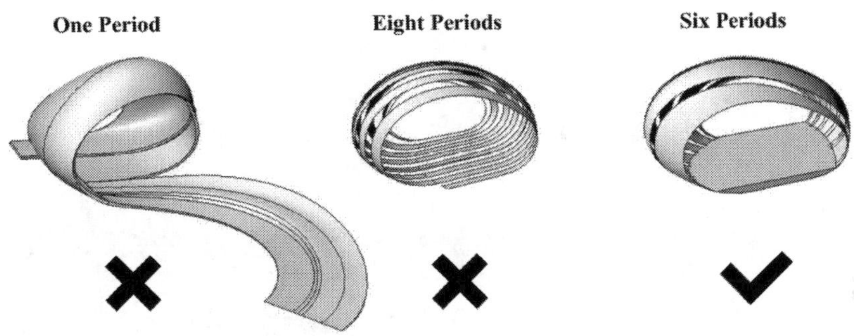

Figure 5. Wave with different frequencies generate different evaluation of architecture

3.2 The schema generation of architecture functional logic

The architecture form logic prototype is a schematization explanation of the wave. By translating the perceptual schema into the algorithm schema, period, frequency and amplitude have become the parameters that affect the generation of architecture mass. The original prototype of wave could not meet the functional requirement of the broadcasting studio. To make the architecture form more reasonable,we should make a topological transformation of the architecture prototype which is generated by the wave schema.

According to the requirements of design, the broadcasting studio should be divided into three main function areas: live broadcasting area, guided propagation area and leisure area. The live broadcasting area should be easy for lighting and have a good visibility. Moreover, it also could be divided into three areas respectively for radio host, guest and brief interactive performance. There should be a shoe changing area at the entrance of the studio. In addition, a small outdoor stage outside the studio is needed to make it convenient for the host to interact with passers-by when the program is on air. Then, we need to obtain a form logic from topological transformation of the wave schema to orgnize these functions rationally.

When we listen to the radio, a song, a word said by the host or even a story told by the audience may touch our hearts. The song may be a consolation when you are tired of the hard life; A story told by smoeone may strike a responsive chord among us. But there is nothing outside—not the broadcast content or the radio wave--touch our hearts. There is just the radio wave which is inside our hearts meet itself. In other words,we are just moved by ourselves. Accordingly, the schema of wave intersecting itself which is extracted from the perceptual image we depicted above formates a topological transformation structure of the architecture prototype. Extending the head and the tail of the wave and intersecting them inside the space so that it divides the space into two function areas---- live broadcasting and leisure area,guided propagation area. Extending the head and the tail of the wave continusly in order to fomate the outdoor stage and LED wall outside the space. In this way,the functional logic of the architecture's form which is extracted from the prototype of wave that has been established.

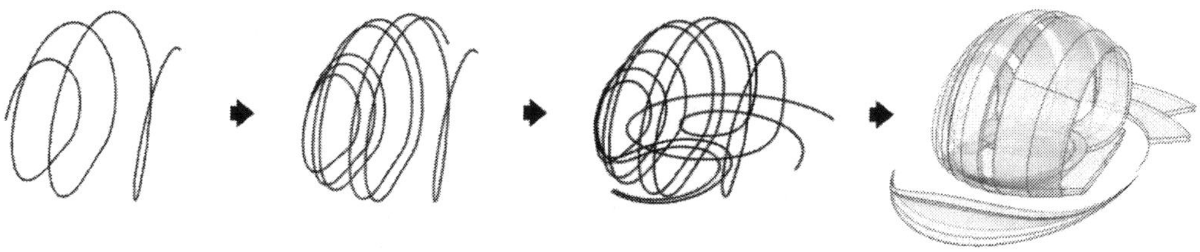

Figure 6. The generation of architecture form logic extraction from the schema of wave intersecting itself

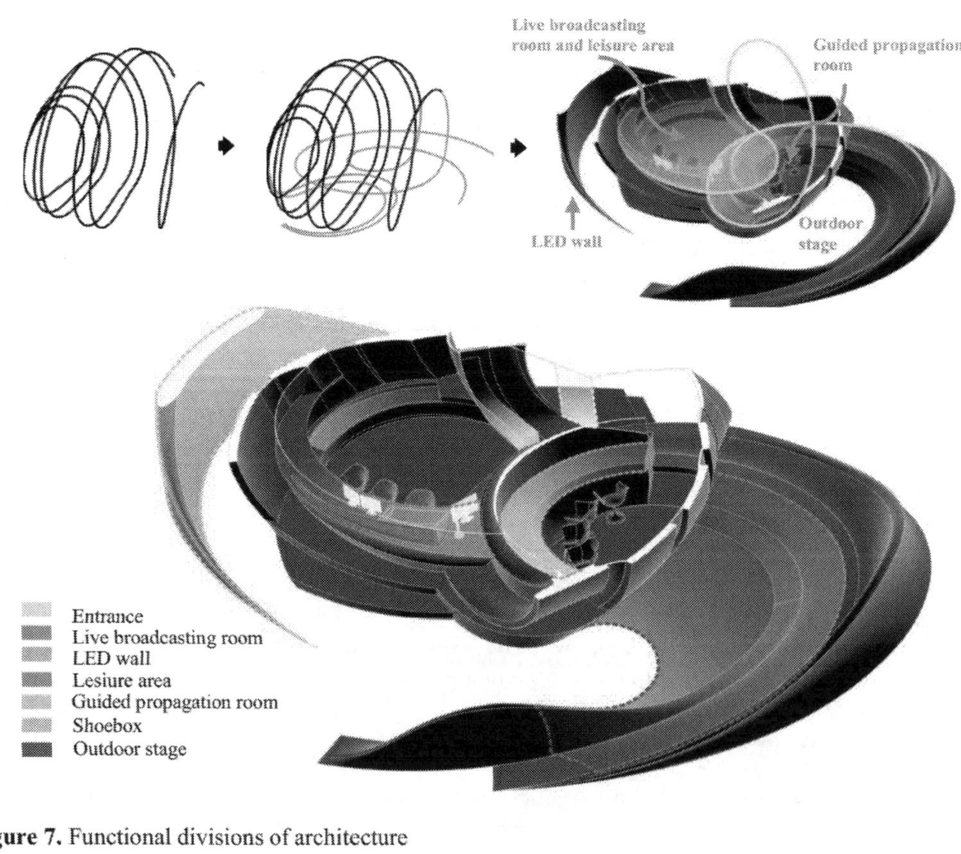

Figure 7. Functional divisions of architecture

DOI: 10.1051/matecconf/20168203001

Figure 8. Rendering image of the outdoor broadcasting studio

4. Conclusion

Schema can help us cognize the architecture form logic generation. The abstract logic which has effects on form generation could be understood through the way of perceptual schema. In addition, the abstract schema structure can be projected to architecture form logic generation, therefore a scheme logic could be established to generate rational architecture form in the way of parametric design. This paper explores a new method to use the schema theory combining the parametric design through the case of outdoor broadcasting studio. The purpose of this paper is to investigate the relevance of scheme logic in scheme cognition and parametric design, so as to seek a new way of architecture design method under the new environment of architecture design.

References

1. J. Piaget, *The Principles of Genetic Epistemology*(Rout ledge and K. Paul, London, 1972)

2. C.Alexander, *A Pattern Language*(Oxford University Press, London,1977)
3. Weiguo Xu, Architectural Creation and Thinking of the Parametric Design in China: Interview with Professor Xu Weiguo & Xu Feng of School of Architecture of Tsinghua University, Urban Environment Design, **35**,108 (2009)
4. Tianyu Zhang, Yukun Zhang, Di Wang, Space Organization of Architecture and the Psychological Schemata, Huazhong Architecture, **24**,65(2006)
5. Li Zhu, Nan Zhang, Xiaoying Qing, Research on Algorithm Schema of Parametric Architecture Design Based on Schema Theory, Advanced in Intelligent Systems Research,**126**,576(2015)
6. Xiong Wang, Guoping Pang,Schema Theory and Its Teaching Significance, Journal of Yulin Teachers College,**25**,98(2004)

MATEC Web of Conferences **82**, 03002 (2016) DOI: 10.1051/matecconf/20168203002
D2ME 2016

The Research on Color Grading of Green Jade Images Based on HSL Chromaticity Analysis

Lili Zhang [a] and Xinqiang Yuan
Gemmological Institute, China University of Geosciences, Wuhan, 430074, China

Abstract: This study aims to establish a simple and convenient method suitable for the market to grade the jade colors. Based on the HSL color model and with the assistance of the HSL analysis on green jade images acquired by camera flashlights, this study obtains the following six kinds of jade colors: melon green, emerald green, yellow green, bean green, light bean green, and light green, based on which the law governing the values and the value range can be determined, and the respective HSL value ranges ,together with their threshold values, for the seven green jades can be finally found, thus laying an important foundation for further researches on computer-based quick grading. It is also hoped that this method will be widely applied and promoted in the marketplace and realizes our ultimate goal of setting guidelines to commercial jade prices.
Keywords: Green jade; color evaluation; image analysis method; HSL value

1 Introduction

Image analysis method, also known as digital color camera colorimetric analysis method, will be applied in this research. Digital color cameras are equipped with CCD color chips full of light-sensitive photosensitive diodes that contain the so-called pixels. Each pixel in the CCD color chip is individually covered with red, green and blue filters so as to only sense the light in the corresponding color. For instance, a jade imaged to the CCD color chip through a digital camera's lens will be sensed by the pixel of the chip, after which electrical signals will be formed. The strength of the electrical signals will be expressed within the range of 0-255 and saved to the memorizer. These signals can be read on a computer, and the color images displayed on the screen. If chromaticity analysis is adopted for reading the signals, quantitative data describing the image colors, such as the R, G and B data, can be obtained. In spite of the fact that RGB data (also referred to as tristimulus values) are the most basic colorimetric parameters, namely, any color can be reproduced via the mixture of the three colors red (R), green (G) and blue (B), and they stand for the quantity of the RGB contained in these colors, such a color description differs from the color description practices of the author and is inconvenient to use. Therefore, RGB values are converted to the more widely-used values of hue (H), saturation (S), and lightness (L).

2 HSL Color Model

[a] Corresponding author : lilili29@sina.com

This research focusing on acquiring the color image of ordinary green jade on the market has its basis on the HSL color model, a model catering to human senses of colors. This model, as a color system established by the scholar Smith, is subject to artists' visual sense on color and lustre, light and shade and hue[2]. In the HSL model, all colors are described with the three basic parameters: hue, saturation and lightness. Therefore, users without knowledge of color mixture can specify colors as well. The difference between the HSL and the common RGB color models lies in that the latter is a psychophysical color system, while the former perception one[3].

3 HSL Colorimetric Test for Green Jade

3.1 Introduction of HSL Colorimetric Test for Green Jade

To better understand the colorimetric characteristics of green jade, the author has acquired a sample of 1590 green jade photos taken by cameras with flashlights, and selected 70 effective photos among the sample for this research. Next, obtain the HSL chromatic value for each photo with the help of relevant software by setting to the appropriate value on the HSL color-block coordinate axis, as shown in Figure 1 (H=120).

Though green jade exemplifies a wide range of color changes, a few such change series can still be observed and distinguished via naked eyes. The experiments divided the jades into various types, and a unanimous result of 6 distinctive color types is obtained. According to the color depth sequence from visual sense, the colors well-recognized in the jade industry are: melon green, emerald green, yellow green, bean green, light bean green, light green and pale green. In this experiment, HSL test was carried out on the 6 colors[4][5].

In addition, since jade has a complex color distribution, with some evenly distributed, while others are not, a color analysis solely dependent on the HSL values will with no doubt lead to bias. Therefore, for a more objective result, the author has in particular applied the software to get the average value of the jade image colors and selected the single point values for representative colors on the jade image. Besides, coordinate points indicating different colors for each kind of jade are used to mark on the color block coordinates (see Figure 1). Specific HSL analysis data are shown in Table 1 below, where data in the first row under the figure represent the average green values. For convenience, data recording is simplified, only data are kept here with their units omitted. For instance, in the table the first photo' HSL values are 128° , 25% and 52%, which are simplified as 128, 25 and 52.

3.2 Test data results as well as the distribution of the corresponding color block coordinate chart

Table 1. Images acquired and HSL parameters of green jade

	1	2	3	4	5	6	7
	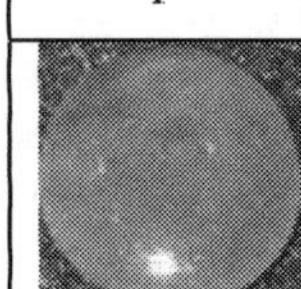	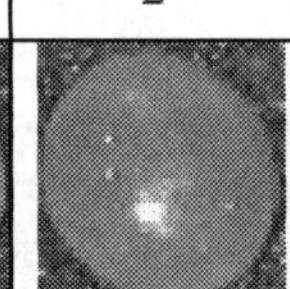	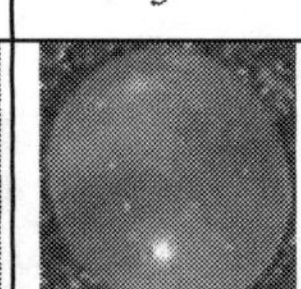	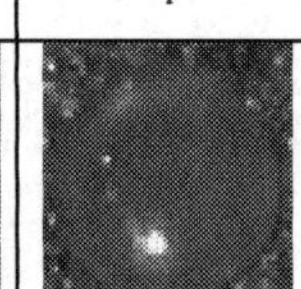	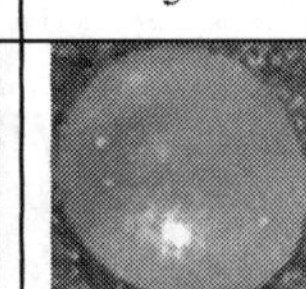	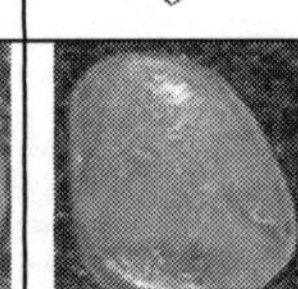	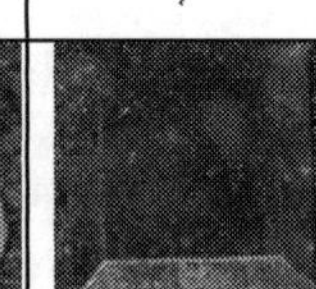
Melon green	128,25,52	135,28,47	124,32,41	131,55,25	128,31,48	126,43,47	141,30,18
	8	9	10	11			
	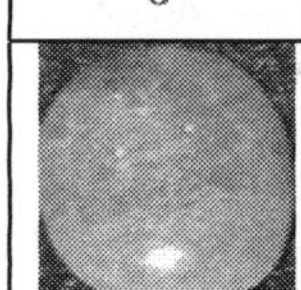	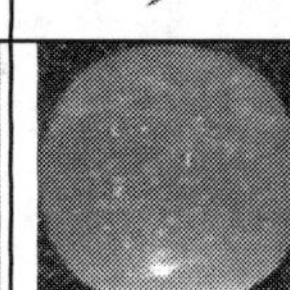	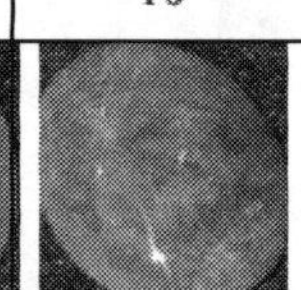	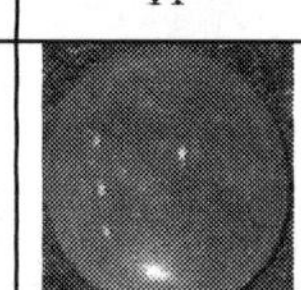			

	127,33,48	130,37,37	131,36,35	133,25,29			
	12	13	14	15	16	17	18
Emerald green	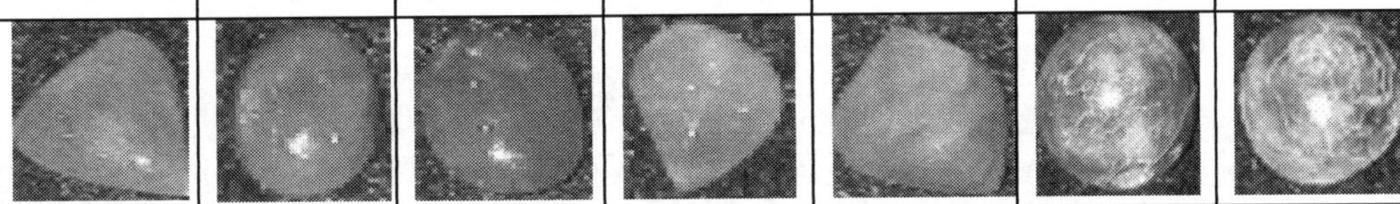						
	128,61,41	137,76,34	132,89,28	133,80,35	132,70,42	131,73,34	122,77,39
	19	20					
	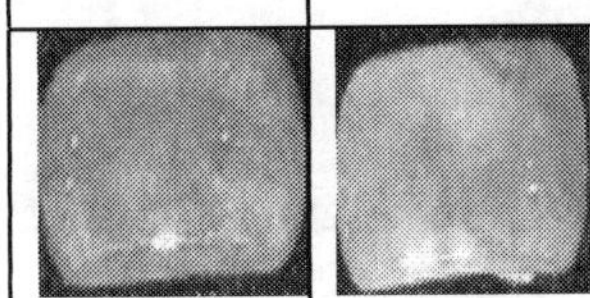						
	138,72,42	129,90,37					
Yellow green	21	22	23	24	25	26	27
	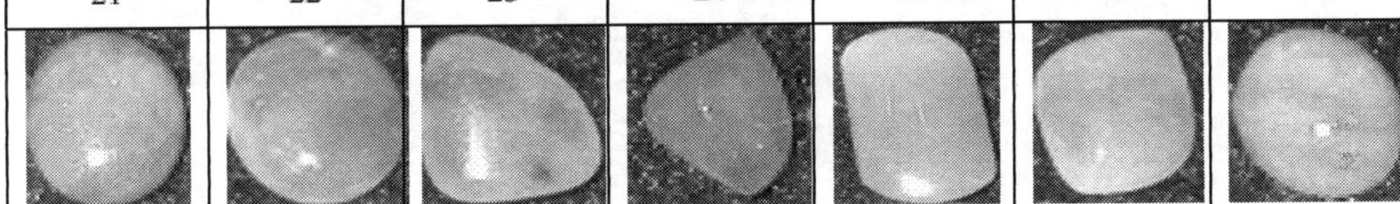						
	133,72,63	132,70,45	133,58,47	125,57,49	129,67,58	135,71,58	134,81,64
	28	29	30	31	32	33	34
	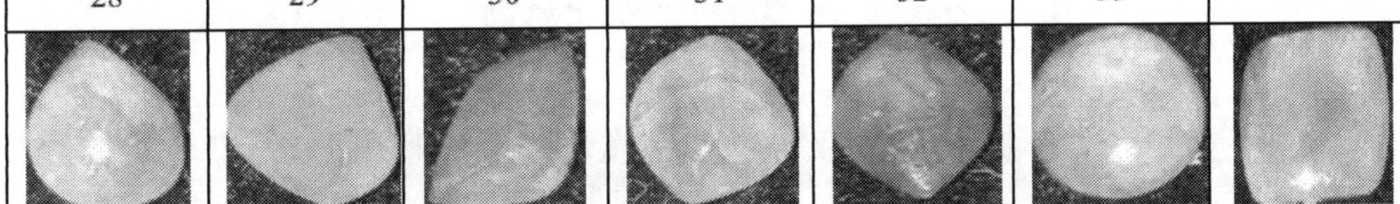						
	134,71,53	132,77,64	127,59,54	138,70,56	127,58,42	133,67,51	133,66,50
Bean green	35	36	37	38	39	40	41
	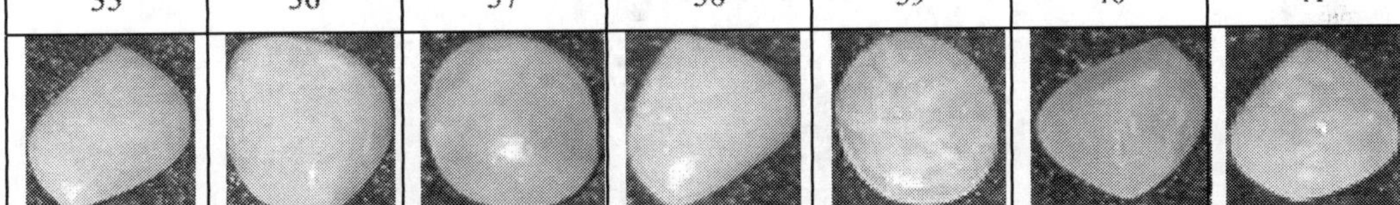						
	133,77,71	134,88,70	129,48,55	133,95,76	130,77,71	135,51,61	126,58,62
	42	43	44	45	46	47	48
	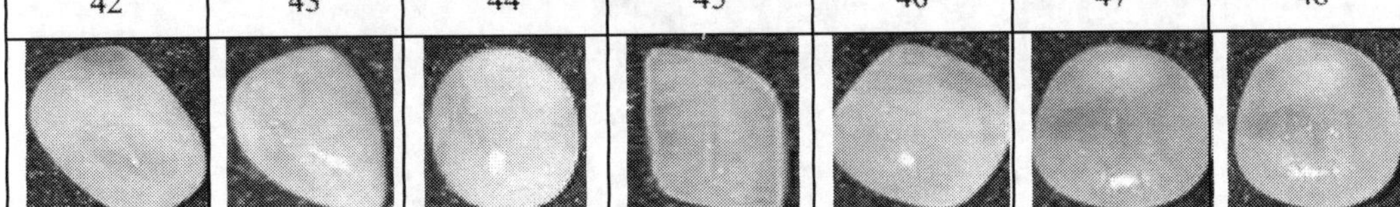						
	132,55,58	144,71,66	134,96,69	134,61,70	131,77,75	145,45,57	146,49,56
	49	50	51	52			
	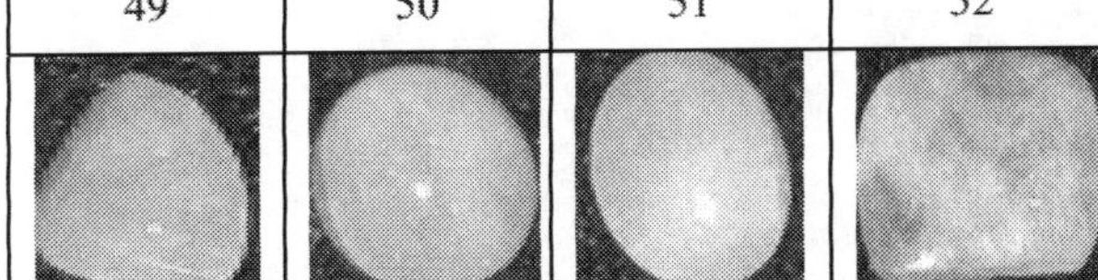						
	130,63,67	141,79,67	141,73,68	131,51,53			

Light bean green	53	54	55	56	57	58	59
	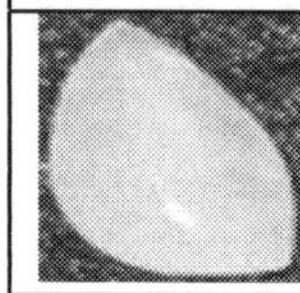	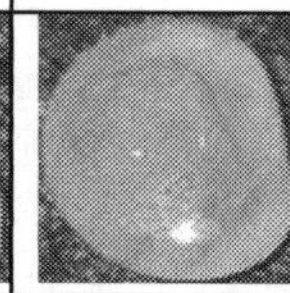	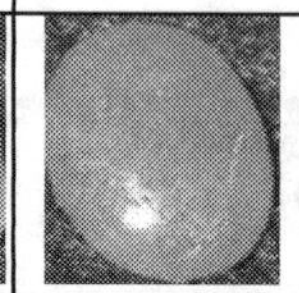	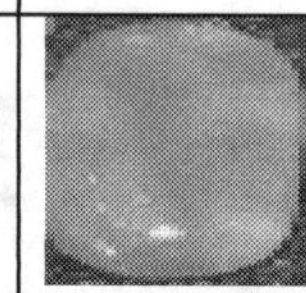	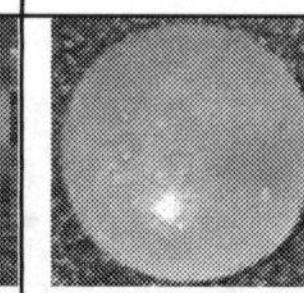	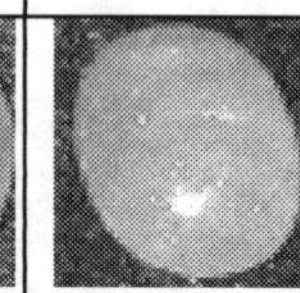	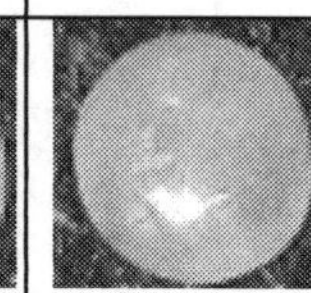
	136,88,80	137,31,61	123,29,57	126,28,56	128,34,59	131,43,65	129,48,70
	60	61	62	63	64	65	
	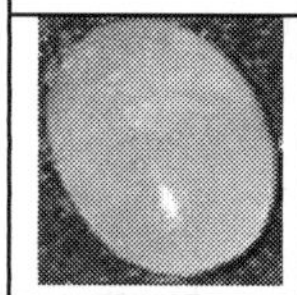	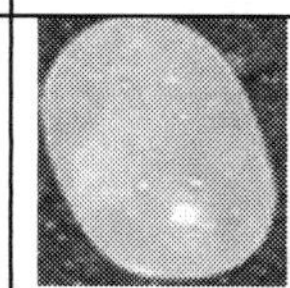	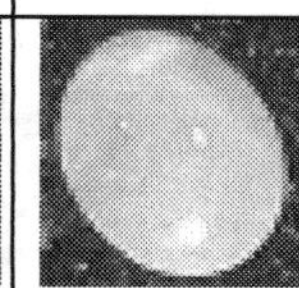	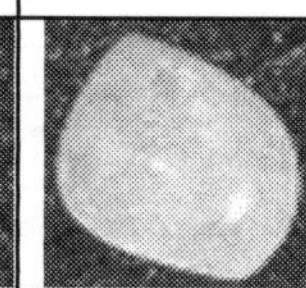	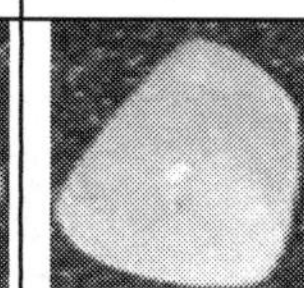		
	131,52,63	138,58,68	119,64,75	127,61,78	131,67,72	138,47,63	
Light green	66	67	68	69	70		
	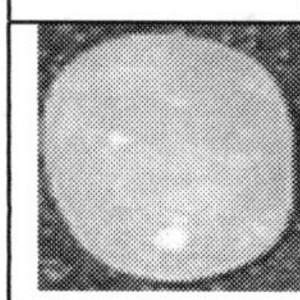	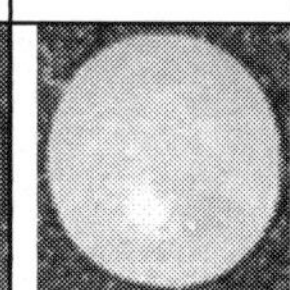	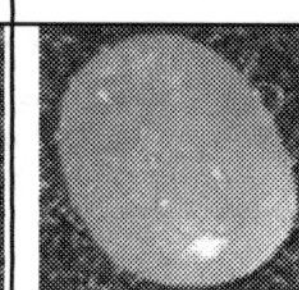	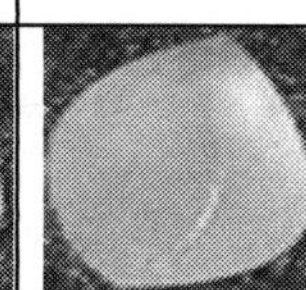	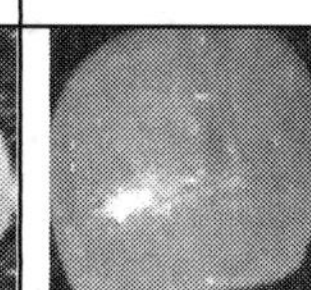		
	137,63,72	124,64,80	121,33,65	141,39,62	146,29,63		

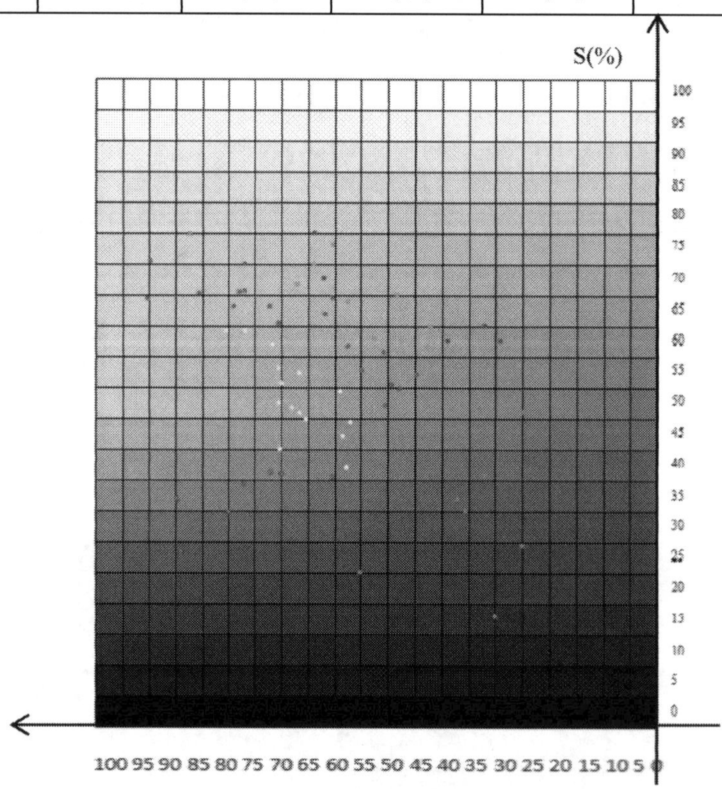

*S (saturation)*L (lightness)

*Melon green is marked with brown points, emerald green is marked with bright-red points, yellow green is marked with yellow points, bean green is marked with purple points, light bean green is marked with reddish orange points, light green is marked with blue points and pale green is marked with white points.

Figure 1. HSL color analysis color block coordinate of green jade-- average value

 DOI: 10.1051/matecconf/20168203002

4 Test analysis and discussion

The above HSL value test result shows that: The 6 colors of green jade are different, and based on such difference, various colors can be differentiated. HSL values of each color have their own scope of law, which can be seen obviously from the distribution characteristics analysis of the HSL values of green jade in the color block coordinates shown above. Jades with bright color and uniform color are characterized in high saturation of above 80%; moderate lightness, mostly of 20-30; hue of 120-149; which are green. Those whose hue is above 150 will have blue tone, affecting the green beautiful degree of the jade. Those whose lightness is below 10 are too dark, which can be classified into black jadeite. Those whose lightness is above 80, and saturation is below 20 can be classified into white jade due to light color, which is close to white (or colorless). The green beautiful degree of jade is mainly affected by saturation and lightness, definite relevance lies between the color saturation and lightness, while hue is secondary.

Problems emerged during the test: Since jade color is featured by hierarchy and complexity, the most common jades in the market often do not have the relatively single color change like diamond, lack of progressive color gradation. Therefore, even if differentiated with naked eyes, under the circumstance where samples are uniform, such jade with close color grade will have inconsistent color grading. The authors find from this test that, when analyzing the HSL values, for green jades with close color grade, boundary overlap or confusion of the color values will occur.

5 CONCLUSIONS

Through the above test and analysis of the HSL values of the green jade, HSL values of the seven kinds of green jade can be obtained, and through the analysis of these values, the seven kinds of color of the green jade can be numeralized. Besides, numerical range and threshold interval of each color of jade can be pinpointed (see table 2). This research makes the fast digitization of jade color possible, which can be served as the colorimetric indicators for color grading of green jade, thus, creating a new way for the fast and effective grading of jade color. Namely, color grade of the jade can be obtained quickly and precisely by image analysis of the HSL values of each jade. Meanwhile, it has laid an important foundation for the late stage research of computer jade color automatic grading system.

Table 2. HSL Parameter Distribution Scope of All Kinds of Colors of Green Jade

Color grade	Color name	Hue (H)	Saturation (S)	Lightness (L)
1	Melon green	120-150	20-50	15-40
2	Emerald green	120-135	60-90	28-40
3	Yellow green	125-138	60-80	45-60
4	Bean green	130-140	50-80	60-70
5	Light bean green	125-135	20-60	50-70
6	Light green	120-140	20-60	60-80

References

1. Mo Site, Liu Tianqi, Li Bixiong., HSL Color Space-Based Automatic White Balance Algorithm. *Journal of Sichuan University (Engineering Science Edition),* 6 (45): 95-99(2013)
2. Liu Xiao, Jiang Gnagyi, Wu Xunwei., HSL Space-based Color Morphological Transformation. *Journal of Circuits and Systems,* 3 (7): 52-56(2002)
3. Hen Zhang: Optimization of Green Jade color grading standard (MS., Kunming University of Science and Technology, China,14-24(2010)
4. Yuan Xinqiang., *Application Jade Gemology*. Wuhan: China University of Geosciences Press Co., Ltd.:91(2009)
5. Hui Zhang, Beili Zhang, Manjun Wang: Application of Method of Colour Measurement in Colour Appraisement of Jadeite Jade, *Journal of Gems and Gemmology*, Vol.**8** No.3:16-20(2006)

MATEC Web of Conferences **82**, 03003 (2016) DOI: 10.1051/matecconf/20168203003
D2ME 2016

Comprehensive Bridge Health Evaluation Method Based on Information Fusion

Li-ping Li[1,a], Wen-xia Ding[2] and Xiao-li Lu[3]

[1]*Faculty of Engineering, China University of Geoscience, 430074 Wuhan, China*
[2]*Hubei Communications Technical College, 430079 Wuhan, China*
[3]*Wuhan University of Engineering Science, 430200 Wuhan, China*

Abstract. In the bridge maintenance management decision, the health evaluation on bridge structure is far more than determination of abnormal index exceed fixed threshold, or evaluation on damage status and carrying capacity, the on-site human inspection should also be paid attention. In order to achieve more accurate and reliable health status evaluation result, it needs to take full advantage of various information and parameter. The D-S evidence information fusion was used in the paper to present comprehensive bridge health evaluation method. It fuses bridge operation status using abnormal status identification, bridge durability with residual life estimation, as well as human inspection rating, to arrive at health evaluation. The actual application examples to verify the method is also provided.

1 Introduction

The safety evaluation in bridge structural health monitoring is always one of the research focuses. Due to professional difference, the work of two main parts as information acquisition and safety assessment in structural health monitoring is completed by personnel of information science and mechanics science respectively. As a result, the existing structural safety assessment theory is limited to the mechanics field. Most methods depend on accurate structural model and known excitation system heavily. The safety assessment is achieved with reduction degree of structural stiffness [1, 2]. The health safety assessment of bridge always regards structural damage identification as premise, namely firstly determine existence of damage and locate it, then evaluate safety status of structure with stiffness reduction.

In the existing bridge structure safety assessment methods, most depend on single parameter as the evaluating index, such as bridge damage or durability. As to complexity of bridge structure and environment, as well as randomness of measurement result, the evaluation evidence is always inadequate. It is impossible to accurately describe overall status of bridge structure, or even arrive at wrong evaluation result, so it has not high confidence for evaluating with single parameter [3, 4]. Some researchers proposed multi-parameter evaluation method and achieved satisfactory effect. Therefore, it is a valuable focus to perform overall evaluation on bridge status in accordance with acquired data from sensors as well as various parameter using information fusion method [5].

[a] Corresponding author: llpok795@163.com.

The paper combines various parameter of bridge health status with information fusion method to arrive at overall assessment on bridge. The rest of the paper is organized as follows: Section 2 introduces D-S evidence theory of information fusion; Section 3 gives comprehensive health evaluation method based on multiple information fusion; Section 4 and 5 present the computation example and the conclusion.

2 D-S Evidence Theory

Suppose Θ as identification framework. If the set function $m:2^{\Theta}\rightarrow[0,1]$ meet $m(\phi)=0$ and $\sum_{A\subset\Theta} m(A)=1$, m is called the basic confidence distribution on framework Θ. As to $\forall A\subset\Theta$, $m(A)$ is the basic confidence distribution value.

In order to obtain total confidence of A, it needs to add the basic confidence distribution value of all subsets B in A, and then represent it with confidence function. Assume Θ as an identification framework, the set function $B_{el}:2^{\Theta}\rightarrow[0,1]$ is confidence function if and only if it meets conditions as following:

(a) $B_{el}(\phi)=0$;

(b) $B_{el}(\Theta)=1$;

(c) To $\forall A_1, A_2, \cdots, A_n\subset\Theta$, there is

$$B_{el}=\left(\overset{n}{\underset{i=1}{Y}} A_i\right)\geq\sum_{i=1}^{n}B_{el}(A_i)-\sum_{i<j}B_{el}(A_i\cap A_j)+\cdots+(-1)^{n+1}B_{el}(\bigcap_{i=1}^{n}A_i) \tag{1}$$

Set B_{el1} and B_{el2} as two confidence functions on same identification framework Θ, m_1 and m_2 as corresponding basic confidence distribution. Set $\sum_{A_i\cap B_j=\phi} m_1(A_i)m_2(B_j)<1$.

Then, the function defined by following is the basic confidence distribution of function $m:2^{\Theta}\rightarrow[0,1]$,

$$m(A)=\begin{cases}0, & A=\phi\\ \dfrac{\sum_{A_i\cap B_j=\phi} m_1(A_i)m_2(B_j)}{1-\sum_{A_i\cap B_j=\phi} m_1(A_i)m_2(B_j)}, & A\neq\phi\end{cases} \tag{2}$$

As to synthesis of multiple confidence functions, set $B_{el1},\cdots,B_{eln}$ as confidence function on the unified framework Θ, $m_1,\cdots,m_n$ as corresponding basic confidence distribution, if $B_{el1}\oplus\cdots\oplus B_{eln}$ exists and the basic confidence distribution is m, as to $\forall A\subset\Theta$, $A\neq\phi$, there is:

$$m(A)=k\cdot\sum_{\substack{A_1,\cdots,A_n\subset\Theta\\ A_1\cap\cdots\cap A_n=A}} m_1(A_1)\cdots m_n(A_n), \tag{3}$$

where, $k=\left[\sum_{\substack{A_1,\cdots,A_n\subset\Theta\\ A_1\cap\cdots\cap A_n\neq A}} m_1(A_1)\cdots m_n(A_n)\right]^{-1}$.

It can be seen from (2) and (3) that the combination of multiple evidences has nothing to do with order. The computation of multiple evidence combination can be obtained by recursive computation of two evidences [6].

3 Comprehensive Safety Assessment Based on Multiple Information Fusion

3.1 Evaluation Flow

The evaluation flow of the proposed method is shown in Fig. 1.

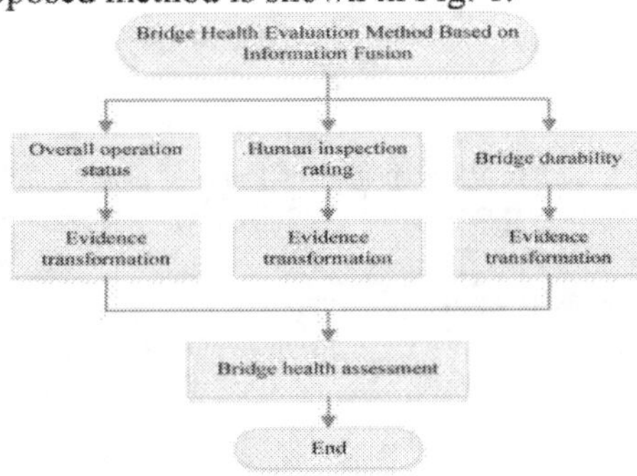

Figure 1. Multiple information fusion evaluation flow.

3.2 Evidence Transformation of Evaluation Information

As to comprehensive bridge health evaluation, its health status is a problem of data fusion, including statuses of health, unhealthy and uncertain. The paper constructs identification framework $\mu = \{A_1, A_2\}$. The $m(A_1)$ shows the structure is in safe state and $m(A_2)$ in unsafety status, so the uncertain status is $m(\mu)$. It is also an important in the information fusion to represent above evaluation results in the probability form and complete confidence distribution of different evaluation information.

3.2.1 Evidence Transformation of Abnormal Status Evaluation

The abnormal status evaluation on bridge usually determines whether some indicator exceed a fixed threshold. The abnormality degree of structure can also be obtained by computing indicator deviation from normal value. Due to complexity of structure itself and surrounding environment, it is quite difficult to determine a fixed evaluation threshold. If the inspection indicator is exactly equal to or close to the threshold, it is difficult to determine abnormal structure. The Exponentially Weighted Moving Average (EWMA) control chart was selected for bridge structure safety evaluation using monitoring information statistical analysis [7], Double control threshold alarming method was presented, namely preliminary alarm and safety alarm.

Referring to the EWMA control chart theory, the paper proposes double threshold abnormal status evaluation method, namely preliminary threshold δ_{min} and alarm threshold δ_{max}.

To address the bridge structure abnormal determination, the system uses threshold as:

$$\delta_{\lambda} = \bar{\lambda} + 4\sigma_{\lambda}. \quad (4)$$

In order to improve accuracy of structural abnormal determination, as well as reduce the difficulty when the detection indicator is close to δ_{λ}, the experiences can be used to set preliminary threshold $\delta_{min} = \bar{\lambda} + 4\sigma_{\lambda}$, alarm threshold $\delta_{max} = \bar{\lambda} + 6\sigma_{\lambda}$.

In accordance with the double threshold evaluation method and bridge management experiences, if the indicator obtained from abnormal diagnosis system is within the range of preliminary threshold, it may be considered $m(A_1) = 0.8$, $m(A_2) = 0.0$ and $m(\mu) = 0.2$. When the detection indicator is beyond the alarm threshold, it can be considered $m(A_1) = 0.0$, $m(A_2) = 0.8$ and $m(\mu) = 0.2$. If the indicator is between above thresholds, it can be considered $m(A_1)=0.33$, $m(A_2)=0.33$ and $m(\mu)=0.34$.

3.2.2 Evidence transformation of bridge durability

The bridge durability analysis can be obtained from durability of main structure. Boom is the key component to connect arch and road in the middle or low arch bridges. Its remaining life can be estimated from daily fatigue damage and cumulative damage degree. In order to evaluate fatigue durability of boom, the cumulative damage as well as relationship between remaining life and theoretical life should be used. The computation method considers not only cumulative damage of boom, but also effect of recent vehicle load on fatigue life. As the boom is force key of long span arch bridge, it directly affects the safety of arch bridge. The boom with the poorest durability can be used as basis to determine durability of bridge booms.

Represent the theoretical life is N and remaining life obtained from fatigue estimation as n. The proportion relationship ϕ between estimated remaining life and theoretical remaining life as shown in (5) can be used to evaluate durability referring to classification boundaries in the bridge level maintenance specifications. The corresponding confidence distribution can also be determined.

$$\phi=\frac{n}{N} \tag{5}$$

In accordance with bridge level classification boundaries, the confidence distribution based on ϕ is shown in Table 1.

Table 1. Boom durability confidence distribution.

ϕ	$m(A_1)$	$m(A_2)$	$m(\mu)$
$\phi\geq 0.88$	0.8	0	0.2
$0.88>\phi\geq 0.60$	0.7	0.2	0.1
$0.60>\phi\geq 0.40$	0.4	0.4	0.2
$\phi<0.4$	0	0.8	0.2

3.2.3 Evidence transformation of human inspection rating

The traditional bridge health monitoring system is overly dependent on data acquisition system, but despise human inspection. As the health monitoring system only deploys sensors on some key components, so the sensor network cannot cover whole bridge. The uncovered sites are usually blind area for safe operation of bridge, while regularly necessary inspection may supplement above corners of health monitoring system. Therefore, it is very important of human inspection to health evaluation.

As to bridge health monitoring in the paper, human inspection is also designed, such as regular testing on linear between rib and deck, settlement of piers, as well as corrosion cracking of components. The confidence distribution of human inspection is always based on experiences. After each time of human inspection, the technical condition rating can be obtained from management department. The corresponding confidence distribution based on bridge rating is shown in Table 2.

Table 2. Confidence distribution of human inspection status evaluation.

Bridge Level	$m(A_1)$	$m(A_2)$	$m(\mu)$
First	0.8	0	0.2
Second	0.7	0.2	0.1
Third	0.4	0.4	0.2
Fourth and fifth	0	0.8	0.2

4 Comprehensive Evaluation Example

Before safety assessment, it should firstly acquire needed data from corresponding database in accordance with evaluation needs. The evaluation amount can be obtained after information pre-processing. The preliminary result of bridge abnormal status, boom durability and human inspection will be transformed into corresponding evidence. The comprehensive assessment result can be achieved after D-S evidence fusion.

As the abnormal status diagnosis system is a near real-time system, the bridge is performed 6 time of abnormal diagnosis per minute. The boom lifetime computation updates once per day. Limited by factors, the update frequency of human inspection is uncertain. The maximum value of bridge status novelty indicator in each minute is used as basis for status scoring. The boom fatigue status rating and human inspection scoring are based on latest assessment.

Taking monitoring data of some bridge as example, the detection indicator data in early one hour is shown in Fig. 2.

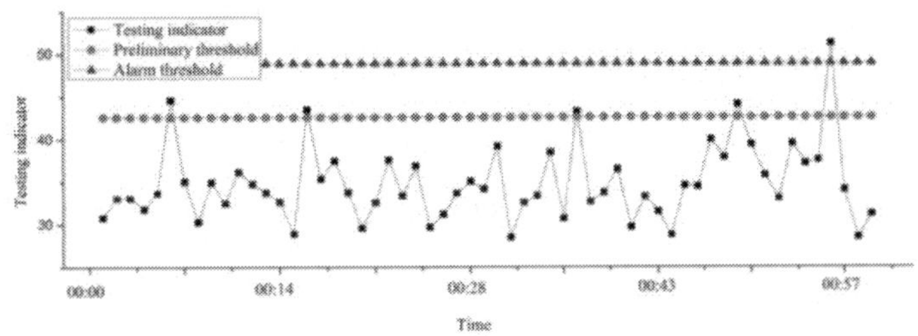

Figure 2. Overall status detection indicator.

As shown in Fig. 2, most detection indicators in the one hour are normal except for 5 points between preliminary threshold and alarm threshold, 1 point larger than the alarm threshold. It indicates that the bridge is normal in most of time. The sensor error or overload vehicles may temporarily impact on bridge structure. The monitoring indicator returns to normal after abnormal time points, which indicates the overall bridge structure is in good condition.

In accordance with computation result of daily damage, cumulative damage and remaining life of boom, the shortest remaining life of booms is 26.5 years. It can be regarded as the basis for boom durability determination. As the theoretical fatigue life of boom is 30 years and it has worked 1.25 years, the theoretical remaining life is 28.75 years. Based on (5),

$$\phi = \frac{n}{N} = 0.92 > 0.88 .$$

It indicates the bridge boom durability is normal and the confidence distribution is $m(A_1) = 0.8$, $m(A_2) = 0.0$ and $m(\mu) = 0.2$.

Based on latest bridge maintenance report and scoring result of management department, the bridge is in the second-class, so the confidence distribution is $m(A_1) = 0.7$, $m(A_2) = 0.2$ and $m(\mu) = 0.1$.

Conduct D-S evidence fusion on abnormal status evaluation combining with boom durability and human inspection result, the overall safety assessment can be arrived as shown in Table 3, Table 4 and Table 5.

Table 3. Overall bridge evaluation when the detection indicator is less than preliminary threshold.

Evidence	$m(A_1)$	$m(A_2)$	$m(\mu)$
Abnormal status evidence	0.8	0	0.2
Durability evidence	0.8	0	0.2
Human inspection evidence	0.7	0.2	0.1

D-S fusion evidence *m*	0.98	0.01	0.01

Table 4. Overall bridge evaluation when the detection indicator is between preliminary threshold and alarm threshold.

Evidence	$m(A_1)$	$m(A_2)$	$m(\mu)$
Abnormal status evidence	0.33	0.33	0.34
Durability evidence	0.8	0	0.2
Human inspection evidence	0.7	0.2	0.1
D-S fusion evidence *m*	0.93	0.26	0.02

Table 5. Overall bridge evaluation when the detection indicator is larger than alarm threshold.

Evidence	$m(A_1)$	$m(A_2)$	$m(\mu)$
Abnormal status evidence	0	0.8	0.2
Durability evidence	0.8	0	0.2
Human inspection evidence	0.7	0.2	0.1
D-S fusion evidence *m*	0.72	0.26	0.02

According to actual siting inspection, the bridge is in good operation, which is consistent with evaluation result obtained from comprehensive bridge safety evaluation based on D-S evidence fusion method.

5 Conclusion

Using D-S evidence fusion method, the paper achieves comprehensive bridge safety assessment fusing operation status, durability and human inspection result. The accuracy of health safety assessment can be improved. However, the paper does not consider weight in the proposed method. In the later research, the D-S evidence theory should be thoroughly studied and weight of each evidence be studied, so as to further improve accuracy of safety assessment.

References

1. Abdelfateh Kerrouche, J. Leighton, W. J. O. Boyle, Y. M. Gebremichael, Strain Measurement on a Rail Bridge Loaded to Failure Using a Fiber Bragg Grating-Based Distributed Sensor System, IEEE Sensors Journal, vol. **8**, no. 12, 2059-2065 (2008)
2. Victoria J. Hodge, Simon O'Keefe, Michael Weeks, Anthony Moulds, Wireless Sensor Networks for Condition Monitoring in the Railway Industry: A Survey, IEEE Transactions on Intelligent Transportation Systems, vol. **16**, no. 3, 1088-1106 (2015)
3. Zhou Jianting, Huang Hanmin, Huang Shanglian, Chen Weiming, Remote Real-time Health Monitoring and Evaluation System for Long Bridge Structure, *Proceedings of IMACS Multiconference on Computational Engineering in Systems Applications*, vol. **2**, 1751-1755 (2006)
4. Yu Chongchong, Zhang Jia, Tan Li, Wang Jinyan, Distortion data research of bridge structure health monitoring based on LS-SVM classification, *Proceedings of 3rd International Conference on Information Sciences and Interaction Sciences (ICIS)*, 118-123 (2010)
5. F. Bruschetta, D. Zonta, C. Cappello, R. Zandonini, Fusion of monitoring data from cable-stayed bridge, *Proceedings of IEEE Workshop on Environmental Energy and Structural Monitoring Systems (EESMS)*, 1-6 (2013)
6. Lan jin-hui, Ma Bao-hua, D-S evidence reasoning and its data fusion application in target recognition, Journal of Tsinghua University, vol. **41**, no. 2, 53-55 (2001)
7. Qin Ping, Condition assessment based on cointegration and multivariate control chart for large-span bridges, Chongqing University, 45-46 (2015)

MATEC Web of Conferences **82**, 03004 (2016) DOI: 10.1051/matecconf/20168203004
D2ME 2016

Design and Simulation of 3-DOF Reconfigurable Planar Parallel Robot

Zhili Hou [1,a], Wenge Wu [1] , Ruiqin Li [1] and Huibin Qin [1]

[1] *School of Mechanical and Power Engineering, North University of China, 030051Taiyuan, China*

Abstract. This paper presents the design approach for 3-DOF (degrees-of-freedom) reconfigurable planar parallel mechanism (RPPM) with modular kinematic limbs and hybrid platform. Using the proposed approach, a series of kinematic models with rotatability are obtained. At first, the main models of PPMs are constructed. Then a series of configuration models of PPM modules are reconstructed. And the new configuration families of 3-DOF PPMs are obtained. Finally, the motion trajectories of three driven layout configurations are simulated, and the results show that the hybrid-loop platform can expand the workspace of PPM. Especially, the configurations of the PPM can transform from the two operations modes to the original mode.

1 Introduction

In the past twenty years, researchers have made extensive research on reconfigurable mechanisms for its potential benefits and good application perspectives. Reconfigurable mechanisms can convert the original configuration from one to several new configurations in order to meet various task requirements [1]. The research involves the key problems such as configuration reorganization and reconstruction, singularity, controllability and so on.

Many application examples of reconfigurable parallel mechanism can be found in the recent studies [2–4] for both fields of machine tools [5–8] and industrial robotics [9–10]. And the end-effector of a manipulator is able to perform different kinds of motion.

A new reconstruction method for kinematic chain is the key step for the reconfigurable mechanism design. At present, the configuration reconstruction of parallel mechanism is mostly based on the diverse complex mathematical calculation, such as the screw theory [11], constraint changes characteristic of synthesis method [12], limb construction method [13], and so on. These calculation methods provide the algorithm rules for the configuration reconstruction.

In fact the concept of reconfigurable technology derived from the concept of modules as robotic cells [9]. For the design of most mechanical system, modular design has the advantages of high reliability and low cost, which is also suitable for the design of parallel mechanism [14].The parallel mechanisms are made of the same kinematic chains, and all the components used in each kinematic chain are the same [15].

3-DOF PPMs have the prominent characteristics, such as high rigidity, high precision and small workspace [16-17]. Because the traditional 3-DOF PPMs are not reconfigurable, the model design and

[a] Corresponding author : whitelily9988@163.com

kinematic analysis of the eight structures have been studied from one type to another. Most researchers focus on characteristics analysis of one typical configuration for the PPMs [18-19]. In fact, the two or more types of the 3-DOF PPMs have the natural basis for reconfigurable configuration design. Moreover the eight configurations of the 3-DOF PPMs proposed in the paper can be reconfigured with some form of manual assistance. It is very important to find a unified method to realize its varying kinematics, and the modular units can quickly reconstruct new configurations.

Since the numbers of configurations are large, the configuration transformation efficiency should be improved. A new integrated modelling method is presented for the 3-DOF PPMs in this paper, which contains scale information and motion information of mechanism.

2 Description of kinematic models

A planar parallel robot consists of a moving platform and a fixed base which are connected to several kinematic limbs. There are two types of active joints used in planar parallel robots, namely, revolute (R) and prismatic (P). All the rotation axes of the revolute joints are parallel to the direction of the actuator. The axes between revolute pair and the prismatic pair are orthogonal to each other.

When the order of the kinematic pair R and P are changed, in accordance with the R and P of different combinations, eight planar kinematic chains having various structures are synthesized, they are RRR, RRP, RPR, RPP, PRR, PRP, PPR, PPP, but 3-PPP kinematic chain structure can't make the moving platform move, so the configuration is removed. Therefore, seven kinds of kinematic chains can be synthesized, the classification of planar kinematic chains are shown in table 1.

Table 1. The classification of planar kinematic chains.

Number of prismatic pair	Types
0	RRR
1	RRP、RPR、PRR
2	RPP、PRP、PPR

The structure of planar kinematic chain is completely symmetrical for the 3-DOF PPM. In order to meet the requirements of the symmetrical workspace, the kinematic pair (R or P) near to the fixed platform is selected as the actuator. In accordance with the joint position of the driven pair, eight planar kinematic chains are classified, as shown in Table 2.

Table 2. Eight kinematic chains of 3-DOF PPM.

Actuator	Types							Remark
Joint 1	RRR	RRP	RPR	PRR		PRP	PPR	The underlined letter indicate the position of actuated joint
Joint 2			RPR		RPP			

3 Design methodology

3.1 Full information modeling method

Variable topology mechanism can realize special motion trajectories by changing topologized structure of the mechanism. Based on top-to-down design process, the full information models of PPMs are established, and the design process as shown in Figure 1.

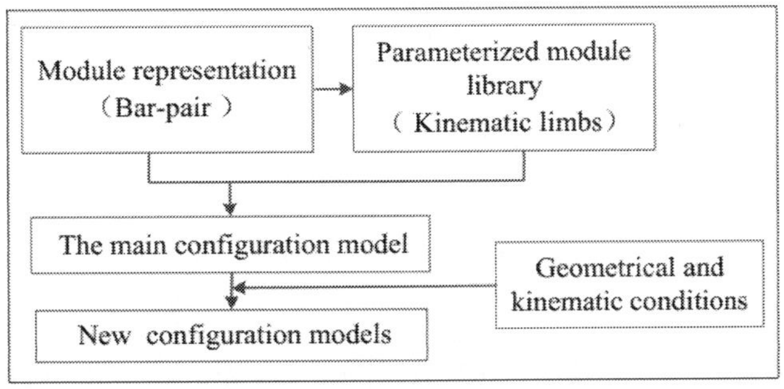

Figure.1 The top-to-down design flowchart

The full information model of a mechanism contains scale information and kinematic information. Scale information refers to the kinematic scale parameter (bar length, twist angle, axial length) and geometric constraint (co-axially, coplanar, parallel, etc.). Kinematic information includes the input and output motion parameters of PPMs. The full information models of the mechanism include main model and configuration rules. The main model contains full parameter model of the standard template.

Selecting a configuration of the PPMs as main model, a new family of PPMs and their configuration models are generated through topological transformation. It can be divided into three aspects to explain the topological transformation.

- The variable driven layout is changing position of the actuated pair through the hybrid platform.
- The variable parameter configuration is using the similar geometric features of configuration models to achieve a series of configuration models of PPMs.
- The variable kinematic chain is changing the modules of kinematic chains to construct new kinematic chains.

These transformation methods in the reconfiguration design reflect the relationship between the scale and motion constraints. The motion information can be mapped to the scale information.

3. 2 Configuration modules

In order to design the RPPM models properly, PPM can divide into five modules, such as the input module, kinematic limb module, connecting layer module, support module and output module. There are several design rules should be required as follows:

When the constraint relationships of the main parameters are established, series of configuration model library are generated for the PPM. Take the bar-pair model as an example, the scale information models of planar bar-pair are constructed, as shown in Figure 2.In order to avoid motion interference phenomenon, the joint between bar-pair models should be designed as layer arrangement assembly.

Three-dimensional planar kinematic chains are constructed, as shown in Figure 3. 3-DOF planar parallel robot can be reconstructed with eight kinematic chains models, as shown in Figure 4.

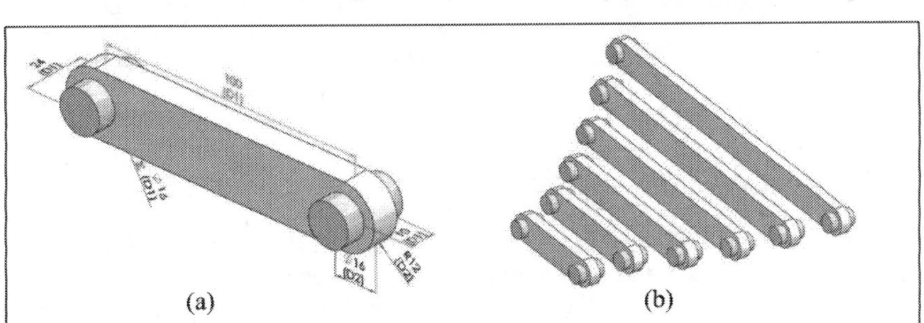

Figure.2 Models of planar bar-pair: (a) bar-pair model, and (b) serial bar-pair configuration models.

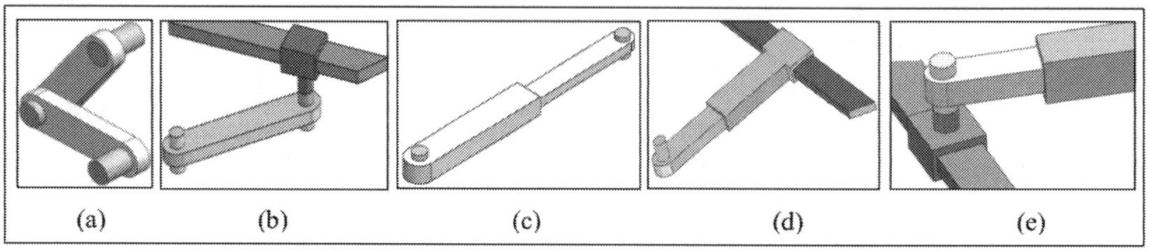

Figure.3 Models of planar kinematic chains: (a)RRR, (b)RRP/PRR, (c)RPR, (d)RPP/PPR, and (e)PRP.

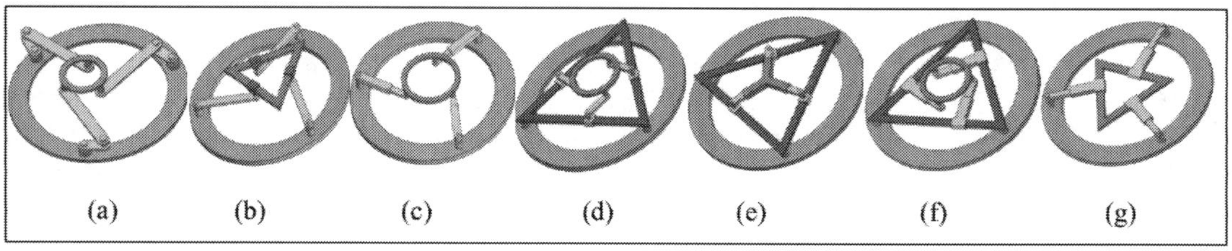

Figure.4 Models of 3-DOF PPM: (a)3-RRR,(b)3-RRP,(c)3-RPR or 3-RPR,(d) 3-PRR,(e) 3-PRP,(f) 3-PPR, and (g) 3-RPP.

4 Design case

When the driven layout of a mechanism is transformed, the different workspace of the mechanism can be obtained. Parallel hybrid platform is a new way to change the driven layout. If the parallel hybrid platform is established, the narrow workspace of parallel mechanism can be expanded.

Take the 3-RRR parallel mechanisms as an example, we establish the mixed orthogonal fixed platform and triangular fixed platform, as shown in Figure 5. When the joint P in the fixed platform is auto-locked in appropriate position, we can achieve different architectures of 3-RRR PPMs. The two modes (Figure 5) can also be swiched to the original mode (Figure 4(a)).When the joint P (Figure 5) is auto-locked in appropriate position, the original architecture of 3-RRR (Figure 4 (a)) are converted. Thus the workspace of the orginal PPM is transformed. The three configurations such as 3-RRP, 3-RPR and 3-RPR (Figure 4(a)-(c)) can also establish the hybrid platform to transform the driven layout.

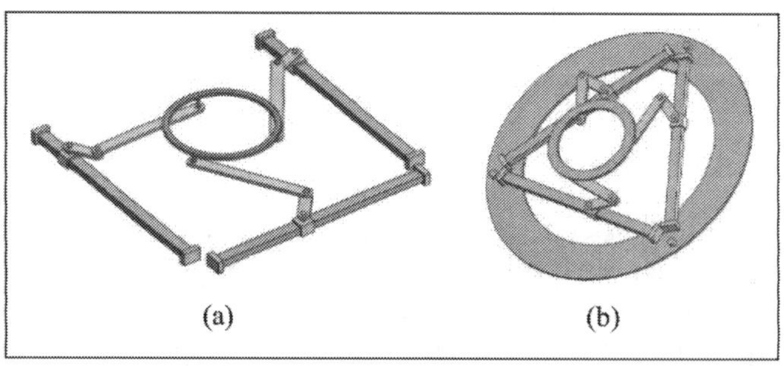

Figure.5 The parallel hybrid platform: (a) orthogonal layout and (b) triangular layout.

5 Workspace analysis and simulation

Workspace is an important characteristic for parallel mechanism. The workspace is defined as the region that consists of a large number of spatial points can be reached by the output point *P*.

Taking the kinematic model of 3-RRR parallel mechanism as representative object, and the motion model is shown in Figure 6. According to Literature [20], if l_4 is the longest link, $l_{4max} > l_2 > l_1$,

$l_2 > l_3$, and $l_{4\max} < \sqrt{3}l_2 + l_3$, we derive the conditions of three-crank 3-RRR RPPM, and the following inequality condition should be hold.

$$2(l_1+l_2) - l_3 < l_{4\max} < 2l_2 + l_3 \quad (1)$$

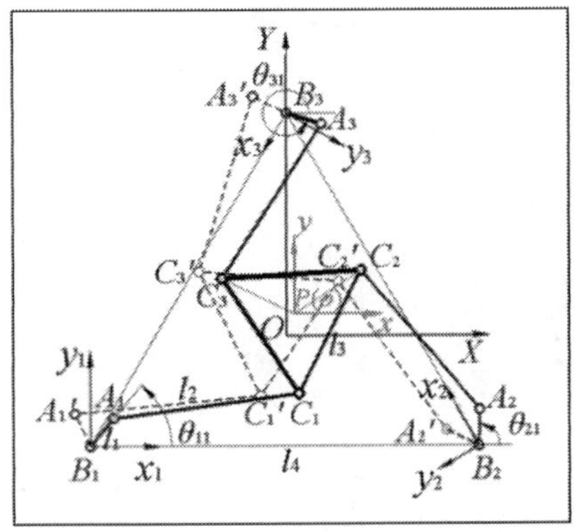

Figure.6 Motion model of a planar 3-DOF parallel mechanism with revolute actuators.

The symmetrical 3-RRR PPM is investigated as in Figure 6, with link lengths $l_1 = 80mm$, $l_2 = 400mm$, $l_3 = 300mm$, and $l_4 = 890mm$.

According to the kinematics analysis of section 3, we solve the reachable workspace with the help of Matlab software, and the results of the reachable workspace for 3-RRR PPM is shown in Figure 7. The workspace is a symmetrical distribution. The search method is based on the kinematic analysis, which requires much computation for more complex configuration.

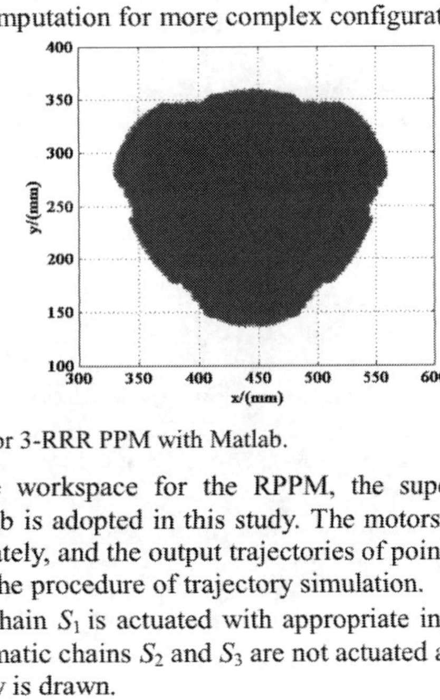

Figure.7 The reachable workspace for 3-RRR PPM with Matlab.

In order to draw the reachable workspace for the RPPM, the superposition method of motion trajectory for each kinematic limb is adopted in this study. The motors of the three kinematic chains (S_1, S_2 and S_3) are actuated separately, and the output trajectories of point P in the moving platform are tracked. There are three steps in the procedure of trajectory simulation.

• The first motor of kinematic chain S_1 is actuated with appropriate input parameters, but the motor parameters of the other two kinematic chains S_2 and S_3 are not actuated and follow the kinematic chain S_1.Thus the first motion trajectory is drawn.

• Refer to the first step, the other two motion trajectories can be drawn.

• The values of three motion trajectories are got in EXCEL CSV file format. And the three group values are superimposed together in one scatter diagram.

The trajectories of three metamorphic driven layouts for 3-RRR PPM have been simulated using the SolidWorks motion software, as shown in Figure 8.

 DOI: 10.1051/matecconf/20168203004

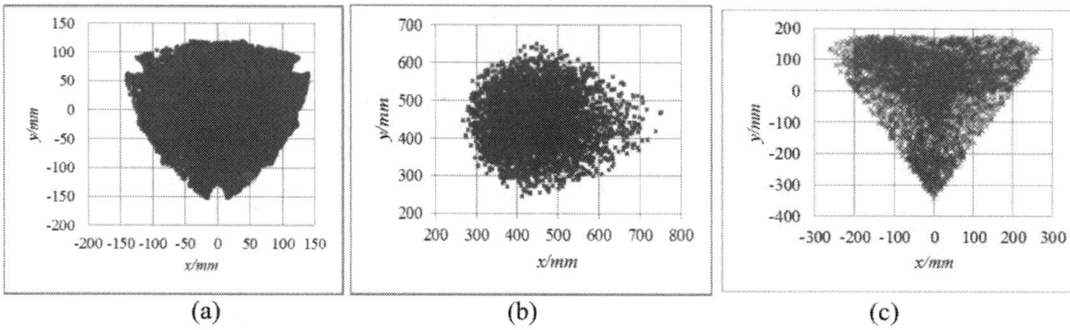

Figure.8 Reachable workspace of 3-RRR configurations by SolidWorks motion software : (a) original layout, (b) orthogonal layout, and(c) triangular layout.

Compared with Figure 7 and Figure 8(a), the almost identical graphics are obtained with the two methods. As can be seen from Figure 8, the motion trajectories are infinitely close to the boundary of the workspace. From Figure 8(a)-(c), three different workspaces by comparing the boundary coordinates can be found. Otherwise, hybrid platform can expand the workspace of the RPPM.

6 Summary

A new full information modeling method has been proposed to the RPPM. By establishing main models of the 3-DOF PPMs, a series of mechanisms configurations have been obtained from the original models. It not only supports the integration parameters modeling between the configuration design and motion analysis but also reconfigurable design of PPMs related to main model.

The superposition method of motion trajectory is adopted in this paper approximate to the real workspace, and avoids complex calculation process. The work has provided a new starting point for further research on reconfigurable mechanisms.

References

1. K.T. Zhang and J.S. Dai, J Mech Design **137**,1 (2015)
2. X. Kong, Mech Mach T 74, 188 (2014)
3. W. Ye, Y. Fang, K. Zhang, S.Guo, Mech Mach T **74**,1 (2014)
4. Q. Zeng, K.F. Ehmann, Mech Mach T **71**,1(2014)
5. L. Carbonari, M. Callegari, G. Palmieri, M.C. Palpacelli, Mech Mach T **79**,173 (2014)
6. D. Zhang, *Parallel robotic machine tools*(Springer, New York,2010)
7. Z. Bi, L. Wang, Robot CIM **25**,591 (2009)
8. Perez R, Molina A, Ramirez-Cadena M, J MANUF SCI **136**,213 (2013)
9. W. Yao , J.S. Dai, IND Robot **37**,542 (2010)
10. W. Yao, F. Cannella, J.S. Dai, Robot CIM **27**,604 (2011)
11. J.S. Dai, *Screw Algebra and Li Groups and Lie Algebras*(Higher Education Press, Beijing, 2014)
12. W.X. Zhang, X.L. Ding, J.S. Dai, J Mech Eng (in Chinese)**49**,601(2013)
13. S. Guo, Z.Y. Sun, H.B. Qu, J Mech Eng (in Chinese)**51**,35(2015)
14. F. Xi, Y. Li, H. Wang,*2010 International Conference on Mechatronics and Automation* (IEEE, New York,2010)
15. L. Carbonari, M. Callegari, Palmieri G, M.C Palpacelli, Mech Mach T **79**,173 (2014)
16. Y.K. Yong, T.F. Lu, Mech Mach T **44**,1156 (2009)
17. Z.F. Shao, X.Q. Tang, X. Chen, L. Wang, Chin. J. Mech Eng **22**,791 (2009)
18. E. Can, H. Stachel, Mech Mach T **79**,29 (2014)
19. B. Bihari, D. Kumar, C. Jha, V.S. Rathorea1, A.K. Dasha, Robotica **34**, 738, (2016)
20. R.Q. Li, J.S. Dai, Scie. China Ser. E **52**, 3601(2009)

MATEC Web of Conferences **82**, 03005 (2016) DOI: 10.1051/matecconf/20168203005
D2ME 2016

Discussion on the Coordinated Development of Ecological Protection and Tourism Development from the Current Situation of Country Park in Shenzhen

Hu yu[1] and Huang Hui[2,a]

[1] *Lecturer of Shenzhen Polytechnic, China*
[2] *Associate Professor of Shenzhen Polytechnic, China*

Abstract. Country parks consist of an important part of the urban green space construction system, which have a dual function of ecological protection and tourism. It is difficult to maintain an equilibrium between ecological protection and tourism development when it comes to the construction of country parks. Shenzhen has obtained great results in the construction of country parks, even thou some problems were encountered. Basing on the principles of ecological protection, this paper puts forward a combination of planning design and tourism planning, while fully considering the fundamental needs of tourism, in order to improve the construction of the country park facilities

1 Country Park Development

With the rapid economical development and the accelerated process of urbanization, the city environment is continuously deteriorating. Due to the lack of new urban development space, the urban sprawl has become increasingly prominent. The green space surrounding the cities have been occupied, the suburban living environment has been destroyed, and the biological diversity has also been threatened in many situations.To solve this problem in cities, building country park is a practical and efficient way.

The concept of country parks originated from the National Parks in the United States, which was set up in order to protect the natural environmental resources, the ecological systems, and the cultural and historical sites. The principle of country parks was first put forward by Britain, whose aim is to make people conveniently the suburb recreation, while reducing the destruction of rural resources, and protecting the rural landscape resources. According to Wikipedia: "A country park is an area designated to visit and enjoy recreation in a countryside environment [1]". In China, Hong Kong is the earliest city that has set up country parks, it is also the city that has the most successful and mature constructions of country parks. The emergence of country parks in Hong Kong came along the rapid economical development and urban expansion of the city. Country parks are established in goal of protecting the local natural environment resources and providing leisure recreation and natural education to the citizens. The Hong Kong country park concepts and systems have a quite deep influence on the construction of country parks in Mainland China.

[a] Corresponding author : Huang Hui, 641602451@qq.com

Because of the particularity in its geographical position, country parks are different from the traditional city parks, forest parks, and natural reserves. The location of country parks are selected mainly for its advantages in ecological protection. Tourism attraction and scientific education are supplemental characteristics. The key point is to protect the natural landscape resources, to improve the ecological conditions, to protect the diversity of the species, and to reduce the destruction of natural ecological environment done by urban expansion. As a new type of park in China, the construction and practice of country parks lack experiences. It is still in exploration and trial stage, a complete set of method system in planning and construction is lacking. Therefore, many problems have emerged in the construction of the existing country parks. Taking the construction of country park in Shenzhen as an example, this paper discusses the coordinated development of ecological protection and tourism planning in the construction of country parks, and hopes to promote the construction and sustainable development of country parks.

2 Shenzhen Country Park Construction Status

2.1 Shenzhen Country Park planning and distribution status

In order to develop the external ecological green space for the city's ecological conservation, Shenzhen is one of cities in China that start early on country park planning and construction. In "Green Shenzhen Urban Development Plan (2010-2020) " published in 2012 , a total delineated plan includes Tang Long Shan Country Park , Ma Luan Shan Country Park , Yang Tai Shan Country Park , Qi niang Shan Country Park and others 21 Country Parks (FIG. 1) , covering the city's six districts , with a total area of 680.7KM^2 [2], accounting for 13 % of the total area of Shenzhen and 27% of the city's total forest area. Until now, 7 country parks have been built.

Shenzhen country parks are concentrated in the eastern coasts and the northern mountainous area of the city, which form an east -to-west oriented ecological corridor. In the recreation space system, country parks are mainly responsible for the ecological conservation, natural resources protection, recreation, and the control of urban spatial culture at the same time [3].

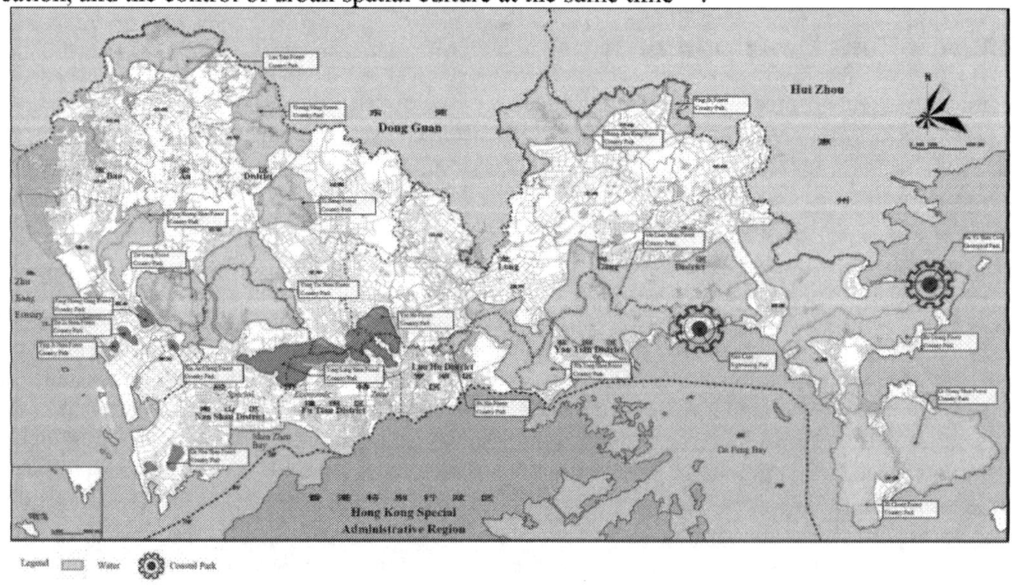

Figure 1. Shenzhen Country Park Map [4]

"Green Shenzhen Urban Development Plan (2010-2020) " mentioned that " Forest Country Park to protect the ecological construction, mainly to provide recreational activities and leisure facilities , supplemented by strict conservation of biological diversity, ecological resources and the countryside to watch value, is a prerequisite for development and construction" . In the beginning of planning, Shenzhen country parks are designed on the basic principles of ecological protection , in terms of natural resources, plant resources , animal resources , landscape and cultural resources to provide a detailed analysis on Shenzhen green space system. Basing on the survey of natural and human resources, evaluation, use of ecological suitability, ecological sensitivity index , slope geological hazard and the ecological carrying capacity of each country parks and other ecological function zonings, a distinction between the country resort and the ecological protection zone has been made, in accordance to the core area of ecology and the ecological restoration zone zoning plan.

Shenzhen Country Park Planning and Construction is a prerequisite for development of ecological protection planning. Ecological protection is the most important and the core aspect. The evaluation of landscape value are carried out in this context , but in terms of country parks , tourism and leisure also are very important aspects . It is also very critical in term of the perspective of tourism , interpretation and understanding of tourism resources with ecological protection needs, to enhance the overall resort.

2.2 Investigation on the tourism demand in Shenzhen country parks

Landscape value, price formulation and location choice of country park are key factors in the planning. Study found that a certain degree of preference to the functional types of the recreational product by the visitors can be expressed by the price, distance, and other factors. However, effective information extracted from the recreation preference survey, as well as the amount of empirical studies made to provide basis and suggestion to the country park planning are very few in the planning and designs of country parks in China [5]. Shang Fengbiao carried out a questionnaire survey for residents in 8 cities in China. After having analyzed the types of country parks, demands, management systems provided, service contents, etc, Shang Fengbiao have put forward 6 elements to construct the development system of country parks [6]. A questionnaire survey on the behavior characteristics of tourists in 2 Shenzhen country parks was carried out by Zhu Jiang. Statistics on tourists travel time, method, time spend, motivation, behaviors, activity type, and other indicators have been collected in percentages [7]. Table 2 provides the description on tourists' demands. Suggestions for the country park recreation facilities have also been put forward.

In seven country parks which have been built in Shenzhen, we conducted a survey of visitors . The survey contains the basic attributes of tourists, behavioral characteristics, satisfaction with three top items, subdivided into ten sub items.

2.2.1 Country Park Visitor basic properties (Table 1)

(1) Gender structure : Male to female ratio close to 2 : 1 ;
(2) Age structure : 88% of young , barely 65 years old ;
(3) Education: mainly High school and college;
(4) Income Characteristics : mainly middle-income;
(5) source distribution : mainly urban residents.

Table 1. Country Park Visitor basic properties.

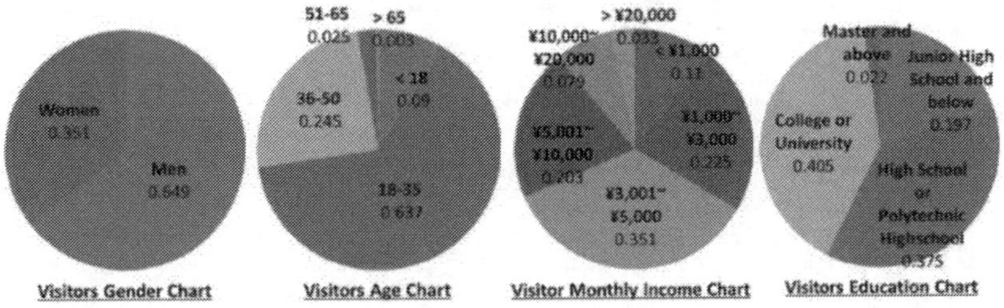

2.2.2 Tourists' Behavior Characteristics (Table 2)

(1) Visiting purpose : visiting and fitness .
(2) Visiting frequency : revisit higher.
(3) Duration of visits : mainly half-day , day trips.
(4) Events feature : mainly mountain climbing , secondly riding.
(5) Consumer features : mainly low consumption and zero consumption.

Table 2. Tourists' Behavior Characteristics.

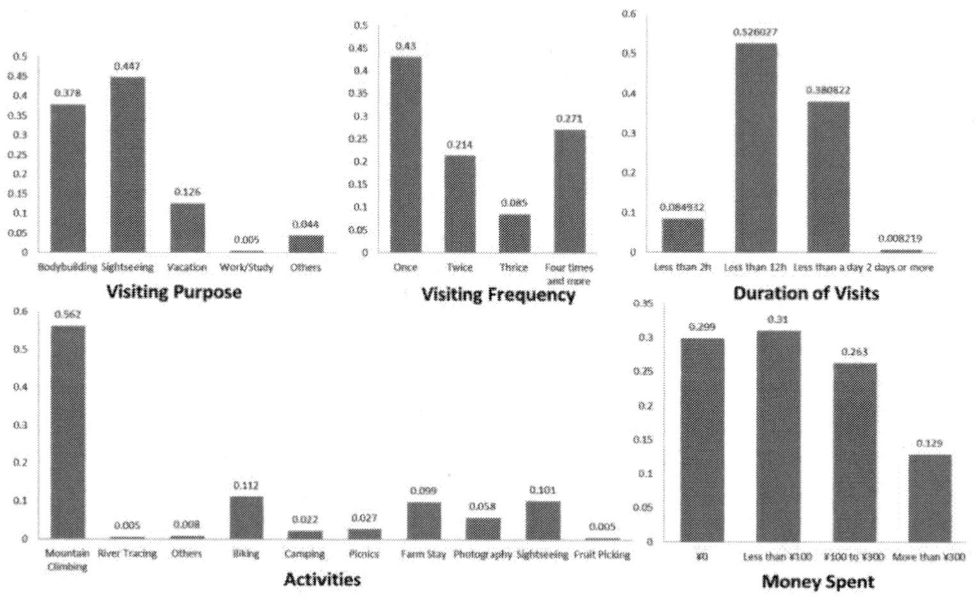

2.2.3 Visitor reception status

(1) Analysis of tourist satisfaction: overall satisfaction : 65% ; dissatisfied focus areas: identification system.

(2) Analysis of the number of visitors : tourists is growing rapidly ; Country Park is very obvious seasonal tourism , tourist busy season is from September to May of the following year; the tourist low season is from June to August each year . Tourism has entered the second stage of the life cycle.

(3) According to the tourism life cycle theories, by Butler , the Shenzhen Country Park is basically in "participatory phase" , the amount of tourists is increasing , locals are offering tourist infrastructure.

2.3 Problems in Shenzhen Country Parks

(1) serious recreation project homogeneity
(2) serious lack of service facilities and safety facilities
(3) imperfect the traffic system
(4) intelligent management system is not yet popularized

2.4 The root of problems

2.4.1 Ignorance of the rigid demands of tourism body

There is a deviation between law of tourism, tourism or biased understanding, including more development and interpretation of landscape, land, and technical point of view, ignoring the leisure and tourism body - the tourists. For tourists, the rustic charm of a country park meets more curious nature adventure of visitors. The number of tourists will not be decreased because of construction imperfections. Country Park should have basic facilities.

2.4.2 Conflicts between ecological conservation and tourism development

First, the overall outcome is the development planning of the premise of ecological protection. Ecological protection is the most important and prominent, including landscape value evaluation, which is carried out in this context. But for country parks, tourism and leisure is one of the very important aspects. It is also very critical in term of the perspective of tourism, interpretation and understanding of tourism resources with ecological protection needs, enhancing the overall resort.

3 Country park planning design ideas and suggestions

3.1 Combination of construction planning and tourism planning

The voices of commercial development and ecological protection have always existed. The concept of tourism project is in line with the current trend of tourism development. Considerations on forest protections fields and self driving camps are in line with the country's current tourism and leisure orientation, which is conductive for new tourism formats. After many years of exploration, ecological protection has become as the consensus of the community, when it comes to selecting between economic development and ecological protection. However touristic services tend to decline in its quality as ecological protection is chosen as the main construction and design criterion. Thus, when choosing between constructing a purely ecological protection park and a country park with tourism planning, the latter tend to be more beneficial.

3.2 Improvements on touristic infrastructures

3.2.1 Traffic facilities

Urban public transport should be enhanced in country park planning. Convenient external traffic can reduce the damage of self driving to the ecological environment. The key point is to focus on the coordination of "walking", "bicycles", and "vehicles" travel routes. In country park area, excluding necessary highway systems, walking and cycling routes should be prioritized. On the walk routes, there should be an exit in connection with public transport after every 4 hours of walk. In Shenzhen country parks, especially on mountains and hills, taking in consideration the age groups of visitors and the activities, trail slopes are designed as a combination of high oxygen level trails and medium oxygen level trails; flat hills can have low oxygen level trails.

3.2.2 Service facilities

Due to the large area of country parks, the number and location of service facilities are worth studying. Environmental facilities and rest facilities should be equipped within a certain radius of service. According to the tourism plan, country tours include half-day tours and full day tours. Catering facilities can be located in the terminal, designed for public consumption. Temporary supply points can be set up in halfway.

3.2.3 Safety facilities

Compared to city parks, more attention should be paid to country parks when it comes to safety hazards. Getting lost, falls, wild animal attacks, and fire are all potential disasters. In addition to the inconveniency in transportation, rescue operations are more difficult to execute. Thus, country parks should have enough necessary guide sign systems and safety relief facilities. If the economic conditions allow, intelligent monitoring systems can also be set up.

3.3 Overall Tourism Experience Structure

It is very important to make the tourism planning and the selection of the tourism projects in the country park planning (Table 3). It is beneficial to the rational use of ecological protection, tourism resources and sustainable development. Tourism implementation strategy (Table 4) is helpful to determine the functional zoning, the environmental capacity, the amount of tourists, and the method, routes, and contents of the visits.

Table 3. Country Tourism Planning

Experience System	Plan
Brand building	The characteristics of country parks and meticulous service are the foundation of brand building.
Atmosphere creation	Sports experience time, forest vacation time, green road and countryside time, culture and art time, business communication time.
Carrier Support	Mountaineering, fitness, education, rest, gathering, landscape viewing, natural education, ecological experience, forest health care, etc.
Spatial distribution	Hiking routes can be set in the buffer zone of the ecological protection zone. Cycling sports leisure area, red cultural leisure area, service area, and other services can be set up in the Kang Le recreation area.

Table 4. Implementation strategy of country tourism

System Analysis		Implementation Strategy
Travel Time		■ Mainly half-day tours and full-day tours. In order to reduce the number of service facilities constructed, two day tours or more should preferably not considered.
Tourism Connotation	Hardware	■ Strengthen the construction of service facilities ■ Perfect the slow green routes and the standard cycling trails.
	Software	■ construction of natural education path; meet the natural experience ■ integration of environmental science education; development of Forest Recreation ■ establishment of ecological protection base; protection of endangered plants ■ Creation of environmental education base; raise awareness on environmental protection ■ Linkage to the community comprehensive development; Highlights on the participation in projects

Technical Support	■ Intelligent Management System/ Ma Luan Shan Intelligent Service System ■ Intelligent Marketing System/ Ma Luan Shan Intelligent Marketing System ■ Intelligent Environmental System

References

1. https://en.wikipedia.org/wiki/Country_park
2. http://www.szpl.gov.cn/main/csgh/zxgh/ldgh/fb.htm
3. Gao Xiangduo， Chen Ban， Hu Zhiliang， Chen Jishen. Recreational Space Planning In Countryside Park, Tianjin[S].Planners，2015（11）：63-66
4. http://www.szpl.gov.cn/main/csgh/zxgh/ldgh/ft77.htm
5. FANG Jia, WANG De, ZHU Wei, SONG Shan, Residents' Recreation Preference of the Country Park in Shanghai: A Stated-preference Analysis, Chinese Landscape Architecture, [J],2016(4):50-55
6. SHANG Feng-biao, ZHOU Wu-zhong . Country Park Development Analysis and System Construction Based on Visitor Demands [J]. Journal of Northwest Forestry University，2009(1)：199-203.
7. Zhu Jiang. Country park planning in China: a case study of four cities in Hongkong, Shenzhen, Beijing and Shanghai [D].Beijing：China Academy Of Urban Planning and Design，2010.

MATEC Web of Conferences **82**, 03006 (2016) DOI: 10.1051/matecconf/20168203006
D2ME 2016

Influence Factors of Sports Bra Evaluation and Design Based on Large Size

Lingxi Zhang[1], Jianping Wang[1,a] and Ruixuan Han[1]

[1]*Fashion and Art Design Institute, Donghua University, 1882 West Yan'an Road, Changning District, China*

Abstract. The purpose of this paper was to find the main influence factors of sports bra evaluation by the subjective assessment of different styles commercial sports bra, and to summarize the design elements of sports bra for large size. 10 women in large size (>C80) were chosen to evaluate 9 different sports bras. The main influence factors were extracted by factor analysis and all the product samples were classified by Q-cluster analysis. The conclusions show that breast stability, wearing comfort and bust modelling are the three key factors for sports bra evaluation. And a classification-positioning model of sports bra products was established. The findings can provide theoretical basis and guidance for the research and design of sports bras both for academic and sports or underwear enterprises, and also provide reference value for women customers.

1 Introduction

The demand of sports bra is increasing due to the popularity of sports and the enhancement of the consciousness of bodybuilding. At the same time, more and more women begin to realize the importance of breast health protection during exercise, and they also give a higher demand to the comfort and functionality of sports bra. The researches on the comfort and evaluation of sports bra were mainly focused on two aspects: pressure comfort and breast displacement. Sports bra is more effective in reducing breast pain and breast displacement than other bra [1]. This protection effect is more obvious in the women with large breast sizes [2]. Zhou et al. [3] evaluated the reduction degree of breast displacement during exercise. The study concluded that the most effective bras had the following features: compression type, short vest style, high neckline, slings, cross back, bound neckline, no center gore, no wire, no cradle, no pad and a non-adjustable wide strap. Although there are many researches on the breast displacement and pressure comfort of sports bra [4-7], however, there is little published information on sports bra evaluation and its influence factors.

Given the lack of research about sports bra evaluation, the purpose of this paper was to find the main influence factors, as well as the design elements of sports bra evaluation by the subjective assessment of different styles commercial sports bra from large size women. A classification-positioning model of sports bra products was established afterwards. It can provide theoretical basis and guidance for the research and development, as well as the products design and evaluation of sports bra both for academic and sports or underwear enterprises, and also provide reference value for women customers.

[a] Corresponding author : wangjp@dhu.edu.cn

2 Methods

2.1 Experiment samples

9 different styles commercial sports bras were chosen as the experiment samples. The detailed information was shown in Table 1. The 9 chosen samples are from different sports or underwear brands with high awareness.

Table 1. Information of experiment samples.

No.	Components（Body）	Style characteristic	Figure
1	Nylon 90.0% Spandex10.0%	Light support, inserted pads, adjustable elastic straps	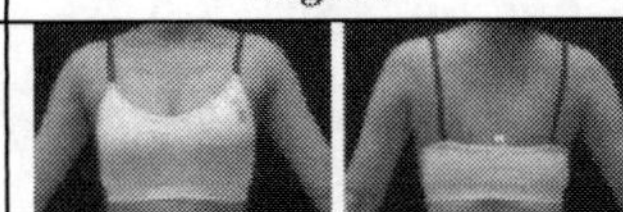
2	Nylon 79.0% Spandex 21.0%	Molded cups, air holes in pads, adjustable elastic straps	
3	Polyester 43.0% Nylon 38.0% Spandex 19.0%	Molded cups, air holes in pads, racerback	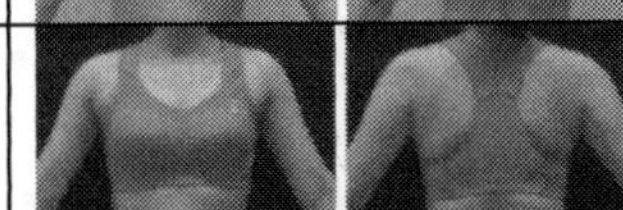
4	Polyester 88.0% Spandex 12.0%	Medium support, molded cups for large size, racerback	
6	Nylon 82.0% Spandex 18.0%	Double layer fabric, with wires, adjustable elastic straps	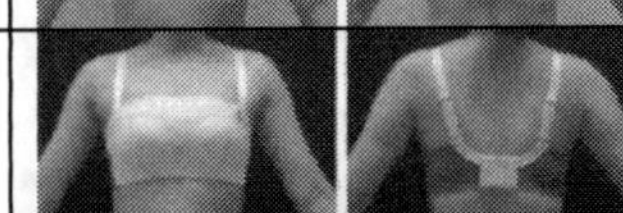
7	Nylon 73.0% Spandex 27.0%	Molded cups, with wires, adjustable elastic straps	
8	Polyester 82.0% Spandex 18.0%	Light support, double layer fabric, molded cups, racerback	
10	Polyester 75.0% Polyurethane 10.0% Nylon 5.0% Spandex 10.0%	High support,inserted pad, racerback	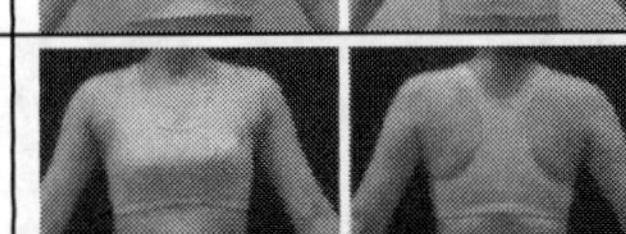
11	Nylon 58.0% Cotton 24.0% Spandex 18.0%	Light support, adjustable elastic straps	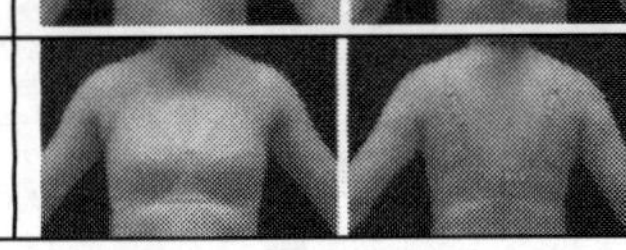

2.2 Participants

Subjects were 10 female (C80, D80, E80) who did sports regularly and were familiar with sports bra, with the age range of 18-30 years. The sizes of sports bra samples corresponded with each breast size.

2.3 Subjective assessment

The experiment was taken by means of subjective assessment. Firstly, a pre-experiment was done to sum up the high mentioned evaluation items and focus points when the subjects were fitting. The results show that the large size pays more attention to the cup inclusiveness and bust gather. The subjects were asked to rate the samples on a scale of 1-5 (1 refers to very uncomfortable, 5 refers to very comfortable), including four stages (before fitting, after fitting, after sports and comprehensive assessment). Table 2 showed the detailed subjective evaluation items.

Every bra would be evaluated by each subject, without time limitation and in a quiet room maintained at a temperature of 20℃ and a relative humidity of 60%. Before fitting, the subjects started to rate the thickness of fabric (1 refers to very thick, 5 refers to very thin) and fabric preference. Then, the subjects began to fit the sample bras in no particular order and were asked to rate the corresponding evaluation items. Afterwards, according to the experiment set, the subjects were required to do five motions under three different exercises intensity (light, medium and high), including arm lifting, body twisting, stand and reach, high knee stepping and jump jack. Each movement repeated three times. After all the motions finishing, the subjects started to rate corresponding evaluation items on a scale of 1 to 5. Finally, the subjects gave a comprehensive preference evaluation for each sample bra.

Table 2. Subjective evaluation items.

Stages	Evaluation items
Before fitting	Thickness of fabric, fabric preference
After fitting	Tightness of under bust, tightness of shoulder straps, cup inclusiveness）, bust modelling（up、protrusion、gather）, outwear, easy to wear
After sports	Breast stability, under bust stability, shoulder straps stability, shoulder straps comfort
Comprehensive assessment	Overall preference

3 Results

3.1 Correlation between products preference and other evaluation items

Pearson correlation was taken. Before fitting was set as state I. After fitting was set as state II. And after sports was set as state III.

As shown in Table 3, before fitting, products preference has a negative correlation with thickness of fabric (P=0.003), while it has a positive correlation with fabric preference (P=0.000). After fitting, products preference is significantly correlated with bust modelling (including protrusion, gather and up) and cup inclusiveness (P=0.000, 0.006, 0.000, 0.004, 0.001). It shows that the large size is more sensitive to cup inclusiveness. After sports, products preference has a positive correlation with breast stability and under bust stability (P=0.000, 0.001).

Table 3. Correlation coefficients of sensory values for each evaluation item.

State	Correlation				
I	Correlation	Thickness	Fabric		
	Preference	-.310**	.380**		
II	Correlation	Under bust tightness	Shoulder straps tightness	Cup inclusiveness	Bust modelling
	Preference	-0.112	0.026	.353**	.597**
	Correlation	Protrusion	Gather	Up	Easy to wear
	Preference	.287**	.463**	.303**	0.182
III	Correlation	Breast stability	Under bust stability	Shoulder straps stability	Shoulder straps comfort
	Preference	.440**	.348**	0.149	0.105

**. Correlation is significant at the 0.01 level (2-tailed).
*. Correlation is significant at the 0.05 level (2-tailed).

3.2 Factor analysis of evaluation items

Based on the results of correlation, factor analysis of evaluation items was done by means of principal component analysis.

In terms of large size, three factors were sufficient to describe the data, given in Table 4. Factor one is highly related to bust modelling (including up, gather and protrusion), defined as modelling factor. Factor two is highly related to breast and under bust stability, defined as stability factor. Factor three is highly related to fabric and cup inclusiveness, defined as inclusiveness and comfort factor.

Table 4. Rotated component matrix[a].

Factors	Items	Component		
		1	2	3
Modelling	Breast modeling	0.779	0.261	0.104
	Protrusion	0.735	-0.245	0.259
	Gather	0.756	0.301	-0.015
	Up	0.816	0.253	-0.074
Stability	Breast stability	0.147	0.88	-0.125
	Acceptance degree of breast motion	-0.268	-0.816	-0.041
	Under bust stability	0.046	0.628	0.407
Inclusiveness and comfort	Fabric	-0.017	-0.06	0.889
	Cup inclusiveness	0.292	0.291	0.408

3.3 Cluster analysis of products based on evaluation factors

Based on the results of factor analysis, factor assignment was done for each product sample before cluster analysis. This paper aims to sum up the classification positioning of sports bra. The method used Q-cluster in hierarchical cluster analysis.

Nine products of the large size were divided into three groups (Table 5). Taking reference to average score of evaluation items (Table 6), key findings show that products no.1, 6, 8, 11 with high scores in cup inclusiveness, along with shoulder strap pressure comfort and easy to wear, belong to the product with good inclusiveness and high comfort; Class two are products no.3 and 7, which gets high scores in breast modelling (including protrusion, gather and up), defined as the product with good modelling; Products no.2, 4, 10 with good performance on breast and under bust stability belong to the product with good motion stability. However, these products get low scores in breast modelling, especially no.10. Product no.4 shows poor performance on shoulder strap pressure comfort. Many subjects felt uncomfortable on the shoulder points.

Table 5. Agglomeration Schedule.

Stage	Cluster Combined		Coefficients	Stage Cluster First Appears		Next Stage
	Cluster 1	Cluster 2		Cluster 1	Cluster 2	
1	8	11	0.052	0	0	2
2	1	8	0.322	0	1	6
3	2	10	0.419	0	0	5
4	3	7	0.567	0	0	7
5	2	4	0.661	3	0	8
6	1	6	1.132	2	0	7
7	1	3	1.664	6	4	8
8	1	2	2.739	7	5	0

Table 6. Average score of evaluation items.

Class	No.	Fabric	Cup inclusiveness	Breast modelling	Protrusion	Gather
Inclusiveness and comfort	1	3.5	3.8	2.8	2.9	2.1
	6	4.1	3.4	3	4.4	2.9
	8	3.5	4.1	3.2	3.2	2.7
	11	3.8	4.2	2.9	3.6	2.4
Modelling	3	4.2	4.7	4.1	4	3.9
	7	3.6	4.2	3.6	3.9	3.4
Stability	2	3.6	4.1	3.2	3.7	2.7
	4	3.8	4.4	3.1	2.3	3.1
	10	3.2	3.1	2.7	2.8	3.4
Class	**No.**	**Up**	**Breast Stability**	**Under bust stability**	**Shoulder strap comfort**	**Easy to wear**
Inclusiveness and comfort	1	2.6	1.8	2.9	4.1	4.5
	6	3.6	1.5	3.3	4.4	4.2
	8	3	2.2	3.6	4.2	3.1
	11	3	2.4	3.4	4.8	4.7
Modelling	3	4.1	3.4	4.6	3.9	3.5
	7	4	2.4	4.4	4.2	4.7
Stability	2	3.9	4.4	4.3	4.3	4.5
	4	2.9	3.8	4.7	2.6	2.8
	10	3.4	3.7	4.7	3.4	2.8

4 Discussions

4.1 Influence factors of sports bra evaluation

The focus of large size women is relatively concentrated. Taken together, the main evaluation factors are breast modelling, motion stability and comfort and inclusiveness.

Related research [8,9] showed that 56% of the women had breast pain and discomfort during exercise. The prime reason is that the breast is doing long-term reciprocating motion during exercise, resulting in breast tissue stretching deformation, even breast ligament injury. Therefore, good breast stability in movement process is one of the key factors in the evaluation of sports bra.

Pressure comfort has been one of the focuses on sports bra comfort. Bowles et al. [10] found that cross-back shoulder strap produced more pressure than traditional shoulder strap, but the difference in pressure produced of two shoulder straps did not reflect on the difference of shoulder comfort. In this evaluation experiment, the results demonstrate that the pressure comfort of cross-back shoulder strap is lower than that of traditional one.

Although sports bra is used in sports, most experimenters put forward the requirement of breast modelling in the evaluation process. Breast modelling effect directly affects the final results of the evaluation. The sports bra with wires had better breast modelling, but the wearing comfort was reduced instead. Therefore, not only the motion function and wearing comfort need to be improved, breast modelling should also be taken into account in sports bra design and research.

4.2 Establishment of classification-positioning model of sports bra products

This paper summed up the key evaluation factors of sports bra products by factor analysis. Namely, modelling factor, stability factor, inclusiveness and comfort factor for large size. Combined with the factor value assignment results of products, a three-dimensional classification-positioning model of sports bra products was established, as shown in Figure 1. From the coordinate model, the classification positioning of different products and the features of each classification group can be seen directly, providing reference for enterprises in the future commercial positioning.

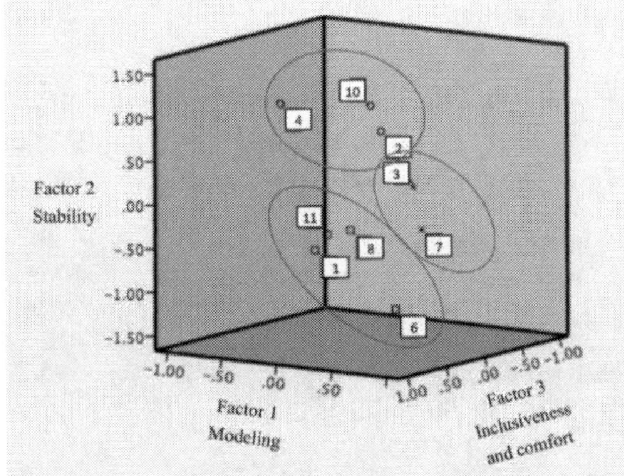

Figure 1. Classification-positioning model of sports bra products.

5 Conclusion

This paper summed up the main influence factors of sports bra evaluation by the subjective assessment of different styles commercial sports bra based on large size, and summarized the design elements of sports bra for large size. Breast stability, wearing comfort and bust modelling are the three important factors for sports bra evaluation. Wearing comfort refers to fabric comfort and pressure comfort. And pressure comfort mostly reflects on the shoulder strap pressure. A three-dimensional classification-positioning model of sports bra products was established based on the influence factors, providing a guiding value in the future commercial positioning and design of sports bra for enterprises.

References

1. K.A. Bowles, J.R. Steele, B. Munro. Brit. J. Sport. Med. **42**,670 (2008).
2. L. Lawson, D Lorentzen. Cloth. & Textiles. Res. J. **8**,55 (1999).
3. J. Zhou, W. Yu, S. Ng. Text. Res. J. **83**,1500 (2013).
4. L. LaJean, L. Deana. Cloth. & Textiles Res. J. **8**, 55(1990).
5. K.A. Page, J.R. Steele. Sports Med. **27**, 205(1999).
6. K.A. Bowles, J.R. Steele, R. Chaunchaiyakul. Med. Sci. Sport. Exer. **37**, 1633(2005).
7. J.C. Scurr, J.L. White, W. Hedger. J Appl. Biomech. **25**, 322(2009).
8. B.R. Mason, K.A. Page, K Fallon. J Sci. Med. Sport **2**,134(1999).
9. D Lorentzen, L Lawson. Phys. Sportsmed. **15**,128 (1987).
10. K.A. Bowles, J.R. Steele. Med. Sci. Sports. Exerc. **45**,1113 (2013).

MATEC Web of Conferences **82**, 03007 (2016) DOI: 10.1051/matecconf/20168203007
D2ME 2016

Healthy Modern Landscape Inherited from Hakka Culture

Huang Hui[1], Gao Ruofei[2] and Kang Xiaoyong[3,a]

[1] *Associate Professor of Shenzhen Polytechnic, China*
[2] *Doctor of Chiba University, Japan*
[3] *Lecturer of Shenzhen Polytechnic, China*

Abstract. The Hakka modern landscape is guided by its spirit of ambition, optimism, tolerance, and enterprising. The Meizhou Ruishan Agricultural and Ecological Garden Project (phase 1) combines the attributes of colors, materials and forms from the traditional Hakka architecture, Hakka culture, and the healthy landscape as a general concept and study. Thus, a new concept of modern Hakka landscape has been put forward through practice.

Introduction

Hakka culture is not only diverse but also unique [1]. After several ethnic migrations, Hakka people in the present day comprises mainly of Han nationality branches. Settled in the southern environment, through cultural interaction and permeability, Hakka people posses several characteristics of nearby local southern ethnics such as Baiyue and Yaoshe, while still retaining the traditional central plains culture as their main practice.

Practical activities in Meizhou Ruishan Agricultural and Ecological Garden (phase 1) have been carried out in order to promote the prosperity of regional culture, as well as the concept of healthy landscape. Attempting to introduce a new concept of healthy modern landscape based on the cultural and spiritual heritage of Hakka culture and the application of healthy landscape, Meizhou, recognized as the capital of all Hakkas, has been visited and studied. The colors, materials, and forms inherited from traditional Hakka architecture have been analyzed on a large scale.

1 Typical Hakka culture characteristics

In 2013, the worldwide Hakka population is 55 million, including 45 million found in China. The rest of the population reside mostly in Malaysia, Indonesia, Thailand, Singapore, and Burma.

The impact of cultural transfer during population migration and the raw nature of Southern mountainous cultures are brought together to form an intercultural Hakka community. However, despite its cultural diversity, Hakkas still root deeply in Confusinism, which preaches that human beings cannot be separate from nature. Mr. Ge Jianxiong said: "Since the purpose of migration is to settle down, people have to make a choice in culture. They can accept the culture which they migrate to so that they can melt into it; or they have to adhere to their own culture so as to carry it out among

---

[a] Corresponding author : 44744093 @qq.com

the local people." [2] The following characteristic features are manifested by Hakkas when it comes to architecture, food, and custom:

- Central Plain cultural heritage
- Extensive collection of diversified cultures
- Unique migration culture

2 Traditional Hakka architecture: colors, materials and forms

Based on the investigation, the authors have extracted the colors, the surroundings, architectural forms, and material language of typical Hakkas buildings, and their application in modern Hakkas landscapes.

- Blue mountains and green rivers are assessed as environmental setting, integral colors with oxford ochre, gray and brown as dominant hues, entranceway with emphasis on gray and white, and interior decoration embellished with the gorgeous colors like red, yellow, and cyan. Application in landscapes should look after both sides of limited use (background colors and decoration colors) and widespread use (entranceway colors and integral colors of hard landscape), thereby laying the foundation for further correlated researches on the materials embodying Hakkas culture in landscapes.
- The spatial structure with plants enclosing hard landscape are prioritized. The quadratic form in the spatial structure in the facade of entranceway and hard space are also prioritized.
- Definition of soft and hard landscape is accomplished by grey granite and tile. Hard pavement gives priority to river gravel which is good in water permeability. China fir and black bricks are used to shape different kinds of facilities and structures.

3 Healthy landscape

Health refers not only to physical health of human body, but also to mental health and the health of social relations. As the WHO (World Health Organization) definition of health goes, "Health is a state of complete physical, mental and social well-being and not merely the absence of disease or infirmity." Based on research findings on health-related landscapes since 1990s, the authors outlined three basic functions of healthy landscapes.

3.1 Improving body functions

Environments featuring plant richness can promote children's and adults' enthusiasm for exercise [3][4], and make them voluntarily extend the time of exercise[5]. Strolling through forests can substantially increase the activity of natural killer cells and the quantity of anti-cancer active proteins, and decrease the secretion of adrenal hormones. While in urban environments dominated by hard landscapes, walking can produce no such effect [6].

3.2 Shaking off weariness and tedium

Studies have found that green landscape can make people relaxed. In accordance with relevant principles of physiology and psychology, in particular in medical premises, patients more frequently exposed to green landscape can expect shorter hospital stays and a relevant normal blood pressure [7], as well as a more positive attitude [8].

3.3 Stress-relieving

Studies have found that in places with green landscape or water, only non-subjective attention is consumed, and the direct attention can be recovered. In contrast, in environments lack of green and full with artificial stimulations, people have to consume large amount of active attention. Enjoying the beauty of green landscape, or resting or exercising in green landscape can relieve stress and recover attention [9]. Children will express higher creativity, self-discipline and attentiveness in green landscapes [10][11][12].

In general, healthy landscapes can be summarized as follows: under the guideline of medical science, environmental psychology, behavioral psychology, horticultural therapy, and traditional science of health preservation, with healthy people and sub-healthy people, as well as people of different age groups, physical conditions and psychological health as the target groups, utilize the health care factors of different landscape elements to the full extent, and establish and maintain an environment good for physical and psychological health, so as to promote the interaction between users and landscapes, comprehensively adjust the physical and psychological status of users, resolve somatopsychic disorders to a possible extent, enhance the awareness of self-care and life cultivation and health preservation, and make the users comfortable and healthy both physically and psychologically.

Healthy landscapes take the form, in particular, of the applications of landscape plants with volatile chemical components in landscapes. Based on the functions of volatile chemical components, the landscape plants can be classified into three groups: plants for health and medical care [13][14][15], plants for sterilization, insecticide and desinsectization [16], and plants for absorbing toxic gases [17].

4 Application practice in Ruishan Agricultural and Ecological Garden

Known as "the capital of Hakkas in the world", Meizhou in Guangdong Province is the hometown of Hakkas. It witnesses the formation and development of the Hakka subgroup and Hakka culture.

Ruishan Agricultural and Ecological Garden is located in Zhourui Town at the south-west boundaries of Dabu County in Meizhou, and bounded to the northwest by Yinjiang Town. As a key node along the world-famous tourist route of Hakka culture in the intersection of the Economic Zone on the west side and the Pearl River Delta Economic Zone, the Garden is within 40 minutes' drive from downtown Meizhou, and only 30 minutes drive to Dabu County. Famous tourist destinations such as Dama Town, Songkou Town and Yanyang Town are all within the "one hour sphere of life".

4.1 Design objective

The landscape of the ecological garden focuses on health and adopts concepts of Hakka culture and healthy landscape based on the quality groundwater resources in the field and the original favorable ecological environment, to develop the ecological garden into a healthy camp with Hakka characteristics that is famous in Meizhou, Guangdong, and South China.

4.2 Design process

The landscape design of the project took about one year, with many on-site design exchanges conducted. Designers had reached agreement on the design objective, Party A's intention, existing problems about the site, the expression of Hakka factors, and users. They also made a 1:100 scale model of the innovative waterscape to verify the result.

4.3 Features of the site

The garden was transformed from a tree farm with a fantastic natural ecological environment. The purpose of landscape design is to make visitors feel comfortable while enjoying more health, tranquility and mildness in a natural oasis. The test by a professional organization shows that the groundwater source of the ecological garden is of metasilicic acid. The water quality is similar to that in the Bama village in Guangxi, where most people live long. The metasilicic acid mineral water is very beneficial to the human body. That's why it is well known as the most high-end and valuable mineral water. It has three characteristics as below:

- Beneficial microelements.
- Natural alkalescence.
- Balance containing of various mineral elements.

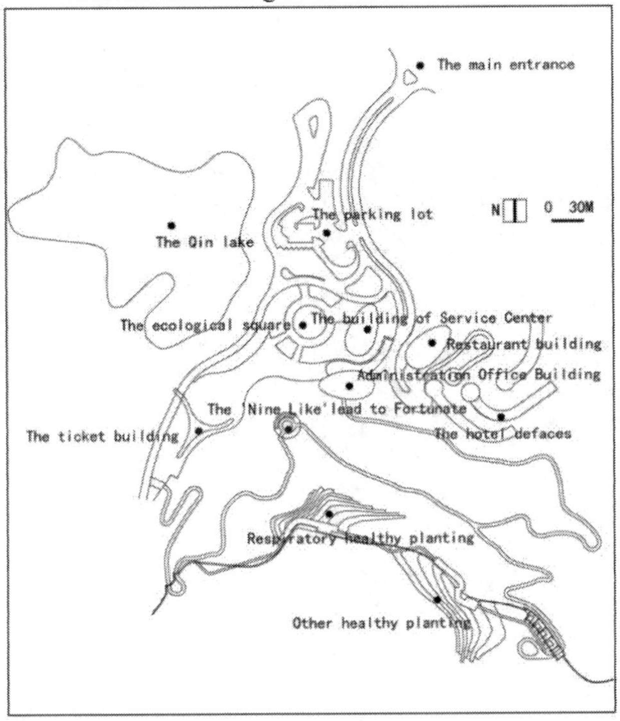

Figure 1 Plan of Ruishan Agricultural and Ecological Garden (phase I)

4.4 Design elements

Phase I of the ecological garden refers to the "Qin Lake Zone" and "Ruixiang Valley" of the "Yungu Zone" which are expected to open in October 2017 (Figure 1). The "Lyre Lake Zone" is located at the main entrance of the ecological garden. With focus on the "Lyre Lake", the form of surrounding eucalyptuses and pinus massoniana forests are improved to form a perfect pattern of sightseeing: mountains (natural landscape) - buildings (artificial landscape) - lakes (natural artificial landscape).

The "Ruixiang Valley" of the "Yungu Zone" are mainly decorated with healthy plants. A wooden trestle is built along the stream. Areas used for growing respiratory care plants and other health care plants are also provided. In addition, a new road for electromobiles is created based on the existing land form. It goes from the entrance at the west side of the valley to the hotel buildings at the east side and through the administrative office buildings. The waterscape across the road for electromobiles presents the auspicious sign of longevity. It makes the road more attractive while achieving a bridge for the "Ruixiang Valley" and "Qin Lake".

Modern Hakka architectures which are built based on the design of typical round earthen building bring benefits of preventing cold air from valley, increased use of space, and less sunshine dead zone, which are more suitable for the settlement pattern of hotels. Hotel buildings in the "Qin Lake Zone" are built around the mountain. Together with the oval service center, dining hall and office buildings, they constitute a shape of Ruyi, which symbolizes good luck and longevity. The GRC stainless steel mosaic externally veneered roofs and glass walls are adopted for three office buildings. In addition, metal mosaic veneered roofs are also adopted for the hotel buildings. In the green background, the buildings shine like dews on a lotus leaf.

Hakka people's traditional settlement areas were located in hilly areas. The hilly area of Meizhou accounts for more than 50% of that of the whole city. Such special geological formation enables the abundance in stones including landscape stones and granites. Local stones are extensively applied in the foundation of traditional Hakka architectures [16]. Hard landscapes of the "Qin Lake Zone" and "Ruixiang Valley" take full advantage of natural stone resources in Ruishan. With less artifacts applied, a more attractive landscape space is formed. Meanwhile, colors with territorial nature and Ruishan characteristics are exploited to match with the nature and Hakka culture. Local stones are widely used in the circular plaza and landscape along the lake in the "Qin Lake Zone". For traditional Hakka architectures, bamboos, woods, red clays, stones and other natural materials are mainly used. However, artificial materials are mainly exploited in modern Hakka architectures for the purpose of modeling and cost control. The landscape structures made of natural materials embody the correspondence between modern Hakka landscapes and traditional culture as well as nature. The combined application of these materials reflects the traditional Hakka culture to some extent. (Figure 2)

Figure 2 Ruishan Agricultural and Ecological Garden (phase I)

The abundant use of healthy plants in the "Ruixiang Valley" of the "Yungu Zone" agrees with the garden's theme - health . In the context that health is gradually incorporated into state policies [17], adoption of more healthy plants in landscape design adapts to needs of social development in China, especially in southern China. It also goes well with the diversity and pioneering spirits of Hakka culture as well as embodies the existence and improvement of Hakka. In the light of prevalent respiratory diseases suffered by people live in modern cities, the area used for growing respiratory care plants mainly consists of the jasmine, Chinese rose, rosemary, etc. The area used for growing other healthy plants mainly consists of the sage, lavender etc. As it is dominated by the existing eucalyptuses and aromatic plants, the valley is called "Ruixiang Valley". Table 1 below listed the effects of health plants in the "Ruixiang Valley".

Table 1: Healthy Plants of "Ruixiang Valley"

Respiratory Health-Types				**Other Health-Types**			
No.	**English**	**Latin**	**Effect**	**No.**	**English**	**Latin**	**Effect**
1	Prynne	*Michelia alba*	Ceasing cough Suppressing asthma Expectorant	1	Mason pine	*Pinus massoniana*	Distributing fine Pythoncidere
2	Manglietia	*Manglietia fordiana*	Antitussive Expectorant	2	Chinese fir	*Cunninghamia lanceolata*	Distributing fine Pythoncidere
3	Velvet	*Magnolia coco*	Antitussive Expectorant Suppressing asthma	3	Lemon eucalyptus	*Eucalyptus citriodora*	Antibacterial
4	Osmanthus	*Osmanthus fragrans*	Suppressing asthma	4	Nagi	*Nageia nagi*	Calming Relaxing and the regulation of mood

5	Tianzhu laurel	*Cinnamomum japonicum*	Suppressing asthma Antitussive Expectorant	5	Camphor	*Cinnamomum camphora*	Antibacterial Absorbing (SO2, Cl2, HF, HCL, benzene, O3, dust)
6	Burmannii	*Cinnamomum burmannii*	Suppressing asthma, Antitussive Expectorant	6	Tallow	*Sapium sebiferum*	Relieving headaches Refreshing
7	Sweetgum	*Liquidambar formosana*	Suppressing asthma	7	Magnolia	*Magnolia denudata*	Ceasing cough Antibacterial Relieving headaches
8	Schefflera	*Schefflera heptaphylla*	Antitussive Expectorant	8	Callistemon	*Callistemon rigidus*	Antitussive Expectorant Absorbing (SO2, Cl2, HCL, dust)
9	Pittosporum	*Pittosporum tobira*	Ceasing cough Suppressing asthma Expectorant	9	Aglaia	*Aglaia odorata*	Calming Relaxing Regulation of mood, Hypnotizing Absorbing SO2
10	Murraya	*Murraya exotica*	Ceasing cough Suppressing asthma Expectorant	10	Chinese rose	*Rosa* cvs.	Refreshing Eliminating fatigue, Hypnotizing
11	Gardenia	*Gardenia jasminoides*	Suppressing asthma,	11	lavender	*Lavandula angustifolia*	Relieving headaches Hypnotizing and Quieting down the mind Antibacterial Antitussive Sterilizing
12	Jasmine	*Jasminum sambac*	Suppressing asthma	12	Sage	*Salvia japonica*	Bacteriostatic Bactericidal Anti-inflammatory Detoxification
13	Basil	*Perilla frutescens*	Antitussive Expectorant Suppressing asthma	13	Rosemary	*Rosmarinus officinalis*	Refreshing Driving mosquitoes
14	Thyme	*Thymus mongolicus*	Suppressing asthma Antibacterial	14	Mint	*Mentha haplocalyx*	Antibacterial Exciting hub Absorbing O3

4.5 Maintenance and management

Phase I of the ecological garden is maintained by the management personnel of the garden. Usually, the management personnel spend two hours a week in maintaining phase I of the garden. In the period when medicinal herbs need to be replaced, the management personnel spend more time as required. Maintenance of the waterscape is entrusted to professionals. To protect the natural environment of the garden, natural methods are taken for pest control.

5 Conclusion

Hakka culture, especially its pioneering spirit, is a key factor for Hakka people to survive and keep vitality. Under the guidance and impact of the essence of Hakka culture and combined with the idea of healthy landscape, Hakka modern landscapes embody a new concept of healthy modern landscape led by Hakka culture. With the development goal of healthy landscape, Hakka modern landscapes will definitely experience a longer life.

The ecological environment of Hakka people's traditional settlement areas lays a foundation for developing Hakka healthy landscape, Hakka culture and spirit provide driving forces, and the abundant rainfall and plant materials in southern China bring application opportunities and various types of materials.

References

1. Hegel, Wang Zaoshi. *Philosophy of History*[M]. Shanghai: Shanghai Bookstore Publishing House, 2001, P572
2. Ge Jianxiong. *History of Chinese Population Six Volumes* [M]. Shanghai: Fudan University Press, 2002
3. Fjortoft, I.,Sageie, J. The natural environment as a playground for children - landscape description and analyses of a natural playscape [J].Landscape and Urban Planning, 2000, **48**(1-2), 83-97.
4. Sullivan, W. C., Kuo, F. E., DePooter, S. F. The fruit of urban nature vital neighborhood spaces[J]. Environment and Behavior, 2004, **36**(5), 678-700.
5. Pennebaker, J. W., & Lightner, J. M. Competition of internal and external information in an exercise setting[J]. Journal of Personality and Social Psychology,1980,**39**(1), 165-174.
6. Li, Q., Morimoto, K., Kobayashi, M., Inagaki, H., Katsumata, M., Hirata, Y. Krensky, A. M. Visiting a forest, but not a city, increases human natural killer activity and expression of anti-cancer proteins[J]. International Journal of Immunopathology and Pharmacology, 2008, **21**(1), 117-127.
7. Ulrich, R. S., Simons, R. F., & Miles, M. A. Effects of environmental simulations and television on blood donor stress[J]. Journal of Architectural and Planning Research, 2003, **20**(1), 38-47.
8. Raanaas, R. K., Patil, G. G., & Hartig, T. Effects of an indoor foliage plant intervention onpatient well-being during a residential rehabilitation program[J]. Hortscience, 2010, **45**(3), 387-392.
9. Pretty, J., Peacock, J., Sellens, M., & Griffin, M. The mental and physical health outcomes of green exercise[J]. International Journal of Environmental Health Research, 2005,15(5), 319-337.
10. Taylor, A. F., Wiley, A., Kuo, F. E., & Sullivan, W. C. Growing up in the inner city - green spaces as places to grow[J]. Environment and Behavior,1998, **30**(1), 3-27.
11. Taylor, A. F., Kuo, F. E., & Sullivan, W. C. Coping with add – the surprising connection togreen play settings[J]. Environment and Behavior, 2001, 33(1), 54-77.
12. Taylor, A. F., Kuo, F. E., & Sullivan, W. C. Views of nature and selfdiscipline: Evidence from inner city children[J]. Journal of Environmental Psychology,2002,22(1-2), 49-63.
13. Yuan Jiamo. *Aromatic Plants in Guizhou* [M]. Guiyang: Guizhou Science and Technology Publishing House, 1990
14. Zhou Bo, Su Weiwei. Temperament Analysis for Healthy Volatile Substances of Landscape Plants in South of the Five Ridges[D]. Guangzhou: Sun Yat-sen University, 2005
15. Chen Li, Qin Hua, etc. Study on Phytocoenosium Design of Healthy Landscape Plants [J]. Heilongjiang, Northern Horticulture, 2011, (05):152-154
16. Zheng Linsen, Pang Mingyu, Jiang Yihua, etc. *Determination of 47 Kinds Healthy Volatile Substances of Landscape Plants* [C]. Selected Works of The Fourth National Congress Papers, Collected Works (Volume I). Chinese Society of Landscape Architecture, 2008:189-193
17. Su Xuehen. *Plant Landscape* [M]. Beijing: China Forestry Publishing House, 1998: 150

Design Study of 200MWth Gas Cooled Fast Reactor with Nitride (UN-PuN) Fuel Long Life without Refueling

Ratna Dewi Syarifah[1,a], Yacobus Yulianto[1], Zaki Su'ud[1], Khairul Basar[1] and Dwi Irwanto[1]

[1]*Nuclear Physics and Biophysics Research Group, Department of Physics, Faculty of Mathematics and Natural Sciences, Bandung Institute of Technology, Bandung, Indonesia.*

Abstract. Design study of 200 MWth Gas Cooled Fast Reactor with UN-PuN fuel long life without refueling has been done. GFR is one type reactor in Generation IV reactor system. It uses helium coolant and fast neutron spectrum. Helium is chemical inert, single phase and low neutron moderation. In this study the calculations are performed by using SRAC code with PIJ calculation for the fuel pin cell calculation and CITATION calculation for core calculation. The data libraries use JENDL 3.2. The variation fuel fractions are 50% until 60%. The diameter active core is 150 cm and the height active core is 100 cm. The reflector radial-axial width is 50 cm. The variation of the powers are 100 MWth up to 500 MWth. The high power causes the high k-eff value. The optimum design is reached when the power is 200 MWth, variation percentage Plutonium for fuel F1:F2:F3=9%:11%:13%. The comparation of fuel:cladding:coolant fraction = 55%:10%:35%. The cooling down time of Plutonium is nine months. The optimum k-eff value is 1.0142 with excess reactivity value 1.403%. The decay of Plutonium decrease k-eff value in the beginning of burn up.

1 Introduction

Nuclear power plants, which produce low-carbon electricity at stable and competitive costs, constitute an element of the solution to global warming and means of delivering power to emerging and developed countries. Further development of nuclear technology is needed to meet future energy demand [1]. The Fukushima Daiichi nuclear power plant accident has emphasized the important of designing nuclear systems with highest levels of safety. Lessons learnt from the accident will benefit the current operating fleet, as well as future nuclear system, including Generation IV systems [1].

Challenging technology goals for Generation IV nuclear energy system are defined in four areas. They are sustainability, economics, safety and reliability, and proliferation resistance and physical protection. There are six systems of Generation IV technologies, i.e. Gas Cooled Fast Reactor (GFR), Lead Cooled Fast Reactor (LFR), Molten Salt Reactor (MSR), Sodium Cooled Fast Reactor (SFR), Supercritical Water Cooled Reactor (SCWR), and Very High Temperature Reactor (VHTR) [2]. Beforehand, neutronics analysis of GFR with SRAC Code and modified CANDLE burn up scheme with uranium natural has been analyzed [3, 4, 5, 6].

The GFR uses gas helium coolant and it uses fast neutron spectrum. The advantages of the gas coolant are it is chemical inert (allowing high temperature operation without corrosion and coolant

[a] Corresponding author : syarifah.physics@gmail.com

radio-toxicity) and single phase (eliminating boiling), and it is low neutron moderation (the void coefficient of reactivity is small) [1]. In this research design study of 200 MWth GFR with nitride fuel long life without refueling has been done.

2 Design Concept and Calculation Method

Burn up. During reactor operation, neutron interactions with fuel give rise to various nuclear reactions such as fission of fissile nuclides, conversion of fertile nuclides into fissile ones, and production of FPs. The burn up equation is to determine atomic number densities of fissile and fertile nuclides in fuel, and considers the changes with the fuel burn up [7].

Recalling the volumetric reaction rate (reaction rate per unit time and volume) expressed by $\sigma\phi N$ (λN in case of decay) and making a balanced relation between the production and destruction rates of a target nuclide, the burn up equation can be given by the following equations, where nuclides are identified by a two-digit superscript indicating the last digit in the atomic number and the last digit in the mass number [3]. Eq. (1) up to eq. (4) shows the equation for burn up Pu-239, Pu-240, Pu-241, and Pu-242. Plutonium isotope is very important in the reactor, because it is fissile material and we use natural uranium.

$$^{239}Pu: \frac{\partial N^{39}}{\partial t} = \lambda^{39} N^{39} - \sigma_a^{49} \phi N^{49} \quad (1)$$

$$^{240}Pu: \frac{\partial N^{40}}{\partial t} = \sigma_c^{49} \phi N^{49} - \sigma_a^{40} \phi N^{40} \quad (2)$$

$$^{241}Pu: \frac{\partial N^{41}}{\partial t} = \sigma_c^{40} \phi N^{40} - \left(\lambda^{41} + \sigma_a^{41} \phi\right) N^{41} \quad (3)$$

$$^{242}Pu: \frac{\partial N^{42}}{\partial t} = \sigma_c^{41} \phi N^{41} - \sigma_a^{42} \phi N^{42} \quad (4)$$

Table 1 show the parameter design of the reactor which has been designed. Determination of fuel, cladding, and coolant material in the reactor influence safety and economic factor of the reactor. In this study, the pin design geometry use hexagonal cell geometry which is presented in Figure 1. But it is divided by six regions, the first until third region are fuel regions, fourth is cladding, fifth and sixth are coolant. The diameter active core is 150 cm and height active core is 100 cm which is divided by three variation fuel (F1, F2, and F3) which presented in Table 2. The half of core geometry design in axial view is showed in Figure 2.

The calculation method use SRAC (Standard Thermal Reactor Analysis Code) System. It is designed to permit neutronics calculation for various type reactors. This research use JENDL 3.2 as a neutron cross section library. SRAC covers production of effective microscopic and macroscopic group cross section, and static cell and core calculation including burn-up analysis. First, it calculate fuel pin cell PIJ with Collision Probability Method (CPM). In PIJ calculation, we get k-inf, burn-up analysis and more. After that, it is continued to calculate core reactor with CITATION, with various core configuration [8].

Table 1. Parameter design of the reactor.

Parameter	Specification
Power (MWth)	100 -500

DOI: 10.1051/matecconf/20168203008

Fuel material	UN-PuN
Cladding material	Stainless stell
Coolant material	Helium
Fuel volume fraction	50% - 60%
Cladding volume fraction	10%
Coolant volume fraction	30% - 40%
Active core diameter and height	150 & 100 cm
Reflector radial & axial width	50 cm
Pin pitch	1.45 cm
Reactor life	> 10 years

Table 2. Percentage and volume Plutonium in each type of fuel design of the reactor.

Fuel Type	Percentage Plutonium	Volume in the reactor core
F1	9%	0,19250 m^3
F2	11%	0,71264 m^3
F3	13%	0,86271 m^3

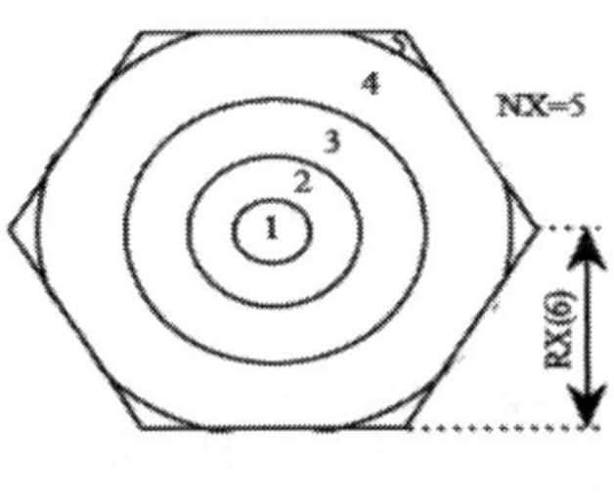

Figure 1. Hexagonal geometries for PIJ [4]

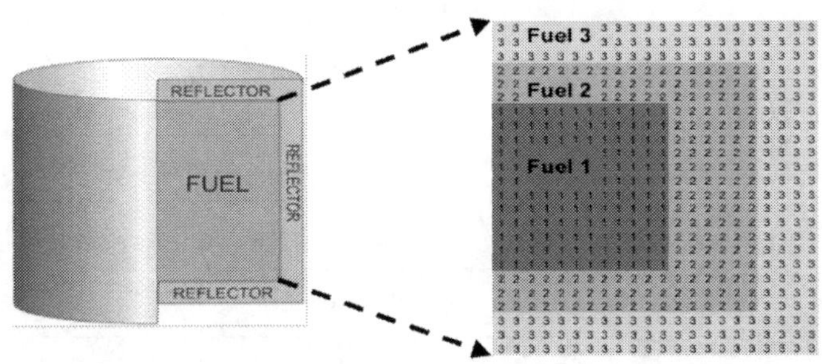

Figure 2. Half heterogen core configuration with different volume in variation fuel.

3 Results and Discussion

The design study of 200 MWth GFR with UN-PuN fuel has been done. The first step calculation, k-inf value has been calculate for percentage Plutonium 9% up to 13% with 60% fuel fraction, 10% cladding, and 30% coolant fraction. Figure 3 show that the percentage Plutonium 11% has flatness value of k-inf. The k-inf value is around 1.35 and its constant up to burn up time ten years. From this parameter survey, 11% Pu become a referable to the next calculation. The k-eff value of 11% Pu is also flat, but for power distribution, it is not good, because the power distribution is too high in center of the core. So, the design must be varied the percentage Plutonium. There are three variations, the first fuel (F1), the second fuel (F2) and the third fuel (F3). It shows in Table 2. The F1 is located in the center of the core, followed by the F2 and F3 (see Figure 2). It is used to make the power in the reactor to be flat.

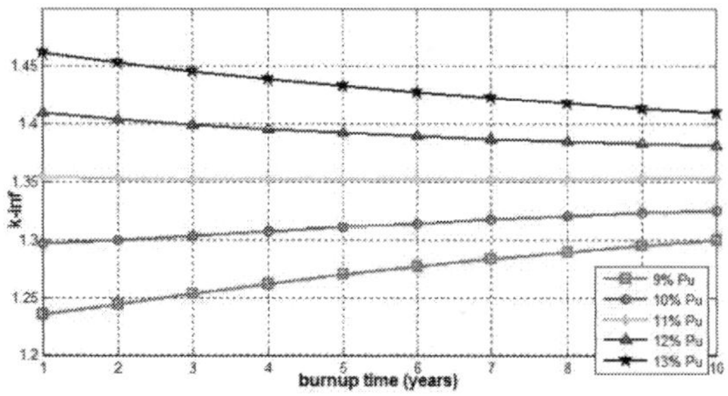

Figure 3. K-inf value with various percentage of Plutonium.

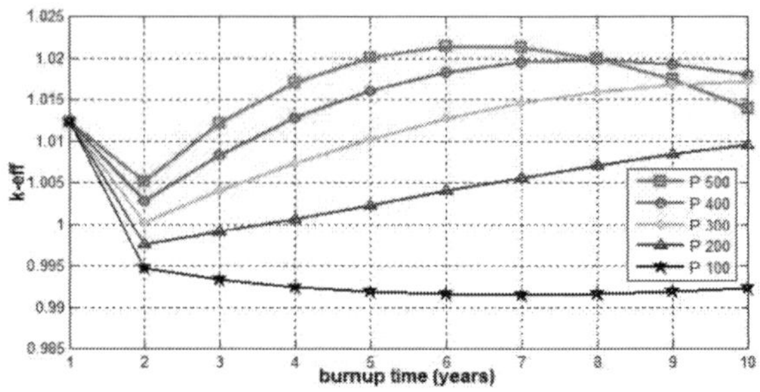

Figure 4. K-eff value with various power from 100MWth – 500MWth with three variation fuel.

Figure 4 presented the k-eff value with various powers from 100MWth-500MWth with three variation fuel. From the Figure 4, when the power is 500 MWth, the k-eff value is highest than the others. And the optimum value is 1.0215 in the step burn up six years. When it decreases the power, the optimum k-eff value is decrease too. Because the flux is reduced, the power also decreases. But in the first step burn up, the k-eff value is same with various powers. When the power is 200 MWth and 100 MWth, the optimum k-eff value are in the first step burn up. So, we want to decrease the k-eff value in the first step burn-up with cooling down the Plutonium before it use in fuel reactor. Plutonium can be found from waste LWR. We are cooling down the Plutonium, so it decays the amount of Pu-241. The Pu-241 has cross section absorption bigger than Pu-239. It cause there are peak k-eff value in the first step burn up. So, cooling down the Plutonium is good choice to decrease the amount of Pu-241. The half-life plutonium can be seen in the Table 3. Table 3 shows the half-life of Pu-241 is 14.4 years. So, when it cooling down the Plutonium from LWR waste, it just decrease the amount of Pu-241.

Table 3. Half-life Plutonium.

Isotope	Half-life (years)
Pu-238	87,74
Pu-239	24100
Pu-240	6560
Pu-241	14,4
Pu-242	376000

Figure 5 presented the k-eff value of 500Mth with Plutonium decay from 0 years up to 2 years. From Figure 5, the k-eff value in first step of burn up is decrease when it cooling down or it decay half years (six month) up to 2 years. The k-eff value in the tenth step burn up disposed the same value. So, the decay just influence in the beginning of burn up. From Figure 5 the excess reactivity value is still high. In this case, the smallest value of excess reactivity is 1.98%. We must decrease the excess reactivity value for safety reactor. The optimum excess reactivity can be reach when the k-eff value is around 1 (see Eq. 5).

$$excess\ reactivity = \frac{\left(k_{eff} - 1\right)}{k_{eff}} \times 100\% \qquad (5)$$

 DOI: 10.1051/matecconf/20168203008

Figure 6 presented the k-eff value of 200Mth with Plutonium decay from 0 years up to 2 years. It does the same case with Figure 5. But, it reduces the power until 200MWth, due to decrease the excess reactivity value. From Figure 6, the optimum k-eff value when the Plutonium decay 0.75 years (nine month). The fuel fraction of this case is 55%. The optimum k-eff value is 1.0142 with excess reactivity value 1.403%.

Figure 7 and Figure 8 show the atomic density of some isotope i.e. U-238, Pu-239, Pu-240, Pu-241 and Pu-242. From Figure 7, the reduction of atomic density U-238 cause increasing Pu-239. It means the reactor is breeding. GFR is one type of fast reactor, so it must breed the fuel. Figure 8 shows the reduction of Pu-241 significantly than Pu-240 and Pu-242. It is because Pu-241 has the biggest cross section absorption than the others. Pu-241 has a larger cross section than Pu-239, this cause a decline in k-eff value in the second burn up time.

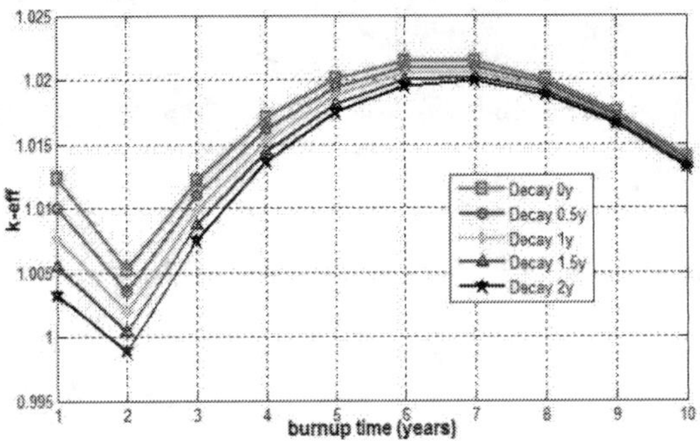

Figure 5. K-eff value of 500Mth with Plutonium decay from 0 years up to 2 years

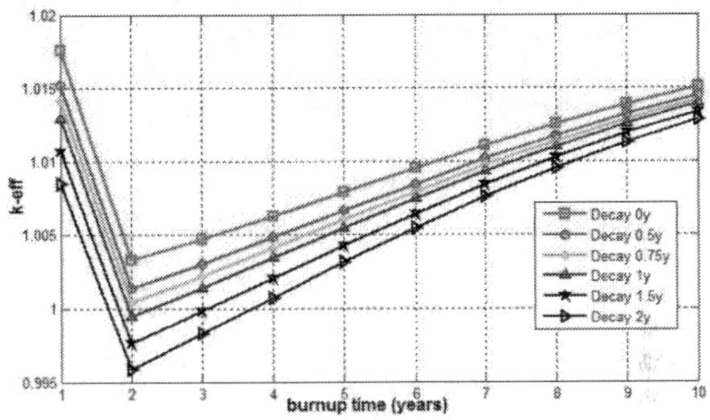

Figure 6. K-eff value of 200Mth with Plutonium decay from 0 years up to 2 years

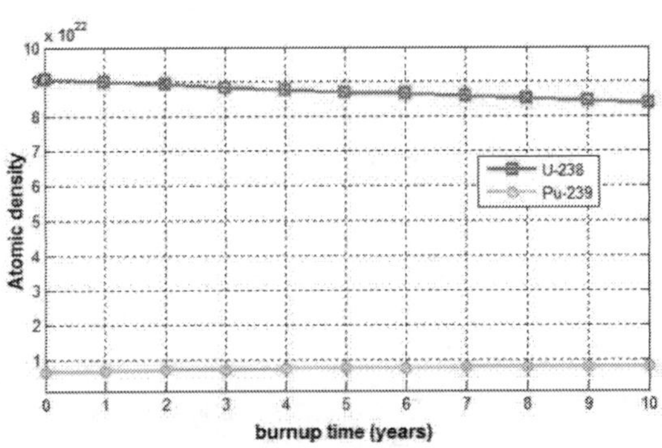

Figure 7. Pu-239 and U-238 atomic density change during burn up time

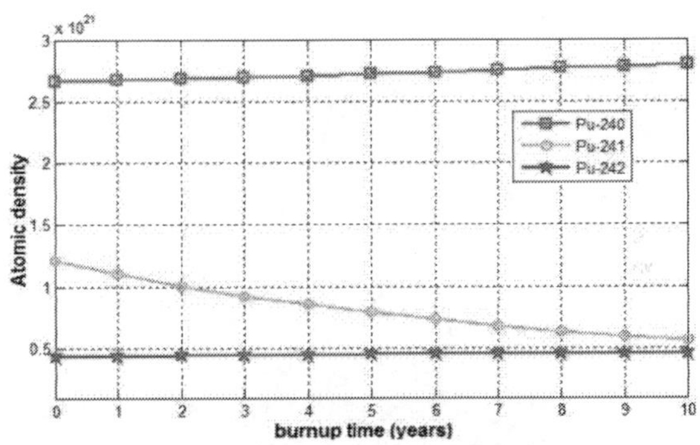

Figure 8. Pu-240, Pu-241 and Pu-242 atomic density change during burn up time

4 Conclusion

The design of 200MWth Gas Cooled Fast Reactor with nitride fuel has been performed. The optimum design reaches when the power is 200 MWth, variation percentage Plutonium for fuel F1:F2:F3 = 9%:11%:13%. The comparation of fuel : cladding: coolant fraction = 55%:10%:35%. The cooling down time of Plutonium is nine months. The optimum k-eff value is 1.0142 with excess reactivity value 1.403%. The decay of Plutonium influence k-eff value in the beginning of burn up.

Acknowledgment

This work was supported by Decentralization Research Grand 2016, DIKTI, Ministry of Research Technology and Higher Education of the Republic of Indonesia which are gratefully acknowledged.

References

1. GIF (The Generation IV International Forum), *Technology Roadmap Update for Generation IV Nuclear Energy System* (the OECD Nuclear Energy Agency, 2014)
2. GIF (The Generation IV International Forum), *A technology Roadmap for Generation IV Nuclear Energy System* (U.S DOE Nuclear Energy Research Advisory Committee, 2002)
3. Z. Su'ud, H. Sekimoto, Annals of Nuclear Energy **54**, 58-66 (2013)
4. F. Monado, Z. Su'ud, A. Waris, K. Basar, M. Ariani, H. Sekimoto, Advanced Materials Research **772**, 501-506 (2013)
5. F. Monado, M. Ariani, Z. Su'ud, A. Waris, K. Basar, F. Aziz, S. Permana, H. Sekimoto, AIP Conference Proceedings **1584**, 105 (2014)
6. F. Monado, Z. Su'ud, A. Waris, K. Basar, M. Ariani, H. Sekimoto, AIP Conference Proceedings **1615**, 47 (2014)
7. Y. Oka, S. Ohki, *An Advanced Course in Nuclear Engineering 2* (Japan, 2014)
8. K. Okumura, SRAC2002: *The Comprehensive Neutronics Calculation Code System* (JAEA, Japan, 2002)

AUTHOR INDEX

AUTHOR INDEX

EDP Sciences – Web of Conferences
17, Avenue du Hoggar
Parc d'Activité de Courtabouf
BP 112
F-91944 Les Ulis Cedex A
France

ISBN 978-1-5108-3169-8

CURRAN ASSOCIATES INC.
proceedings.com

9 781510 831698